Irreversible Phenomena

Kunio Terao

Irreversible Phenomena

Ignitions, Combustion and Detonation Waves

With 363 Figures and 23 Tables

Kunio Terao
Tsuruga-oka 1-8-3
238-0056 Yokosuka
Japan

Library of Congress Control Number: 2006939786

ISBN-10 3-540-49900-8 Springer Berlin Heidelberg New York
ISBN-13 978-3-540-49900-8 Springer Berlin Heidelberg New York

This work is subject to copyright. All rights are reserved, whether the whole or part of the material is concerned, specifically the rights of translation, reprinting, reuse of illustrations, recitation, broadcasting, reproduction on microfilm or in any other way, and storage in data banks. Duplication of this publication or parts thereof is permitted only under the provisions of the German Copyright Law of September 9, 1965, in its current version, and permission for use must always be obtained from Springer. Violations are liable to prosecution under the German Copyright Law.

Springer is a part of Springer Science+Business Media.

springer.com

© Springer-Verlag Berlin Heidelberg 2007

The use of general descriptive names, registered names, trademarks, etc. in this publication does not imply, even in the absence of a specific statement, that such names are exempt from the relevant protective laws and regulations and therefore free for general use.

Typesetting by SPi

Cover design: Erich Kirchner, Heidelberg

Printed on acid-free paper   SPIN 11803218   89/3100/SPi   5 4 3 2 1 0

Preface

Ideals are simple and able to be easily understood, but never exist in reality.

In this book a theory based on the second law of thermodynamics and its applications are described. In thermodynamics there is a concept of an ideal gas which satisfies a mathematical formula $PV = RT$. This formula can approximately be applied to the real gas, so far as the gas has not an especially high pressure and low temperature. In connection with the second law of thermodynamics there is also a concept of reversible and irreversible processes. The reversible process is a phenomenon proceeding at an infinitely low velocity, while the irreversible process is that proceeding with a finite velocity. Such a process with an infinitely slow velocity can really never take place, and all processes observed are always irreversible, therefore, the reversible process is an ideal process, while the irreversible process is a real process.

According to the first law of thermodynamics the energy increase dU of the thermodynamic system is a sum of the heat dQ added to the system and work dW done in the system. Practically, however, the mathematical formula of the law is often expressed by the equation , or some similar equations derived from this formula, is applied to many phenomena. Such formulae are, however, theoretically only applicable to phenomena proceeding at an infinitely low velocity, that is, reversible processes or ideal processes. The question arrives whether or not such mathematical formulae which are only applicable to ideal processes can also approximately to real processes.

Since Jost wrote the book on combustion "Explosions-und Verbrennungsvorgänge in Gasen," a lot of book on ignition, combustion, flames, and detonation waves have been published. In these books, the mathematical formulae which are applicable only to ideal processes are applied to all phenomena of ignition, combustion, and explosion, assuming that the mathematical formulae introduced for the reversible processes can approximately be applied to the irreversible phenomena, too.

Nevertheless there are still many phenomena, for example, fluctuating phenomena in ignition and detonation waves, or anomalous high temperature of free electrons and ions in the flame and behind detonation waves which cannot be explained by the theories introduced and applied in those books.

In an irreversible process at least once for a short time a nonequilibrium and heterogeneous state appears which causes some stochastic phenomena. In this book the author tries to explain that the reversible and irreversible processes are quite different from one another and that the mathematical formulae which are applicable to the reversible process cannot be applied to the irreversible process, not even approximately, in so far the process has a large entropy increase, i.e., a large irreversibility, and further tries to explain many phenomena in the combustion science applying the characteristics of the irreversible phenomena, since ignition, combustion, explosion, detonation, and other phenomena in the combustion are distinctly irreversible processes accompanied by a large entropy increase. By observing the phenomena in combustion as irreversible, we find quite a different world from that described according to classical concepts and theories.

In this book the theory and its applications are explained mainly according to the experimental results obtained by the author and his coworkers, because there are very little experiments carried out under the concept of an irreversible process. Therefore, only a few references are made of books on combustion science, since these books, using the classical concept, are written in a philosophy which is quite different from that used in this book.

The dimensions of the experimental apparatus illustrated in this book are shown as exactly as possible, because the irreversible phenomena proceed accompanied by some stochastic phenomena and the probabilities also depend on the quantity of materials used in the phenomena.

The theory is very simple and can easily be applied to the practical calculation and industrial purposes. The author is, however, only afraid, if he had preached Buddha, or carried owls to Athens.

CONTENTS

1 **Introduction** 1

2 **Classical Ignition Theories** 3
   2.1 Thermal Explosion Theory 3
   2.2 Chain-Branching Kinetics 8
   2.3 Induction Period of Ignition and Ignition Mechanism 14

3 **Stochastic Theory of Irreversible Phenomena** 17
   3.1 Irreversible Process 18
   3.2 Fluctuation in Irreversible Process 21
   3.3 Stochastic Theory for Irreversible Phenomena 23

4 **Nucleation in Phase Transition** 27
   4.1 Classical Nucleation Theory 27
   4.2 Stochastic Nucleation Theory at Ebullition of Liquids 29
       4.2.1 Experiments for measuring the ebullition
             induction period 30
       4.2.2 Probability of ebullition 31
       4.2.3 Activation energies of ebullition 32
       4.2.4 Frequency factor 34
       4.2.5 Ebullition mechanism 35
   4.3 Stochastic Theory for Ice Formation in Water 36
       4.3.1 Experiments 36
       4.3.2 Probability of ice formation 37
       4.3.3 Activation energies of ice formation 40
       4.3.4 Ice formation under a radiation of high energy 42

5 **Shock Tubes** 45
   5.1 Shock Waves 45
   5.2 Simple Shock Tube 49

Contents

6 Stochastic Ignition Theory 57
   6.1 Probability of Ignition Behind Shock Waves 57
   6.2 Spontaneous Ignition in a Hydrogen–Oxygen Mixture
        Behind Shock Waves 61
        6.2.1 Experimental method 62
        6.2.2 Experimental results 63
        6.2.3 Explosion limits 68
        6.2.4 Quantity effect on the ignition 71
        6.2.5 Reaction mechanism 72
   6.3 Spontaneous Ignition in Hydrocarbon–Air Mixtures 74
        6.3.1 Ignition probability and mechanism
                in paraffin–fuel–air mixtures 74
        6.3.2 Ignition limits of hydrocarbon fuel-air mixtures 84
        6.3.3 Influences of tetraethyl lead on the ignition 87
   6.4 The Chain-Branching Kinetics and Stochastic
        Ignition Theory 97

7 Ignition in a Fuel Spray 99
   7.1 The Most Inflammable State of a Fuel–Air Mixture 99
        7.1.1 Ignition probability in lean n-octane–air mixtures 100
        7.1.2 The most inflammable state of the mixture 102
   7.2 Ignition Probability in a Fuel Spray 103
        7.2.1 Partial ignition probability 103
        7.2.2 Experiments using a shock tube 104
        7.2.3 Induction period of ignition in the fuel spray 107
        7.2.4 Distribution of the ignition probability 108
        7.2.5 Ignition and combustion in a fuel spray 110
        7.2.6 Conclusions 112

8 Ignition by Electric Sparks 113
   8.1 Igniter Using Induction Coils 114
        8.1.1 Spark ignition and characteristics of electric discharge . . 115
        8.1.2 Ignition and gap distance between the electrodes 115
   8.2 Application of the Stochastic Ignition Theory
        to Spark Ignition 116
        8.2.1 Stochastic theory of ignition by external energies 117
        8.2.2 Experiments 118
        8.2.3 Action of capacity and inductance components 120
        8.2.4 Gap distance and ignition 125

9 Nonequilibrium State 137
   9.1 Adiabatic Combustion Temperature at Equilibrium 139
        9.1.1 Reaction process and dissociation 139
        9.1.2 Reaction heat and adiabatic combustion temperature . . . 141
        9.1.3 Adiabatic combustion temperature of
                propane–oxygen mixture 142

	9.2	Investigation of Flame by Probe Method	146
		9.2.1 Electron temperature and ion temperature	146
		9.2.2 Langmuir probe method	148
		9.2.3 Double probe method	149
		9.2.4 Investigation of ionization in a standing flame by a double probe method	152
		9.2.5 Investigation of ionization in a propagating flame by a double probe method	153
	9.3	Investigation of Flame by a Laser Light Scattering Method	160
		9.3.1 Laser light scattering method	161
		9.3.2 Notices to be considered at the measurement	164
		9.3.3 Investigation of the ionization in a propagating flame using the laser light scattering method	165
	9.4	Nonequilibrium and Heterogeneous State behind Shock Waves	174
		9.4.1 Spectroscopic temperature measurement method	174
		9.4.2 Temperature in argon gas behind shock waves	175
		9.4.3 A stochastic phenomenon behind shock waves	182
		9.4.4 Nonequlibrium and heterogeneous state behind shock waves	183
10	**Interaction Between Combustion and Pressure or Shock Waves**		187
	10.1	Propagation of Combustion Waves	187
	10.2	Flame Propagation as an Irreversible Phenomenon	191
		10.2.1 Theoretical treatment	191
		10.2.2 Experiments of flame propagation	192
		10.2.3 Transition from a laminar flame to a turbulent flame	198
	10.3	Interaction between Combustion and Shock Waves	199
		10.3.1 Interaction modes in the experiments	199
		10.3.2 Experimental results	202
	10.4	Resonance Pulse Jet Engine (Schmidtrohr)	216
		10.4.1 Construction and action mechanism	216
		10.4.2 Performance	218
		10.4.3 Ignition and combustion	220
11	**Gaseous Detonation Waves**		223
	11.1	The Classical Theories of Detonation Waves	223
		11.1.1 Macroscopic structure of detonation waves	224
		11.1.2 Microscopic structure of detonation waves	228
		11.1.3 Transition from deflagration to detonation waves	232
	11.2	Detonation Waves as Irreversible Phenomena	233
		11.2.1 Interaction between converging shock and combustion waves	233
		11.2.2 Cellular structure formation as a stochastic phenomenon	241
		11.2.3 Interaction between shock and detonation waves	251

11.3 Initiation of Detonation Waves.................................... 255
    11.3.1 Transition from combustion to detonation............ 257
    11.3.2 Transition from shock to detonation waves.......... 270
11.4 Propagation of Detonation Waves................................. 281
    11.4.1 Detonation propagating in mixtures having different temperatures............................. 282
    11.4.2 Deceleration of the detonation propagation.......... 289
    11.4.3 Concluding remarks on the propagation velocity of the detonation wave..................... 295
11.5 Ionization of Gases Behind Detonation Wave..................... 296
    11.5.1 Investigation applying a double probe method...... 297
    11.5.2 Investigation by a laser light scattering method..... 302

12 Industrial Applications of Detonation Waves 307
12.1 Imploding Shock Waves 308
    12.1.1 Theoretical calculation of imploding shock waves ... 308
    12.1.2 Radially divergent detonation wave 314
    12.1.3 Cylindrically and spherically imploding detonation waves of small size 319
    12.1.4 Spherically imploding detonation waves of large size.. 336
    12.1.5 Spherically imploding detonation waves initiated by two-step divergent detonation 343
    12.1.6 Nuclear fusion applying spherically imploding detonation waves 354
    12.1.7 Nuclear fusion rocket engines applying imploding detonation waves 358
12.2 Hypersonic Combustion for RAM Jet Engine................... 361
    12.2.1 Standing detonation waves........................ 362
    12.2.2 Advantage of the ram jet using standing detonation waves 362
    12.2.3 Experimental apparatus........................... 364
    12.2.4 Experiments..................................... 365
12.3 Shock Tubes Driven by Detonation Waves 371
    12.3.1 Rarefaction waves behind detonation waves (Taylor expansion).............................. 372
    12.3.2 Shock tube directly driven by detonation waves..... 373
    12.3.3 Shock tube using free piston driven by detonation waves............................. 376

References.. 389

Author Index.. 397

Subject Index... 401

1

Introduction

According to the basic concepts of thermodynamics[1], a reversible process is that in which a phenomenon proceeds with an infinitely slow velocity. Practically such a process can never take place and we observe only irreversible processes in which some phenomena proceed with a certain finite velocity.

On the other hand, in order to analyze some physical or chemical phenomena quantitatively, it is convenient to apply some mathematical formula introduced from basic sciences. Such basic mathematical formulae are, however, in general introduced from the basic laws of sciences which can essentially be applied only to reversible processes, namely in equilibrium or quasiequilibrium states.

These mathematical formulae are very often applied also to irreversible phenomena under the assumption that the observed irreversible phenomena can approximately be treated as reversible ones. The formulae can be applied to phenomena proceeding very slowly, but never to distinctly irreversible phenomena, for example, rupture of material, electric discharge, or earthquake.

The combustion is an oxidation of materials being accompanied by light emission, heat release, and pressure or shock waves. The combustion phenomena have long time been investigated by many scientists and engineers not only from their academic interests but also with intentions of the practical and industrial applications and many books as well as reports of the phenomena have been published. In these books and reports the combustion phenomena have mainly been explained according to the theories based on the first law of thermodynamics expressed in mathematical formulae which can be applied only to reversible processes, though the combustion phenomena are distinctly irreversible.

In this book, comparing with the classical theories, ignition, explosion, combustion, and detonation waves are explained according to an irreversible theory based on the second law of thermodynamics. The explanations and results introduced from the irreversible theory are, therefore, quite different from those described in the classical books, but so clear and simple that they can easily be applied to the understanding, investigation, and numerical estimation of the combustion phenomena as well as detonation waves for practical and industrial purposes.

In combustible mixtures or materials some ignitions take place and then flames propagate from the ignition points, being accompanied by light emission, heat release, and shock waves. As the flame propagation is regarded as a succeeding ignition, the ignition phenomena are first to be studied for understanding the combustion phenomena. In this book, after the explanation of explosion according to the classical theories, a stochastic theory for irreversible phenomena, its application to nucleation and then shock tube as the experimental apparatus of ignition in gases are explained. Next the theoretical and experimental results of spontaneous and spark ignitions in gaseous mixtures are explained mainly according to the stochastic ignition theory. Then combustion waves, ionization in flames, detonation waves, interaction between shock and combustion waves are described as irreversible phenomena. Subsequently, some practical and industrial applications of the stochastic ignition theory and detonation waves are also proposed.

2

Classical Ignition Theories

As the basic phenomenon of combustion the ignition in combustible mixtures has been investigated by many scientists.[2-4] There are two types of ignition. One of them is the spontaneous ignition which is also called explosion and the other is the ignition by an external energy. The most basic ignition is the explosion in a homogeneous mixture and for it there have been two well-known theories, i.e., thermal explosion theory and chain-branching kinetics.

After a combustible mixture, a mixture of fuel and air or oxygen, for example, a propane–air mixture, is brought to a state in which an ignition can take place, none of remarkable changes, for example, pressure or temperature rise in the mixture, is observed for a certain short period, and then some phenomena like pressure or temperature rise, light emission, or heat release occur. In general, the ignition is defined by the instant in which pressure or temperature begins to rise, or the first light emission is observed. The period from the instant where the mixture is brought into the state in which an ignition can take place to the ignition instant is called induction period of ignition or ignition delay as shown in Fig. 2.1 with t_{ind}. Either in the thermal explosion theory or in the chain-branching kinetics, it is assumed that all phenomena during the induction period proceed steadily, since the pressure as well as the temperature is kept almost constant. What phenomena proceed during the induction period comes in question.

2.1 Thermal Explosion Theory[5]

It has been believed that the physical state of a combustible mixture must be divided into two regions, namely, a region where a spontaneous ignition is always observed and the other having no spontaneous ignition. The border-line between both the regions is called explosion limit and one has discussed what mechanism defines such explosion limits.

In order to explain the mechanism of spontaneous ignition, i.e., explosion in a gaseous mixture, the thermal explosion theory was proposed by Van't Hoff

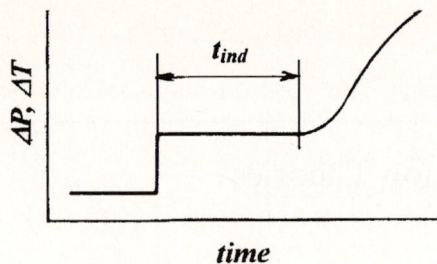

Fig. 2.1. Ignition process

and Semenoff. A combustible mixture is filled in a vessel keeping the same temperature T_0 of the mixture at the initial state. During some chemical reactions proceed in the mixture, increasing the mixture temperature, some of the released heat is transformed through the vessel wall to the outside, keeping the wall temperature T_0 constant. It is also assumed that the mixture keeps a homogeneous state, namely the specific entropy is everywhere equal.

Let us assume

V, the vessel volume (expressed by mole number of the mixture)
T_0, the initial mixture temperature (=the temperature of the vessel wall)
P, the mixture pressure
n, the concentration of reaction product
t, time,

then the reaction rate is expressed by the following equation, where E is called activation energy,

$$\frac{dn}{dt} = f(P)\exp\left(-\frac{E}{RT}\right). \tag{2.1}$$

In a reaction of second order,

$$f(P) \cong \beta_r P^2, \tag{2.2}$$

where β_r is a proportional constant. Further, if

q, reaction heat per unit mole mixture
Q_1, whole reaction heat,

then

$$\frac{dQ_1}{dt} = q\frac{dn}{dt}V = qVf(P)\exp\left(-\frac{E}{RT}\right). \tag{2.3}$$

A part of the reaction heat heats the mixture, while the other heat is transformed to the outside through the vessel wall. If the heat Q_2 transferred outside is assumed to be proportional to the difference between the mixture temperature T and the wall temperature T_0,

$$\frac{dQ_2}{dt} = C(T - T_0), \tag{2.4}$$

where C is the heat conductivity of the mixture depending on the form and size of the vessel, especially proportional to the surface area of the vessel, but in this case we assume it constant during the process. Near the wall the mixture temperature has a gradient, but we take here its mean value. Under these assumptions we discuss the conditions for the explosion. Plotting the released heat per unit time dQ_1/dt in combustible mixture having an initial pressure of P against the mixture temperature T, we obtain a curve C.H. in the diagram of Fig. 2.2.

According to (2.4), the heat dQ_2/dt transferred outside is expressed as a straight line in the same diagram. If the line is expressed by line 1, the mixture temperature rises to the cross point T_c of the curve dQ_1/dt with the straight line 1, but then decreases and no ignition takes place, as the heat transferred outside is larger than the reaction heat. If the heat transferred outside is expressed by the line 2, the reaction heat is always larger than the transferred heat and an explosion takes place. As the limit of both the cases, the straight line of the transferred heat should be tangential to the curve of reaction heat, as shown by line 3. Then the contact point T_1 of the curve with the straight line must be the explosion limit.

Expressing these relations in mathematical formulae

$$\left(\frac{dQ_1}{dt}\right)_{T=T_1} = \left(\frac{dQ_2}{dt}\right)_{T=T_1}, \tag{2.5}$$

$$\left(\frac{\delta}{\delta T}\frac{dQ_1}{dt}\right)_{T=T_1} = \left(\frac{\delta}{\delta T}\frac{dQ_2}{dt}\right)_{T=T_1}. \tag{2.6}$$

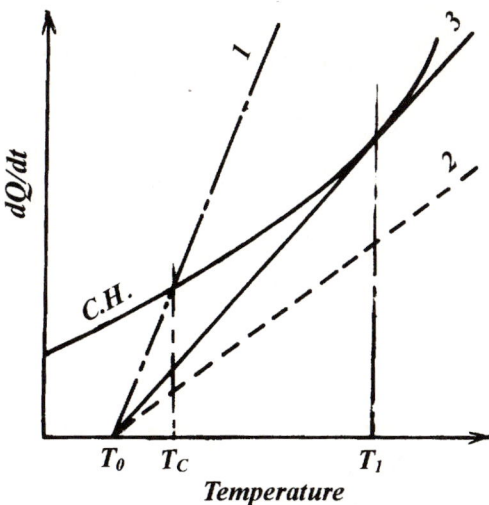

Fig. 2.2. Heat release at the thermal explosion. (C.H.) Heat release by the reactions, (1, 2, and 3) heat loss by conduction

2 Classical Ignition Theories

From (2.3)–(2.5) we obtain

$$qVf \exp\left(-\frac{E}{RT}\right) = C(T_1 - T_0), \qquad (2.7)$$

then from (2.6)

$$\frac{E}{RT_1^2} qVf \exp\left(-\frac{E}{RT}\right) = C. \qquad (2.8)$$

By division of (2.7) by (2.8)

$$\frac{RT_1^2}{E} = T_1 - T_0,$$

then

$$T_1 = \frac{E}{2R} \pm \left(-T_0 \frac{E}{R} + \frac{E^2}{4R^2}\right)^{1/2}.$$

As the equation having plus sign means higher temperature and meaningless in this case, we take only that with minus sign, namely

$$T_1 = \frac{E}{2R} - \left(-T_0 \frac{E}{R} + \frac{E^2}{4R^2}\right)^{1/2} = \frac{E}{2R} - \frac{E}{2R}\left(1 - T_0 \frac{4R}{E}\right)^{1/2}$$

$$\approx \frac{E}{2R} - \frac{E}{2R}\left\{1 - \frac{2RT_0}{E} - \frac{2R^2 T_0^2}{E^2}\right\} \quad \text{(as } T_0 \text{ is much smaller than } E/R\text{)}$$

$$= T_0 + \frac{T_0^2 R}{E}.$$

Substituting this relation into (2.7)

$$qVf \exp\left\{-\frac{E}{RT_0(1 + RT_0/E)}\right\} = C\frac{RT_0^2}{E}.$$

As RT_0/E is much smaller than 1, we approximately obtain the following relation:

$$1 \bigg/ \left(1 + \frac{RT_0}{E}\right) \approx 1 - \frac{RT_0}{E}.$$

Therefore

$$qVf \exp\left\{-\frac{E}{RT_0}\left(1 - \frac{RT_0}{E}\right)\right\} = eqVf \exp\left(-\frac{E}{RT_0}\right), \qquad (2.9)$$

$$eqVf \exp\left(-\frac{E}{RT_0}\right) = \frac{CRT_0^2}{E}.$$

Substituting the relation $f = \beta_r P^2$ of the second-order reaction into the equation and expressing in a logarithmic formula:

2.1 Thermal Explosion Theory

$$\ln eqV\beta_r + 2\ln P - \frac{E}{RT_0} = \ln \frac{CR}{E} + 2\ln T_0. \qquad (2.10)$$

Using the pressure P_{cr} at the explosion limit, we obtain the following relation:

$$\ln \frac{P_{cr}}{T_0} = \frac{E}{2RT_0} - \frac{1}{2}\ln\left(\frac{eqV\beta_r E}{CR}\right) = \frac{E}{2RT_0} + \text{const.} \qquad (2.11)$$

From this equation we can obtain the relation among the pressure P_{cr} and temperature T_0 at the explosion limit and the activation energy E. In other reactions like single molecule or three molecule reaction the formula should be somewhat reformed, but in general we observe mostly the second-order reaction. If we plot the values of $\ln (P_{cr}/T_0)$ against $1/T_0$, we obtain a straight line, from whose gradient the activation energy E is calculated.

According to the thermal explosion theory the explosion limit is thus expressed in a pressure–temperature diagram with a monotone curve as shown in Fig. 2.3. In order to suppress the explosion, some gases having large specific heat and higher heat conductivity should be mixed into the combustible mixture.

In most combustible mixtures such an explosion limit shown in Fig. 2.3 is observed, but in someones like hydrogen–oxygen mixture we observe a curve having an inverse S-form as shown in Fig. 2.4. As such explosion limit having an inverse S-form curve could not be explained by the thermal explosion theory. Semenoff, Hinshelwood, and others proposed the chain-branching kinetics.

The activation energy is now to be explained. In a substance, although in an equilibrium state having a certain finite physical state, a certain definite temperature and pressure, the particles composing the substance have different motion and never keep their velocity and direction constant, namely have never constant energy. In a substance having a constant equilibrium temperature T, for example, it has an energy distribution, that is, so-called Boltzmann's distribution as shown in Fig. 2.5. The particles having energies higher than a certain energy E_0 can participate in the reaction like a combustion, whereas the other particles play no role for the reaction. The reaction

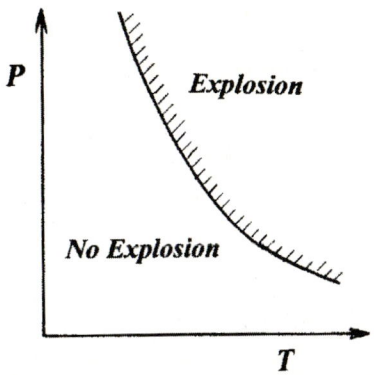

Fig. 2.3. Explosion limit at a thermal explosion

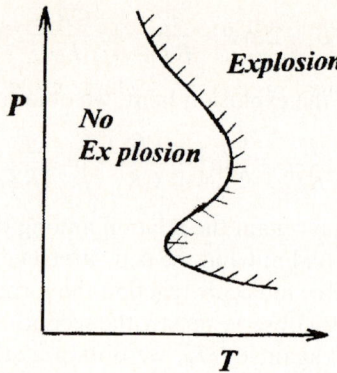

Fig. 2.4. Explosion limit at a chain branching

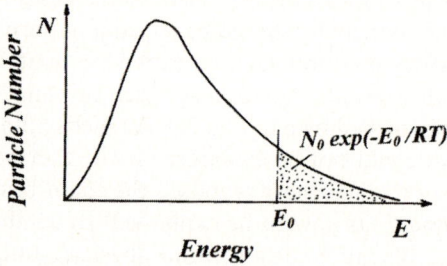

Fig. 2.5. Maxwell–Boltzmann distribution

rate is thus proportional having an energy higher than E_0 as shown in Fig. 2.5, while the probability that a particle has an energy higher than E_0 is proportional to $\exp(-E_0/RT)$, i.e.,

$$\frac{dn}{dt} = A \exp\left(-\frac{E_0}{RT}\right), \tag{2.12}$$

which is called Arrhenius' equation.

2.2 Chain-Branching Kinetics[6–9]

As mentioned above, we observe an explosion limit having an inverse S-form curve with a so-called explosion peninsula in a hydrogen–oxygen mixture. In order to explain the limit, the chain-branching kinetics has been developed. Namely, the chemical kinetics has been applied to explain the phenomenon, assuming the explosion as a chemical reaction composed of several simple elementary single-, two-, or three-molecule reactions.

2.2 Chain-Branching Kinetics

The chemical kinetics is based on the law of mass action, in which the reaction rate is proportional to the mass of each element. This theory, however, can be applied only to reversible processes.[6] A single molecule reaction (first-order reaction) is expressed by

$$(\mathrm{I})\ A \rightarrow x + y \ldots \ldots$$

and a two molecule reaction (second-order reaction) by

$$(\mathrm{II})\ A + B \rightarrow x + y \ldots \ldots$$

The decrease rate of one element in the left-side or increase of one element in the right-side means the reaction rate. Expressing the concentration of each element by square bracket and rate coefficient of reaction by k_n, the rate of the single molecule (first order) reaction is expressed by the following equation:

$$-\frac{d[A]}{dt} = k_1 [A]. \tag{2.13}$$

And that of two molecule (second order) reaction by

$$-\frac{d[A]}{dt} = k_2 [A][B]. \tag{2.14}$$

Assuming the initial concentration of each element to be $[A_0]$ and $[B_0]$, we obtain the following equation for the first-order reaction,

$$-\frac{d[A]}{[A]} = k_1 dt. \tag{2.15}$$

As $[A] = [A_0]$ at $t = 0$,

$$[A] = [A_0] \exp(-k_1 t). \tag{2.16}$$

For the second-order reaction

$$-\frac{d[A]}{dt} = k_2 \{[A_0] - [A]\} \{[B_0] - [B]\}. \tag{2.17}$$

As $[A_0] = [B_0]$

$$-\frac{d[A]}{dt} = k_2 \{[A_0] - [A]\}^2,$$

$$-\frac{[A]}{\{[A_0] - [A]\}[A_0]} = k_2 t. \tag{2.18}$$

We can approximately take the average concentration $\overline{[A]}$ or $\overline{[B]}$ of each element during the reaction:

$$-\frac{d[A]}{dt} \propto k_2 \overline{[A]}\overline{[B]}. \tag{2.19}$$

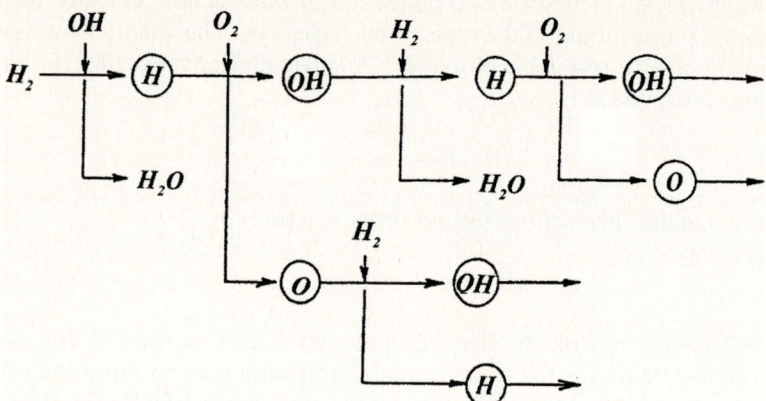

Fig. 2.6. Explosion reaction model in a mixture of hydrogen–oxygen

If only the first stage of reaction comes in question, we can here approximately take the initial concentration of the elements except that calculated as the reaction rate.

Assuming that the reaction of hydrogen with oxygen $2H_2 + O_2 \rightarrow 2H_2O$ is composed of several elementary reactions supported by some activated atoms or radicals, we can assume the following reaction system as an example, whose scheme is shown in Fig. 2.6.

(1) Chain-branching reaction:

$$[1]\ OH + H_2 \rightarrow H_2O + H + 12\ kcal \qquad (k_1)$$
$$[2]\ H + O_2 \rightarrow OH + O - 14\ kcal \qquad (k_2)$$
$$[3]\ O + H_2 \rightarrow OH + H + 0.5\ kcal \qquad (k_3)$$

(2) Chain-breaking reaction:

$$[4]\ H + OH + M \rightarrow H_2O + M, \qquad (k_4)$$

where M is the third element, for example, a metal molecule.

In reactions [2] and [3], not only activated atoms H and O appear in the right-hand side, but also OH is formed together with O or H. When the activated radical OH is formed more than that decreased in the reaction [4], the reaction accelerated to the explosion by increasing the concentration OH. When the activated atoms and radicals are formed in the mixture more than a certain definite concentration, an explosion takes place, after which the chain-branching reactions do not play any role for the reaction, we should only consider the balance of the increase of the activated atoms and radicals in the reactions [2] and [3] with those decreased in the reaction [4], so long as the explosion limit comes in question.

2.2 Chain-Branching Kinetics

Assuming steady reactions, the reaction rates can be calculated as follows:

[I] $\dfrac{d[H]}{dt} = k_1[OH][H_2] - k_2[H][O_2] + k_3[O][H_2] - k_4[H][OH][M]$

[II] $\dfrac{d[O]}{dt} = k_2[H][O_2] - k_3[O][H_2]$

[III] $\dfrac{d[OH]}{dt} = -k_1[OH][H_2] + k_2[H][O_2] + k_3[O][H_2] - k_4[H][OH][M]$.

Applying the conditions of steady reactions $\dfrac{d[H]}{dt} = 0, \dfrac{d[O]}{dt} = 0$, and $\dfrac{d[OH]}{dt} = 0$,

$[O] = \dfrac{k_2[H][O_2]}{k_3[H_2]}$ from [II]

$2k_1[OH][H_2] - 2k_2[H][O_2] = 0$ from [I] and [III], then

$[OH] = \dfrac{k_2[H][O_2]}{k_1[H_2]}$.

Substituting the relations above into [I], we obtain the following relations:

$[H] = \dfrac{k_1[H_2]}{k_4[M]}$

$[O] = \dfrac{k_1 k_2 [H_2][O_2]}{k_3 k_4 [H_2][M]} = \dfrac{k_1 k_2 [O_2]}{k_3 k_4 [M]}$

$[OH] = k_2 \dfrac{[O_2]}{k_4[M]}$.

From [1], [4], and the equations described above the formation of H_2O is

$\dfrac{d[H_2O]}{dt} = k_1[OH][H_2] + k_4[H][OH][M] = \dfrac{2k_1 k_2 [H_2][O_2]}{k_4[M]}$.

Thus, we obtain the rate of overall reaction, but still know nothing of the explosion limit.

Considering the repeating of chain-branching reactions [1], [2], and [3], a reaction scheme as shown in Fig. 2.6 can be proposed. Putting H_2 A or C, H_2O B and activated radicals or atoms OH, O, H R, the reaction can be expressed by the following schemes:

$$R + A \xrightarrow{k_1} B + R \qquad (2.20)$$

$$R + C \xrightarrow{k_2} 2R + D \qquad (2.21)$$

$$R + E \xrightarrow{k_3} E' \qquad (2.22)$$

Besides, considering an active dissociation or first-order reaction

$$U \xrightarrow{k_0} nR. \qquad (2.23)$$

In a steady reaction

$$\frac{d[R]}{dt} = k_0[U] + k_2[R][C] - k_3[R][E] = 0.$$

Thus

$$[R] = \frac{k_0[U]}{k_3[E] - k_2[C]} \quad (2.24)$$

The reaction rate of B is obtained from (2.20)

$$\frac{d[B]}{dt} = k_1[R][A] = \frac{k_1 k_0[A][U]}{k_3[E] - k_2[C]}. \quad (2.25)$$

So long as the denominator is positive, the reaction proceeds with a certain finite rate, but if it is zero, that is,

$$k_3[E] - k_2[C] = 0$$

the reaction proceeds with an infinitely large rate, namely it means the explosion limit.

According to the proposal of Semenoff,[8] if
n_0, number of initiating reaction
α, probability of chain breaking
β, probability of chain branching
The reaction rate w is expressed by the following equation:

$$w = \frac{n_0}{\beta - \alpha}. \quad (2.26)$$

The condition $\beta = \alpha$ defines the explosion limit. The explosion limit of the hydrogen–oxygen mixture has an inverse S-form in a pressure P vs. temperature T diagram, as shown in Fig. 2.7 having a so-called explosion peninsula in the low-pressure region. Dividing the curve of the explosion limit into three parts by two turning points, one calls them first, second, and third limit corresponding to that in the lowest pressure region, that in the middle pressure region, and that in highest pressure region, respectively.

Semenoff explains as follows:

(1) At the first limit the activated particles (chain carriers) are destructed on the vessel wall
(2) At the second limit they are destructed in the gas phase
(3) The third limit is that of thermal explosion

At the first limit the metal of the vessel wall destructs the chain carriers and breaks the chain-branching reaction. In order to prove the theory, he points out

1. The explosion limit is enlarged to the region having lower pressure and temperature with enlargement of the mixture vessel, as the mixture quantity

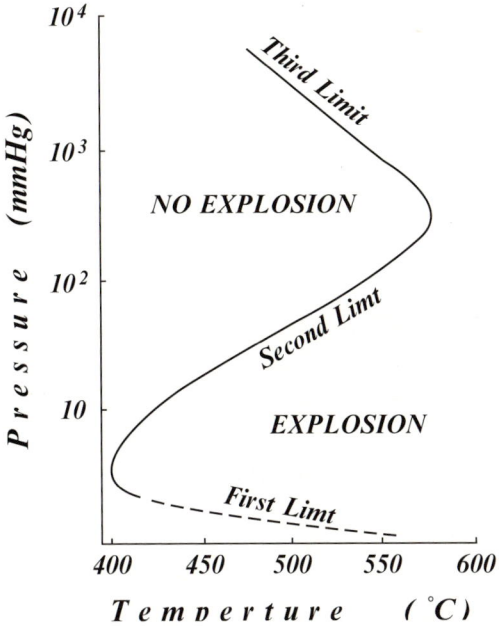

Fig. 2.7. Explosion limits of stoichiometric hydrogen–oxygen mixture[10, 11]

increases against the surface area of the vessel and the collision probability of the chain carriers with the surface decreases
2. The explosion limit is also enlarged, if the vessel surface is coated with KCl to avoid the direct contact of the mixture with the wall
3. The knocking phenomenon in spark ignition engines believed to be a kind of spontaneous ignition is suppressed by adding tetra-ethyl-lead to the fuel, as the metal lead hinders the spontaneous ignition in the engine combustion chamber.

The chain-branching kinetics in the combustion phenomena is still now generally supported by the explanation, experimental and empirical results described above.

This chain-branching kinetics, however, can theoretically be applied only to reversible processes, while the explosion is an irreversible phenomenon. It is, therefore, doubtful to apply the theory to such irreversible phenomena as explosion and combustion, even if approximately. The thermal explosion theory is also introduced under the assumption of equilibrium state and cannot be applied to irreversible phenomena. We can here show an experimental result contradicting the chain-branching kinetics.

Let small balls of platinum having different diameters from 1.0 mm to 6.0 mm heated to different temperatures between 800°C and 1,400 °C fall in a mixture of 3.0% benzene in air or that of 20% hydrogen in air, observing the ignition. From the experimental results the relation between the ignition temperature and the

14   2 Classical Ignition Theories

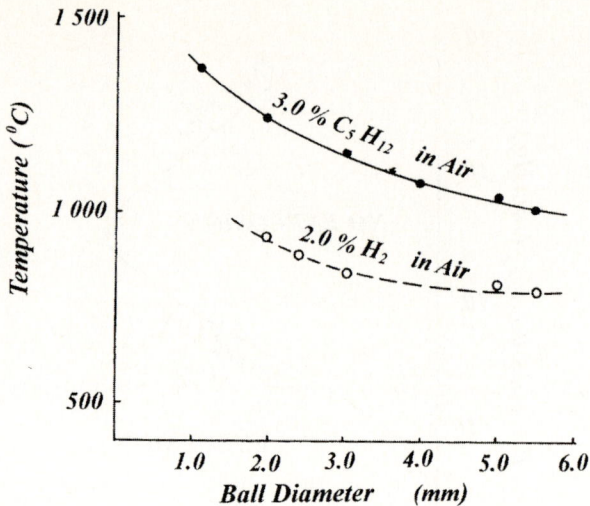

Fig. 2.8. Ignition temperature with respect to the diameter of the platinum balls falling in the mixture[12]

diameter of the ball in both the mixtures are illustrated in Fig. 2.8. The result suggests that the ignition temperature decreases with enlargement of the ball that means that the mixture ignites more easily, when the mixture contacted with metal surface increases. This result contradicts the explanation by the chain-branching kinetics.

2.3 Induction Period of Ignition and Ignition Mechanism

The period from the instant when a combustible substance or mixture is put in a state where it can ignite to that of ignition is called induction period of ignition or ignition delay. When a combustible mixture is compressed adiabatically or by shock waves, it keeps the compressed state for a while, then its pressure or temperature rises first slowly, but then rapidly and it emits light. The instant when the mixture begins to change its state or to emit light, or that when the concentration of some activated radicals like OH increases and reaches a certain definite value, can be defined to be the ignition instant. We can, thus, measure the induction period by observing the light emission, pressure, temperature, or concentration of the activated radicals.

As such induction period of ignition is able to be measured easily; experiments for measuring the ignition induction period are often carried out. In Fig. 2.9 a schematic diagram of the reaction between the logarithm of the induction period $\ln t_i$ and the reciprocal mixture temperature $1/T$ after the compression is illustrated. As the induction period t_i is reciprocally proportional to the reaction rate, we obtain the following relation according to (2.12):

2.3 Induction Period of Ignition and Ignition Mechanism

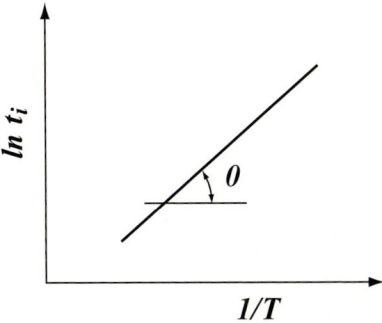

Fig. 2.9. Schematic relation between logarithm of ignition induction period $\ln t_i$ and reciprocal mixture temperature $1/T$

$$t_i = C \exp \frac{E_e}{RT}, \quad (2.27)$$

where C is a proportional constant and E_e the effective activation energy.[13] According to the relation:

$$\tan \theta = \frac{E_e}{R}, \quad (2.28)$$

where θ is the inclination of $\ln t_i$ to the axis of $1/T$, we can obtain the activation energy E_e.

In order to investigate and analyze the ignition mechanism according to the classical theories, a combustible mixture is ignited by compressing and heating through some methods and

1. The ignition induction period is measured to estimate the activation energy
2. While the intermediates are measured during the induction period
3. Considering some elementary reactions having the intermediates observed during the induction period and the same activation energy as that estimated from the above observed ignition induction period, the ignition mechanism is estimated.

It is, however, quite difficult to find all elementary reactions, as all of intermediates cannot be detected.

For example, at the first stage of research on the ignition of methane CH_4, only six elementary reactions were proposed, but as no correct mechanism was found, so many elementary reactions have been proposed as listed in Table 2.1. Using some computers such complicated reactions can be calculated, nevertheless the results do still not agree with the results experimentally observed.

The classical theories can essentially be applied only to the reversible phenomena. For the irreversible phenomena like ignition, explosion, and combustion, other theories must be proposed.

Table 2.1. Methane–oxygen reaction scheme[14] $k = A\, T \exp(-E/T)$ (m mol s)

no.	reaction	A	N	E (K)
1	$CH_4 + M \rightarrow CH_3 + H + M$	1.41	0.0	44,500.0
2	$CH_4 + OH \rightarrow CH_3 + H_2O$	3.47	3.08	1,010.0
3	$CH_4 + H \rightarrow CH_3 + H_2$	2.20	3.0	4,400.0
4	$CH_4 + O \rightarrow CH_3 + OH$	1.20	2.1	3,810.0
5	$CH_3 + HO_2 \rightarrow CH_3O + OH$	3.24	0.0	0.0
6	$CH_3 + O_2 \rightarrow CH_3O + O$	2.51	0.0	14,600.0
7	$CH_3 + OH \rightarrow HCHO + H_2$	3.98	0.0	0.0
8	$CH_3 + O \rightarrow HCHO + H$	7.00	0.0	0.0
9	$CH_2O + O_2 \rightarrow HCHO + HO_2$	1.00	0.0	3,020.0
10	$CH_3O + M \rightarrow HCHO + H + M$	5.0	0.0	10,600.0
11	$HCHO + OH \rightarrow CHO + H_2O$	7.59	0.0	85.5
12	$HCHO + H \rightarrow CHO + H_2$	3.31	0.0	5,280.0
13	$HCHO + O \rightarrow CHO + OH$	5.01	0.0	2,310.0
14	$HCHO + M \rightarrow CHO + H + M$	3.31	0.0	40,800.0
15	$CHO + OH \rightarrow CO + H_2O$	1.00	0.0	0.0
16	$CHO + H \rightarrow CO + H_2$	2.00	0.0	0.0
17	$CHO + O \rightarrow CO + OH$	1.00	0.0	0.0
18	$CHO + O_2 \rightarrow CO + HO_2$	3.98	0.0	3,520.0
19	$CHO + M \rightarrow CO + H + M$	1.42	0.0	8,460.0
20	$CO + OH \rightarrow CO_2 + H$	4.40	1.5	−373.0
21	$CO + O + M \rightarrow CO_2 + M$	5.89	0.0	2,060.0
22	$CO + HO_2 \rightarrow CO_2 + OH$	1.51	0.0	11,900.0
23	$CO + O_2 \rightarrow CO_2 + O$	3.16	0.0	18,900.0
24	$H + O_2 \rightarrow OH + O$	2.20	0.0	8,450.0
25	$O + H_2 \rightarrow OH + H$	1.80	1.0	4,480.0
26	$OH + OH \rightarrow O + H_2O$	6.30	0.0	550.0
27	$OH + H_2 \rightarrow H + H_2O$	2.20	0.0	2,590.0
28	$H + H + M \rightarrow H_2 + M$	2.60	−1.0	0.0
29	$O + O + M \rightarrow O_2 + M$	1.90	0.0	−900.0
30	$H + O + M \rightarrow OH + M$	3.60	−1.0	0.0
31	$OH + H + M \rightarrow H_2O + M$	4.06	−.0	0.0
32	$H + O_2 + M \rightarrow HO_2 + M$	5.00	0.0	−500.0
33	$H + HO_2 \rightarrow H_2 + O_2$	2.50	0.0	350.0
34	$H + HO_2 \rightarrow OH + OH$	2.50	0.0	950.0
35	$H + HO_2 \rightarrow O + H_2O$	9.00	0.5	2,000.0
36	$OH + HO_2 \rightarrow H_2O + O_2$	5.00	0.0	500.0
37	$O + HO_2 \rightarrow OH + O_2$	6.30	0.0	350.0

3
Stochastic Theory of Irreversible Phenomena

According to the second law of thermodynamics, an irreversible process must at least once pass a state of minimum entropy and this means the existence of a heterogeneous and nonequilibrium state which governs the process and causes some fluctuating phenomena. As the fluctuations are governed by a certain probability, a stochastic theory should be applied to investigate and analyze distinctly irreversible phenomena, for which the classical theories developed for a reversible process cannot be used. Applying the stochastic theory developed for irreversible phenomena to experimental results obtained in the spontaneous ignition in combustible mixtures behind shock waves, not only many empirically observed phenomena are well and simply explained, but also some unknown phenomena which have not been found by the classical theories are unveiled.

In different fields of physics, chemistry, and engineering some basic theories which are introduced from the first law of thermodynamics and which are essentially valid only for reversible processes are often applied to investigate irreversible phenomena. As the reversible process is an ideal one and practically never takes place, we observe only irreversible processes. It is, therefore, very important to confirm, if the theories can be applied to irreversible processes, even approximately, or not.

This section first presents the differences between both the reversible and irreversible processes according to the second law of thermodynamics and clarifies that the theories are valid only for reversible processes, for example, chemical kinetics, nucleation theory, etc., cannot be applied to distinctly irreversible processes, even approximately, and then explains a stochastic theory developed for the irreversible process, reporting the results experimentally obtained by application of the stochastic theory to spontaneous ignition in combustible mixtures behind shock waves in a shock tube.

3 Stochastic Theory of Irreversible Phenomena

3.1 Irreversible Process

The definition of a reversible process in thermodynamics is the process proceeding with an infinitely slow velocity, while that of the irreversible process is the process with a finite velocity.[1] Practically there is no reversible process and we observe only irreversible ones.[15, 16]

The first law of thermodynamics applied everywhere in phenomena influenced by heat or temperature is an expression of energy conservation and often expressed as a practical formula with the following equation:

$$dU = dQ - PdV \tag{3.1}$$

or with those modified from this equation, where U is energy, Q the heat, P the pressure, and V the volume of the system concerned. They are used everywhere in the fields of physics, chemistry, and engineering. These formulae are, however, theoretically valid only for the process proceeding with an infinitely slow velocity, that is, only for the reversible process. The question is now whether the mathematical formulae introduced from the first law can also be applied to the irreversible process or not.

According to the second law of thermodynamics the entropy S in a closed system never decreases and therefore an equilibrium state has a maximum entropy.[17, 18] An entropy increase is observed in an irreversible process from one equilibrium state to another. As both the initial and final states have a maximum entropy, the process must pass through a minimum entropy state at least once, as shown in Fig. 3.1. What is the meaning of the minimum entropy state or the entropy decrease ΔS_{mf} from the initial state S_i to the minimum entropy state S_m in the irreversible process?

We can assume an adiabatic process changing from the initial state to the minimum entropy state, in which the macroscopic values of thermodynamic coordinates are kept constant, since no macroscopic change is observed during the process. Considering first a state in a perfect gas as an example, it has an entropy described as follows:[19]

$$S_i = N\left(c_v \ln T_0 + k \ln \frac{V_0}{N} + \sigma\right), \tag{3.2}$$

where N is the number of particles, c_v the specific heat of the gas per particle at constant volume, T_0 the temperature of the gas, k the Boltzmann constant, V_0 the gas volume, and σ is a constant.

Now we divide the system into two parts having the same number of particle $N/2$ and let each part have different temperatures T_1, T_2, different volumes V_1, and V_2, respectively, in adiabatic states, as shown schematically in Fig. 3.2. Then we have the following relations

$$\frac{T_1 + T_2}{2} = T_0, T_1 = T_0 + \Delta T, T_2 = T_0 - \Delta T, V_1 + V_2 = V_0,$$

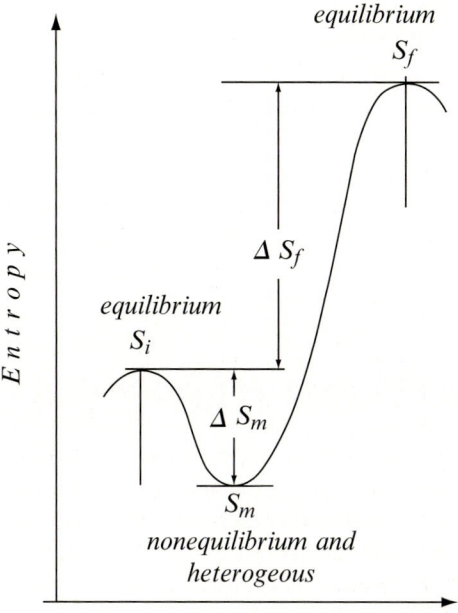

Fig. 3.1. Entropy in an irreversible process

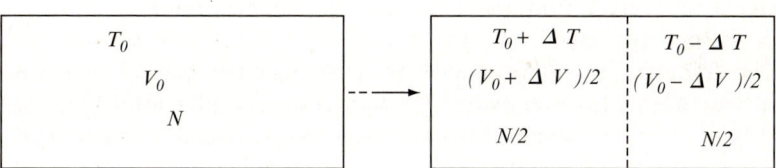

Fig. 3.2. System dividing

$$V_1 = \frac{V_0 + \Delta V}{2}, \quad \text{and} \quad V_2 = \frac{V_0 - \Delta V}{2}.$$

The system after the division has entropy as follows:

$$S_e = \frac{N}{2}\left[c_v \ln(T_0 + \Delta T) + k \ln \frac{V_0 + \Delta V}{N} + \sigma\right]$$
$$+ \frac{N}{2}\left[c_v \ln(T_0 - \Delta T) + k \ln \frac{V_0 - \Delta V}{N} + \sigma\right]. \quad (3.3)$$

The difference of the entropy between both the states before and after the division is:

$$\Delta S = S_e - S_i = \frac{N}{2}\left[c_v \ln\left\{1 - \left(\frac{\Delta T}{T_0}\right)^2\right\} + k\ln\left\{1 - \left(\frac{\Delta V}{T_0}\right)^2\right\}\right]. \quad (3.4)$$

As $0 < \dfrac{\Delta T}{T_0} < 1$

and $0 < \dfrac{\Delta V}{V_0} < 1$,

then $0 < \left\{1 - \left(\dfrac{\Delta T}{T_0}\right)^2\right\} < 1$ and $0 < \left\{1 - \left(\dfrac{\Delta V}{V_0}\right)^2\right\} < 1$,

consequently, $\Delta S < 0$. The entropy decreases with this division.

If we repeat this process in each part and consequently divide the system into many parts having the same number of particles, but different temperatures and volumes, keeping the mean temperature and the whole volume constant, we obtain a much lower entropy state. In the case of condensed states, we need only to consider the temperature.

Thus, in order to decrease the entropy, it is necessary to have a state having different temperatures everywhere in the system like a mosaic. In the irreversible process such a heterogeneous and nonequilibrium state must appear for a short period somehow at the minimum entropy state even though the macroscopic thermodynamic coordinates are kept constant.

The definition of homogeneous state in thermodynamics is the state in which the entropy $S(qU, qV, qN_j) = q\, S(U, V, N_j)$, where q is an arbitrary number, U energy, V volume, and N_j is the particle number.[20] This means that the specific entropy is everywhere the same in the homogeneous system. The quantity of each part composing the mosaic and having the same temperature must, therefore, be so large that the entropy, eventually the temperature can exist in it, that is, each part must be consisted of more than several thousand particles. The heterogeneous state appearing at the entropy minimum state even if in a short time, causes some fluctuating phenomena not only in time but also in space, as the phenomena are initiated in the parts having higher temperature and develop to an observable scale. We can thus conclude that in an irreversible process the following two states appear:

1. A heterogeneous state which causes some fluctuations in time and space.
2. A nonequilibrium state.

The irreversible process is quite different from the reversible process. The mathematical formulae which are introduced from the first law of thermodynamics and valid only for the reversible processes, therefore, cannot be applied to irreversible processes. Only for very slow processes the theories may approximately be available, but never for distinctly irreversible phenomena.

Remark 1. *The first law of thermodynamics for the irreversible process*
The mathematical formulae of first law of thermodynamics generally are introduced under an assumption of reversible process. The first law should naturally be available not only for the reversible process, but also for the irreversible one. The first law is, however, applied to the irreversible process considering an entropy increase from a maximum value at an equilibrium state step by step, keeping an equilibrium state, at each step, but considering never the entropy decrease by an irreversible process. The entropy increase directly from a maximum value, keeping an equilibrium, however, is physically impossible.[21, 22]

Remark 2. *Fluctuating phenomena at an equilibrium state*
As a thermodynamic system is composed of many particles having different energies under a Maxwellian distribution, a phenomenon observed at an equilibrium state has always some fluctuation, but the range of the fluctuation is very small. For example, the broadening of the distribution of the ignition probability density calculated according to the molecular dynamics is roughly estimated to have an order less than 1/10 of those experimentally observed, which are later shown in this book. Besides, there is no experimental result supporting the stochastic theory introduced from the molecular dynamics.[23–25]

3.2 Fluctuation in Irreversible Processes

As an irreversible process must pass a minimum entropy state, the fluctuations observed in the irreversible process take place not at random, but are governed by the probability passing the minimum entropy state. In many irreversible phenomena we really observe some fluctuations. In Fig. 3.3, a histogram of delay of electrical discharge is observed by Zuber[26, 27] and Hirata[28] and in Fig. 3.4 that of rupture of glass pieces is observed by Hirata and Terao.[29] Both histograms show a large fluctuation. Such fluctuations are governed by the probability passing the minimum entropy state.

As the entropy S is expressed as a product of the Boltzmann constant k and the logarithm of the probability W of the state, namely[30, 31]

$$S_i = k \ln W_i, \quad S_m = k \ln W_m \quad \text{and} \quad S_f = k \ln W_f, \tag{3.5}$$

where the subscript i means the initial state, m the minimum entropy state, and f is the final state. Therefore

$$-\Delta S = S_m - S_i = k \ln W_m - k \ln W_i = k \ln\left(\frac{W_m}{W_i}\right). \tag{3.6}$$

Then

$$\frac{W_m}{W_i} = \exp\left(\frac{-\Delta S}{k}\right). \tag{3.7}$$

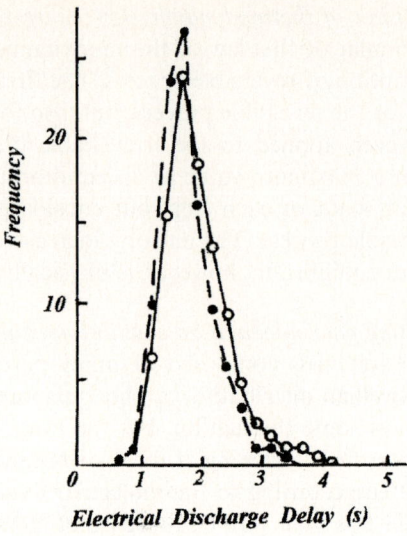

Fig. 3.3. Histogram of an electric discharge delay (Zuber)

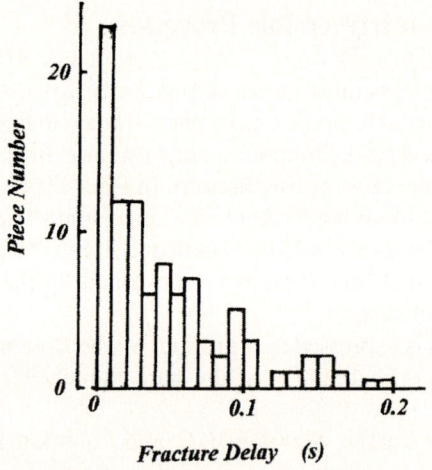

Fig. 3.4. Histogram of a fracture delay of a glass piece (Hirata)

As $\Delta S = \Delta Q/T$, where Q is heat

$$\frac{W_m}{W_i} = \exp\left(-\frac{\Delta Q}{kT}\right). \tag{3.8}$$

Substituting a relation $N \times \Delta Q = E$, where N is Avogadro's number, into (3.8), then for one mole

$$\frac{W_m}{W_i} = \exp\left(-\frac{E}{RT}\right), \tag{3.9}$$

where R is the gas constant. Then the probability W_m which governs the fluctuations is:

$$W_m = W_i \exp\left(-\frac{E}{RT}\right). \tag{3.10}$$

thus, the probability of the irreversible process is expressed by an Arrhenius' formula, where E is called activation energy.

In every irreversible process some stochastic phenomena are always observed.

Remark 3. *Minimum entropy S_m*
At a stable state a thermodynamic system has an entropy S expressed by $S = k \ln W$. In an irreversible process, however, a heterogeneous and nonequilibrium state appears at the entropy minimum state, which seems to be unstable, but is stable even for a very short period. According to the results experimentally observed, the fluctuation of each phenomenon has a Poisson distribution. The particles of the system, therefore, must have also Maxwellian distribution composed of a small number of the element, namely, of several groups having different energy levels. The minimum entropy S_m, thus, can also be expressed by $S_m = k \ln W_m$.[30]

3.3 Stochastic Theory for Irreversible Phenomena

In an irreversible process we observe some fluctuating phenomena. The fluctuations are quite different from the experimental error fluctuating at random, but governed by a probability as described earlier. We can therefore estimate the probability for the occurrence of a phenomenon from the fluctuations. We call such phenomena having some fluctuations governed by a probability "stochastic phenomena."

A model of such a stochastic phenomenon is illustrated in Fig. 3.5.:

1. *Initiation process*. In the heterogeneous state at the entropy minimum some nuclei, origins of the phenomenon, are formed perhaps at the points having higher energy, expressed with white circles.
2. *Development process*. Colliding with other particles having high energy, several of the nuclei develop to those having higher energy state and reach the critical state from which the nuclei develop further only by the reaction energy released in each nucleus itself, expressed with larger white circles, while the other nuclei are destructed by collision with other particles having lower energy.
3. *Growing period of the critical nucleus*. In order to recognize the phenomenon the critical nucleus must further grow to a measurable size. To the measurable size the nucleus grows almost with a certain definite rate.

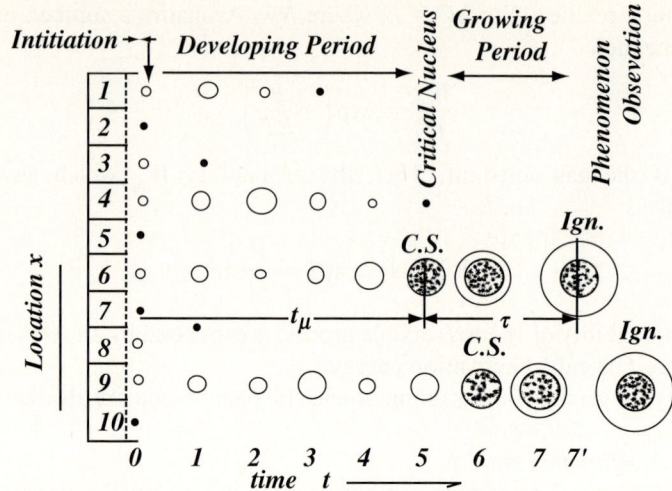

Fig. 3.5. Schematic representation of nucleus development in an irreversible phenomenon. *C.S.*: Nucleus having a critical size, *Ign.*: appearance of the phenomenon

Repeating the same experiment and measurement of a phenomenon, for example, ignition, fracture or ebullition, under the same condition, we can obtain a histogram of the induction period t of the phenomenon, i.e., the period from the instant in which the concerning system is set to a state where the phenomenon can take place to the time when the phenomenon is observed.

If $P(t)$ is the probability for the occurrence of a phenomenon whose induction period is longer than t, $q(t)$ the probability density of the phenomenon occurrence which is obtained by normalizing the histogram of the induction period t and m the molar number of the concerning system, as shown in Fig. 3.6, the probability $\mu(t)$ of the phenomenon occurrence in one mole of the system per unit time can be calculated according to the following equations:[26-31]

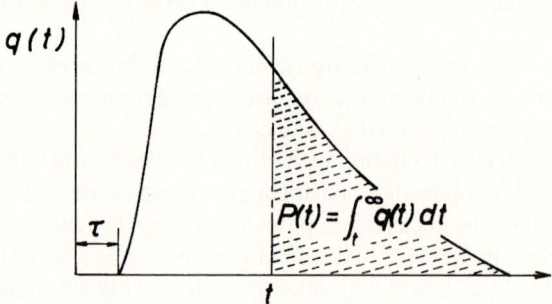

Fig. 3.6. Probability density $q(t)$ and probability $P(t)$ of the phenomenon appearance later than an instant t

3.3 Stochastic Theory for Irreversible Phenomena

$$P(t) = \int_t^\infty q(t) \, dt, \tag{3.11}$$

$$m.\mu(t).P(t) \, dt = -dP, \tag{3.12}$$

$$m.\mu(t) = -\frac{d \ln P(t)}{dt}. \tag{3.13}$$

This probability $\mu(t)$ consists of two components, the probability α of the initiation of the phenomenon and the probability β of the development of the initiation process to a certain critical nucleus which can by itself develop into the phenomenon. That is:

$$\mu(t) = \alpha.\beta. \tag{3.14}$$

As α should be proportional to $\exp(-E_1/RT)$ and β to $\exp\{-(E_2 - W)/RT\}$, where E_1 and E_2 are the activation energies for the initiation and the development of the process, respectively, W the energy released by the initiation and supplied to the development process, R the gas constant, and T is the temperature of the system, the probability

$$\mu(t) = A \exp\left(-\frac{E_1 + E_2 - W}{RT}\right), \tag{3.15}$$

in which A is a frequency factor consisting of the density and the collision probability of each element in the system. The energy process is schematically illustrated in Fig. 3.7.

It is necessary, on the other hand, to consider a certain period τ in which the nucleus grows to a measurable size. This minimum dimension appears in $P(t) - t$ diagram, as shown in Fig. 3.8, but depends on the sensitivity of the applied measurement apparatus. The period τ for the nucleus growth should be inversely proportional to the growing velocity, namely,

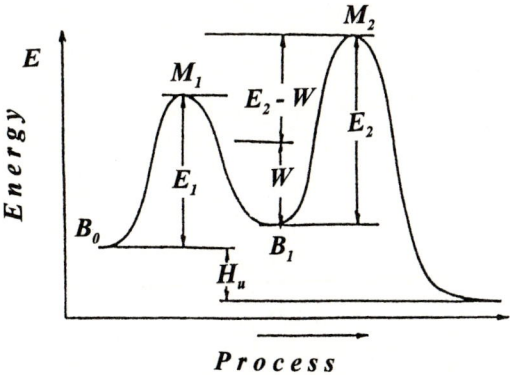

Fig. 3.7. Energy process of a stochastic phenomenon

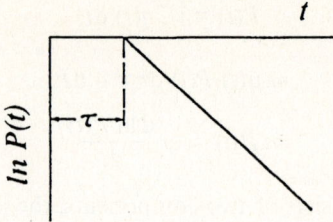

Fig. 3.8. Relation between ln $P(t)$ and time t in a stochastic phenomenon

$$\tau = B\exp\frac{E'_2}{RT}, \qquad (3.16)$$

where E'_2 is the activation energy for the growing process and B is a constant which depends on the minimum measurable size of the phenomenon, the density and the collision probability of each element in the system. E'_2 may be equal to $(E_2 - W)$, as the phenomenon is to grow from the initiation process to the measurable size beyond the critical nucleus. That is

$$\tau = B\cdot\exp\frac{E_2 - W}{RT}. \qquad (3.17)$$

Thus, the mean induction period t_m for the phenomenon occurrence in an m mole system should be expressed as follows:

$$t_m = \tau + \frac{1}{m\mu}. \qquad (3.18)$$

In order to prove our idea experimentally, we present in the following sections applications of the stochastic theory to irreversible phenomena, i.e., first to the nucleation in the phase transition and then to many phenomena in the ignition, combustion, and detonation waves.

4

Nucleation in Phase Transition

In the phase transition under a supersaturated state, some nuclei are first formed in the original phase, which develop to form and enlarge the new phase, for example, ice formation in water, ebullition in liquids, raining, etc. Such phenomena are called nucleation, which has been studied by many scientists, mainly under the assumption of equilibrium state. As the phenomena are irreversible, the stochastic theory can be applied to them.

4.1 Classical Nucleation Theory[32]

Under the assumption of equilibrium and homogeneous state in a phase of a substance, a certain critical size of the nucleus from which the phase can spontaneously be transformed to another phase is theoretically estimated. Assuming that the nucleus having a smaller size than the critical one must be destructed by the outer power, the probability of formation of the nucleus having the critical size is calculated.[32]

As an example we treat the ebullition in a liquid, in which a vapor bubble is formed in the liquid. Halving the bubble by a plane A–A through its center, as shown in Fig. 4.1, we investigate the balance of power acting on the plane. The power acting on the plane by the pressure difference is $\pi r^2(P-P_0)$, while the power compressing the bubble is $2\pi r \sigma$, where r is the radius of the bubble, P the vapor pressure, P_0 the pressure outside the bubble, and σ the surface tension of the bubble. Under the condition $2\pi r \sigma > \pi r^2(P-P_0)$ the bubble must be destructed, while under the condition $2\pi \sigma < \pi r^2(P-P_0)$ the bubble develops, namely at the critical size

$$2\pi r_c \sigma = \pi r_c^2 (P - P_0), \qquad (4.1)$$

where r_c is the critical radius of the bubble. Namely, if the molecules of number N_c in the critical bubble are vaporized, the ebullition takes place in the liquid.

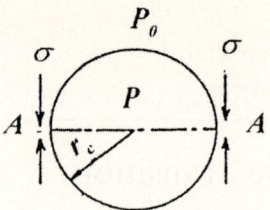

Fig. 4.1. Nucleation model

Also in other nucleation phenomena the critical size of the nucleus is first estimated and the probability of the state in which all N_c molecules at the same instant obtain the energy necessary for forming the critical nucleus is recognized to be the probability of the nucleation. As the phase transition proceeds under a constant temperature and pressure, its work is calculated as the difference of free enthalpy ΔG:

$$G = U - TS + PV \tag{4.2}$$

$$dG = -S\,dT + V\,dP + \Sigma\,\mu_j\,dN_j, \tag{4.3}$$

where U is the energy, T the temperature, V the volume, S the entropy, μ_j the chemical potential, and N_j the particle number.

The work ΔG for the formation of the critical nucleus having a radius of r under the interfacial energy σ is expressed as follows:

$$\Delta G = 4\pi r^2 \sigma + \frac{4}{3}\pi r^3 \Delta G_V, \tag{4.4}$$

where

$$\Delta G_v = -\frac{kT}{V_0}\ln\frac{P}{P_0}. \tag{4.5}$$

In this equation k is the Boltzmann constant, V_0 the volume of one molecule, P the pressure in the nucleus, and P_0 the outside pressure.

As illustrated in Fig. 4.2, $\Delta G = 0$ at $r = 0$, but increases with r to a maximum value ΔG_c, then decreases again. This ΔG_c is the increase of free enthalpy for

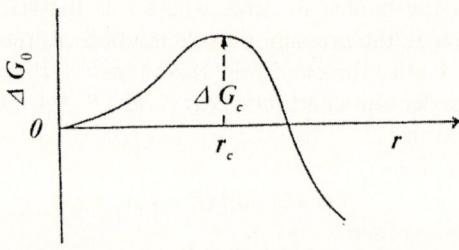

Fig. 4.2. Free enthalpy G in relation to the nucleus radius r. r_c: critical nucleus radius

the development of the nucleus to the size having the critical radius r_c. From the relation $\delta G/\delta r = 0$ at the maximum value of ΔG, ΔG_c can be calculated. In a substance having a particle density of N the density N_c of particle has an energy larger than the critical one ΔG_c is

$$N_c = N \exp\left(-\frac{\Delta G_c}{kT}\right). \tag{4.6}$$

Probability J of nucleation (number of the critical nucleus in the unit time and in 1 mol substance) is expressed as follows:

$$J = \alpha_c (4\pi r_c^2) \frac{P}{(2\pi m_p kT)^{1/2}} N \exp\left(-\frac{\Delta G_c}{kT}\right), \tag{4.7}$$

where m_p is the particle mass, α_c nucleation factor less than 1.0, and $(4\pi r_c^2)\{P/(2\pi m_p kT)^{1/2}\}$ is the particle number passing through the surface of the nucleus in unit time.

Some equations revised by Doering, Volmer, and others have been reported, but are almost the same as that described above. According to the classical theory ΔG_c must be so large, that no nucleation can take place, for example $\Delta G_c = 10^9 - 10^{10}$ kJ mol^{-1} at the ebullition of water, with which the ebullition can never take place. Therefore, the following two ideas have been added to the theory:

(1) Something outsides having a very high energy, for example cosmic rays, comes into the substance, in which they collide with some particles, give them much energy and cause the nucleation according to the classical theory.
(2) Assuming that some solid particles are mixed in the substance as impurity, the nucleation can take place on the surface of the solid particles by ΔG_c much smaller than that in a free space.

The ebullition, however, is observed everywhere, even in a space having little cosmic rays or impurity, the classical nucleation theory, therefore, must be not correct, as it has been introduced under the assumption of a reversible process. It must be treated as an irreversible process.

4.2 Stochastic Nucleation Theory at Ebullition of Liquids

In this section an application of the stochastic theory to the ebullition of liquids is explained.

The induction period of ebullition in distilled pure water as well as in acetone is measured which generally fluctuates in a large range. According to the stochastic theory, the probability of ebullition of both the liquids is calculated from the fluctuations of the ebullition induction period and from the results the ebullition mechanism is investigated.[33, 34]

4.2.1 Experiments for Measuring the Ebullition Induction Period

The experiments of ebullition in liquids are carried out using an apparatus schematically illustrated in Fig. 4.3. Distilled water filled into a cylindrical plexiglas vessel having a thin steel diaphragm at its bottom is set into a thermostat, so that the water temperature can be kept constant at an arbitrary value between 325 and 360 K. The top of the cylindrical vessel is connected with a glass tube having a three-way cock to a glass vessel in which air pressure can be set at an arbitrary value below the equilibrium vapor pressure by a vacuum pump.

Setting the water pressure suddenly from the atmosphere one to that lower than the saturated vapor pressure of the water corresponding to its temperature by turning the three-way cock, the water is put into a supersaturated state and an ebullition takes place in it after an induction period.

Observing the induced current through the coil wound around a permanent magnet set under the steel diaphragm at the bottom of the cylindrical vessel, we can recognize the instant of the pressure drop, while the ebullition beginning can be measured by observing the electrical conductivity of the water, as shown in Fig. 4.4. The period from the pressure drop to the ebullition beginning is taken the induction period of ebullition.

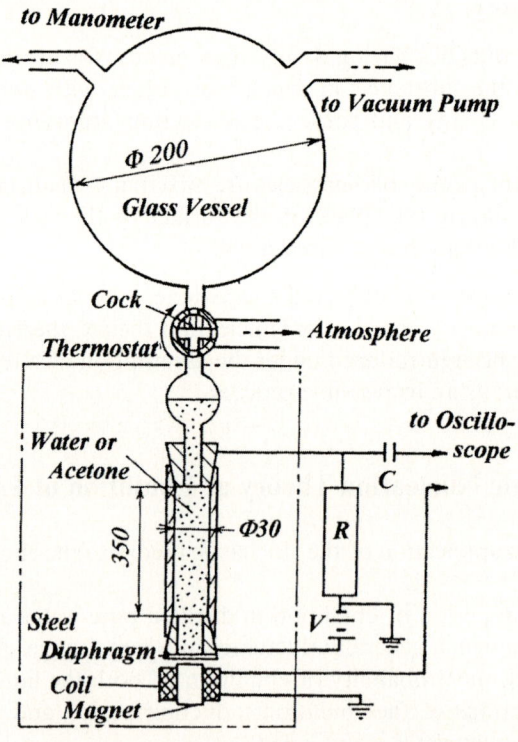

Fig. 4.3. Experimental apparatus for ebullition

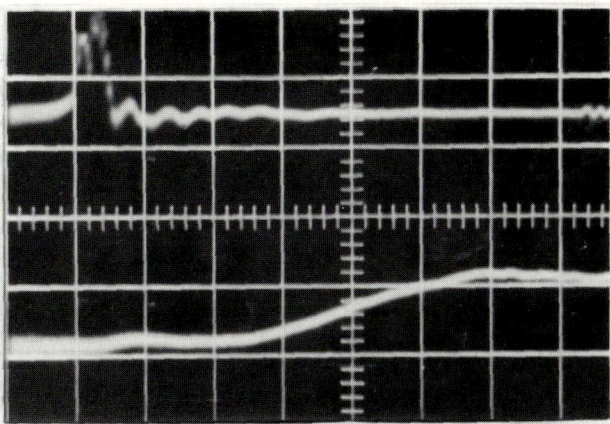

Fig. 4.4. Pressure variation (dP/dt, above) and electrical resistance of water at the ebullition in an arbitrary scale, 20 ms per div.

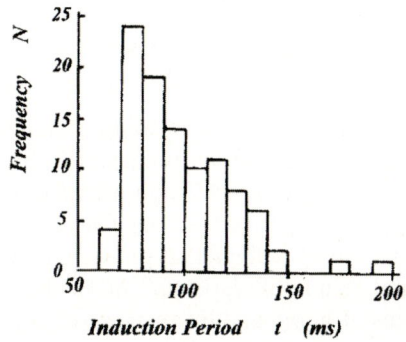

Fig. 4.5. Histogram of the ebullition induction period t in water at 55°C and $\Delta P = 90$ mm Hg (12.2 kPa)

The experiments are carried out at different water temperatures under three different pressures ΔP below the equilibrium vapor pressure. Repeating the experiment many times under the same conditions, a histogram of the ebullition induction period t under each experimental condition can be obtained, as shown in Fig. 4.5. The same experiments of ebullition in acetone are also carried out in the same way as in water.

4.2.2 Probability of Ebullition

According to (3.11)–(3.13), the probability of ebullition can be calculated from the histogram of the ebullition induction period. If

μ, Probability of ebullition occurring in 1 mol liquid during 1 s
m, Mole number of the whole liquid in the state in which the ebullition can take place, namely that of liquid in the cylindrical vessel in this case

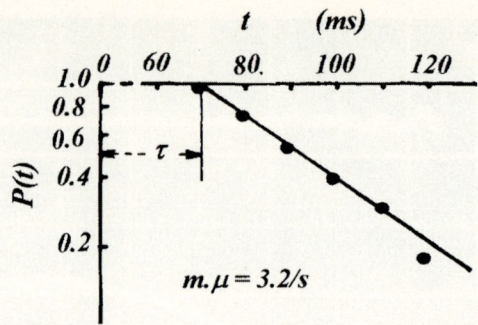

Fig. 4.6. ln $P(t)$ in relation to the induction period of ebullition t

$q(t)$, Probability density obtained from the histogram of ebullition induction period

then

$$P(t) = \int_t^\infty q(t)\, dt, \qquad (3.11)$$

$$m\mu = -\frac{d \ln P(t)}{dt}. \qquad (3.13)$$

An example of the relation between ln $P(t)$ and t is shown in Fig. 4.6 in which we have a straight line. It means that $m\mu$ is constant with time. From the inclination of the straight line the whole ebullition probability $m\mu$ can be calculated according to (3.13), while the period τ of nucleus growth is estimated in the diagram. In the period τ the critical nucleus grows to a measurable size.

All the experimental results of the ebullition in the water and acetone are summarized in diagrams of ln $m\mu$ and ln τ having ΔP as parameter with respect to the reciprocal temperature of the water or acetone at which the phenomena occur, as shown in Figs. 4.7 and 4.8, respectively. At a constant value of ΔP the size of the critical nucleus is the same, i.e., the frequency factor is the same.

4.2.3 Activation Energies of Ebullition

From the slopes of the straight lines of ln $m\mu$ and ln τ at each value of ΔP in these diagrams, the effective activation energies $(E_1 + E_2 - W)$ and $(E_2 - W)$ which are independent of ΔP can be calculated according to (3.15) and (3.17), but in the liquids the collision number of each particles is so large that W must be almost zero, as all energy of the particles after the initiation reaction is lost by the collision. Therefore, μ and τ are expressed by the following equation:

$$\mu = A \exp\left(-\frac{E_1 + E_2}{RT}\right) \qquad (4.8)$$

$$\tau = B \exp\frac{E_2}{RT} \qquad (4.9)$$

4.2 Stochastic Nucleation Theory at Ebullition of Liquids

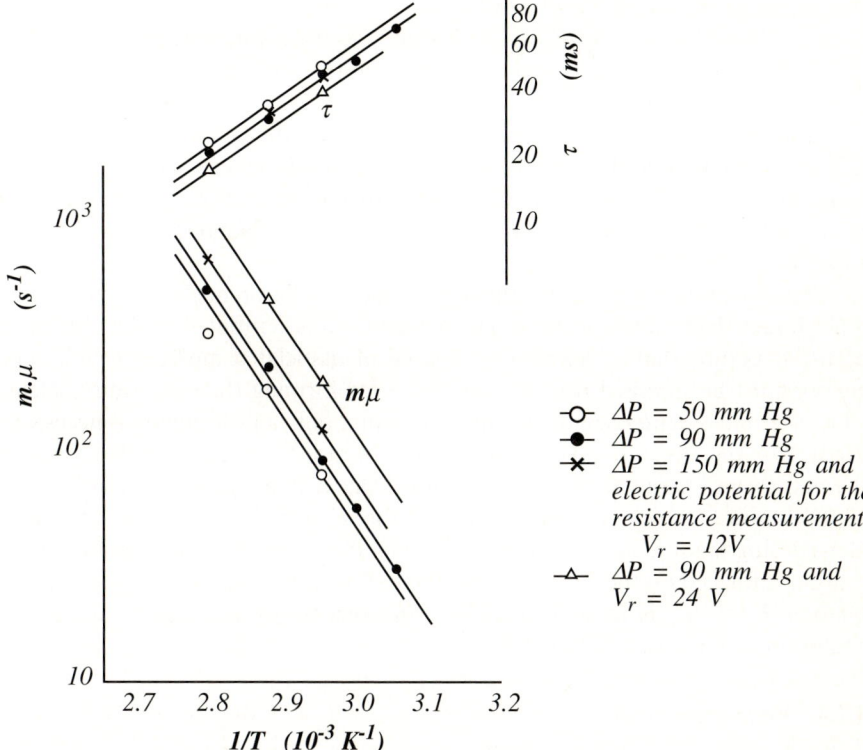

Fig. 4.7. Logarithms of the whole ebullition probability $m\mu$ and nucleus growing period τ in m mole water with respect to the reciprocal water temperature $1/T$

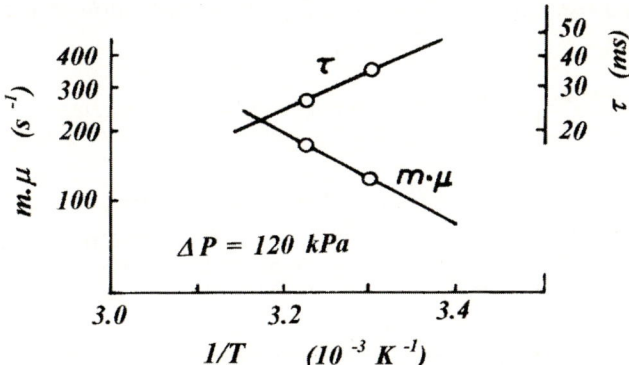

Fig. 4.8. Logarithms of the whole ebullition probability $m\mu$ and nucleus growing period τ in m mole acetone with respect to the reciprocal acetone temperature $1/T$

According to the equations we obtain
$E_1 = 51.0$ kJ mol^{-1} and $E_2 = 42.3$ kJ mol^{-1} in the distilled water and
$E_1 = 3.3$ kJ mol^{-1} and $E_2 = 32.7$ kJ mol^{-1} in the acetone
from the experimental results.

In water as well as acetone E_2 is always equal to the latent heat of evaporation H_v of each liquid. In order to develop the ebullition nuclei, the vaporization heat must be supplied to the nuclei. The activation energy E_2 for the development of the ebullition nucleus to a measurable size is, therefore, equal to the vaporization heat H_v.

The activation energy E_1 for the initiation of ebullition nucleation in water is much larger than that in acetone. The reason for it is attributed to the difference of the structure, that is, water is composed of associated molecules and some energy must be supplied to dissociate them for forming the ebullition nucleus, while acetone is composed of free molecules and only a little energy is necessary for it. On the other hand Vorsanger and Mauret[35] experimentally obtained the activation energy E_v for vaporization from a free surface of water to be 23.0 kJ mol^{-1}, which is almost half of the E_1 for the ebullition of water. Considering that the nucleus of ebullition is formed in water, while that of evaporation is formed on the surface where the half space above the water is free, the relation $E_1 = 2E_v$ is quite reasonable. The scheme of the energy process at ebullition is illustrated in Fig. 4.9.

4.2.4 Frequency Factor

The frequency factor A increases with the pressure difference ΔP between the vapor pressure of the liquid at ebullition and that of the saturated state, while the constant B for the growth of ebullition nucleus decreases. We observe here only the vapor bubble grown from the nuclei of ebullition by the electrical conductivity and never the size of nuclei themselves. Considering that the radius r_c of the critical nucleus $r_c = 2\sigma/\Delta P^{32}$, r_c decreases with increase of ΔP, the number of the nuclei which can be formed in a unit volume of the liquid also increases and consequently the probability of the nucleation increases as well.

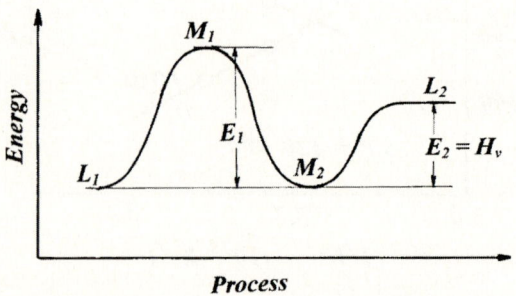

Fig. 4.9. Energy process at the ebullition

With increase of ΔP the rate of vaporization is accelerated, therefore, the growing period τ of nuclei is shortened and B also decreases.

Raising the electrical tension for observing the nucleus formation from 12 V to 24 V, A increases, while B decreases, because we can detect smaller bubbles. From this we can conclude that it is correct to assume τ as the growing period of the nucleus to the measurable size.

4.2.5 Ebullition Mechanism

According to the experimental results carried out and analyzed by the stochastic theory, the ebullition has an energy process schematically shown in Fig. 4.9. From the molecules of liquids at a supersaturated state having an average energy L_1, some molecules having an energy higher than E_1, reach the state M_1, are liberated from the liquid into a gaseous state, giving their energy E_1 to the other ones staying in the liquid and reach M_2 at the same energy state as the initial one but in gaseous and unstable state. In order to have a stable state, the vaporization heat $H_v = E_2$ must be supplied to the unstable gaseous molecules. The formation of nucleus is, thus, initiated and the nucleus develops to the critical size, getting the vaporization energy E_2. The growth of nucleus proceeds further to a measurable size by which the formation of nucleus is recognized.

As the molecules liberated from the liquid can exist in the nucleus even for a short time before growing to the critical size, giving their energy to next ones in the liquid one after another, not so much energy as that estimated by the classical theory is necessary for the nucleation in liquids.

As the frequency factor A depends on the number of nuclei which can be formed in a unit volume of the liquid and the measurable size of the nuclei, the probability of nucleation is

$$\mu = \frac{V_0}{C \frac{4\pi r_c^3}{3}} \exp\left(-\frac{E_1 + E_2}{RT}\right), \qquad (4.10)$$

where V_0 is the volume of 1 mol liquid, r_c radius of the critical nucleus, and C the number of nucleus having the minimum measurable size, depends also on the rate of nucleation. From the experimental results in water we have a relation between C and the pressure difference ΔP below the saturated state shown in the following table:

| ΔP (mm Hg) | 50 | 90 | 150 |
kPa	6.7	12.0	20.0
C	5×10^5	2×10^6	7×10^6

In the experiments of nucleation, the contact of liquid with the vessel wall cannot be avoided and its influence must be considered. The size of the critical nucleus on the wall should be less than half in free space, and the frequency factor A in the liquid contacting with the wall is, therefore, so much larger.

4.3 Stochastic Theory for Ice Formation in Water

In the solidification of a liquid we observe also a nucleation.[36] In this section the ice formation in water is explained according to the stochastic theory, depending on the experimental results. The ebullition is endothermic process, while the ice formation an exothermic one. The question is what influences has the exothermic process on the probability and nucleus growing period of the ice formation.[37]

4.3.1 Experiments

The experiment of ice formation can be carried out under the same principle as the ebullition described in the Sect. 4.2, namely, the state of a distilled water should suddenly be put into a supersaturated state, as illustrated in Fig. 4.10,[33, 34] or shown next:

	temperature	initial pressure	final pressure
1.	−4.62°C	845 kgf cm^{-2} (82.8 MPa)	→ 1.0 kgf cm^{-2} (0.98 MPa)
2.	−4.62°C	845 kgf cm^{-2} (82.8 MPa)	→ 97 kgf cm^{-2} (9.5 MPa)
3.	−4.62°C	845 kgf cm^{-2} (82.8 MPa)	→ 160 kgf cm^{-2} (15.7 MPa)
4.	−6.16°C	1,050 kgf cm^{-2} (102.9 MPa)	→ 215 kgf cm^{-2} (21.1 MPa)
5.	−6.16°C	1,050 kgf cm^{-2} (102.9 MPa)	→ 268 kgf cm^{-2} (26.3 MPa)

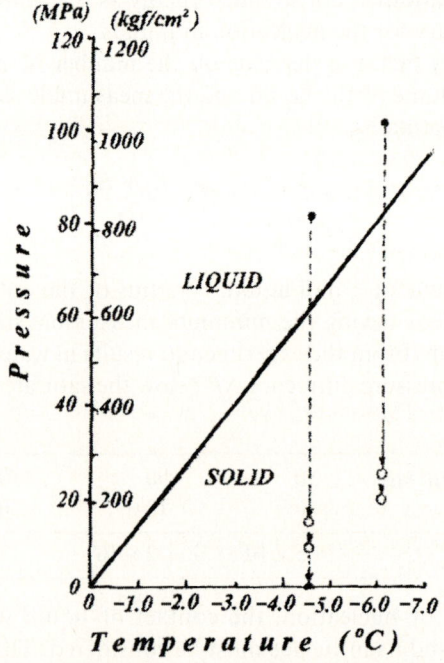

Fig. 4.10. Phase diagram of water[38]

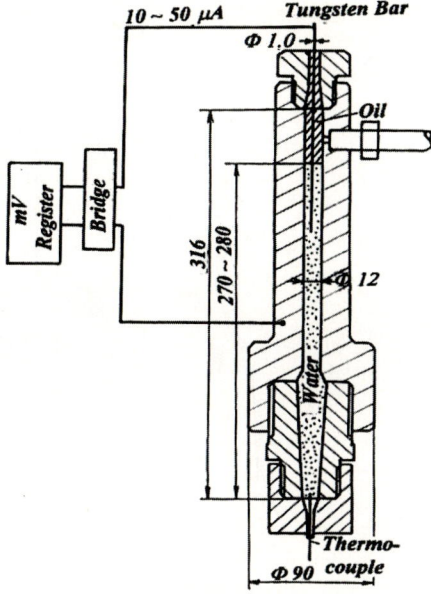

Fig. 4.11. Experimental apparatus for ice formation

The experimental apparatus is illustrated in Fig. 4.11. A deionized, distilled water filled in a vessel having a thick wall of steel is set into a thermostat, so that the water can be kept at an arbitrary temperature and pressure through an oil in a pipe connecting to the vessel. In the top of the water column a tungsten needle having a diameter of 1.0 mm is set, while at the bottom a thermocouple of chromel–alumel is set to measure the water temperature. Observing the electric conductivity between the tungsten needle and steel vessel, and the temperature shown by the thermocouple the ice formation in the water is recognized.

4.3.2 Probability of Ice Formation

An example of the variations of electric conductivity and temperature of the water observed by a pen recorder is illustrated in Fig. 4.12. In this case the difference of initial position between both the pens must be considered, that is, the temperature variation is recorded a little later than that of electric conductivity during the same process. Reducing the pressure from 900 kgf cm^{-2} (88.2 MPa) to 70 kgf cm^{-2} (6.9 MPa), the electric conductivity decreases. Because of the inertia of the recorder, the recorded electric conductivity overshoots once, then returns slowly to the correct value. At the beginning of the ice formation the electric conductivity begins to decrease, the induction period t of ice formation should be that from the pressure reduction to the beginning of ice formation,

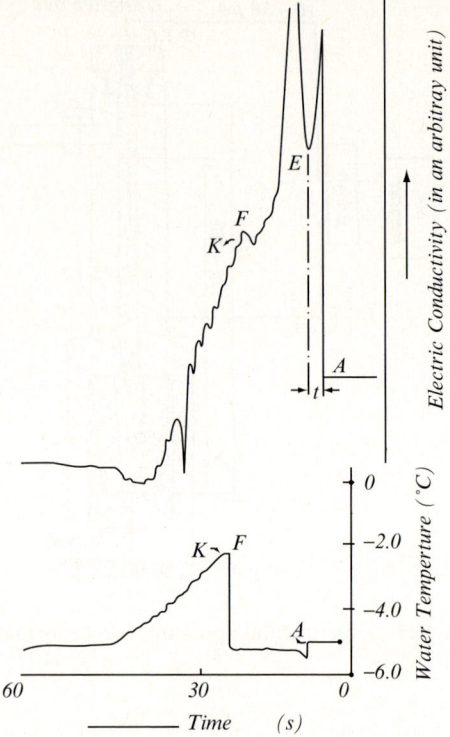

Fig. 4.12. Electric conductivity (above in an arbitrary scale) and temperature of water at ice formation. Pressure drop from 900 kgf cm^{-2} (88.2 MPa) to 70 kgf cm^{-2} (8.9 MPa) at −5.0°C. A: setting of supersaturated state; E: beginning of ice formation; F: freezing around the thermocouple; K: compression; t: induction period of ice formation

i.e., from the instant of the first beginning of the conductivity decrease to the next one. When the pressure decreases, the water temperature decreases too. The temperature decrease ΔT is:

$$\Delta T = \frac{Tv\beta}{C_p}\Delta P, \tag{4.11}$$

where v is the specific volume, β the bulk modulus, and C_p the specific heat under constant pressure of water. According to the equation, $\Delta T = 0.2°C$, while experimentally $\Delta T = 0.15°C$. The temperature decrease by the pressure decrease thus is negligibly small.

Repeating the experiment many times under the same conditions, a histogram of the induction period of ice formation can be obtained, as shown in Fig. 4.13. From such a histogram the probability density $q(t)$ of ice formation and then $P(t) = \int_t^\infty q(t)\mathrm{d}t$ is obtained. In the water having a mole number of m and the

4.3 Stochastic Theory for Ice Formation in Water

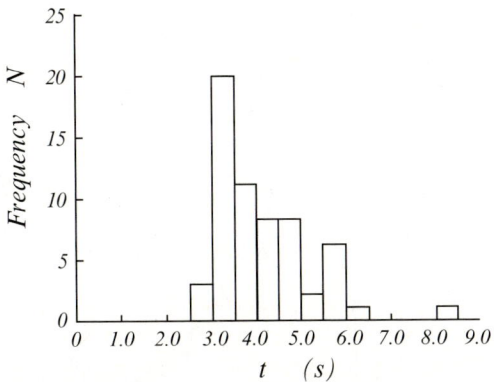

Fig. 4.13. Histogram of induction period t of ice formation. Pressure drop from 845 kgf cm^{-2} (82.8 MPa) to the atmospheric pressure at $-4.62°$C. Supersaturation pressure $\Delta P = 638$ kgf cm^{-2} (62.5 MPa)

probability of ice formation μ(s^{-1} mol^{-1}), the whole ice formation probability of the water is:

$$m\mu = -\frac{d \ln P(t)}{dt} t. \tag{4.12}$$

According to the equation, just like in the ebullition, the whole ice formation probability $m\mu$ as well as τ in which the nucleus grows to a measurable size can be obtained from the relation between $\ln P(t)$ and t as shown in Fig. 4.14.

From the relation between τ and ΔP as shown in Fig. 4.15, we can estimate τ at an arbitrary ΔP by an interpolation. Just like in the case of ebullition the ice formation probability should be reciprocally proportional to the third power of the critical radius r_c which is equal to $2\sigma/\Delta P$, where σ is the interfacial tension. The probability of ice formation is thus proportional to the third power of ΔP,

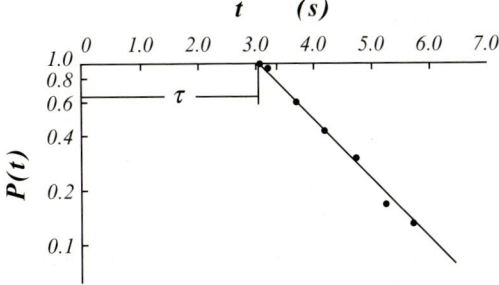

Fig. 4.14. log $P(t)$ with respect to the induction t of ice formation

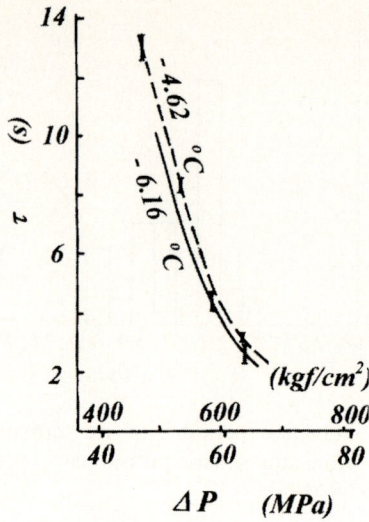

Fig. 4.15. Nucleus growing period τ with respect to the supersaturation pressure ΔP

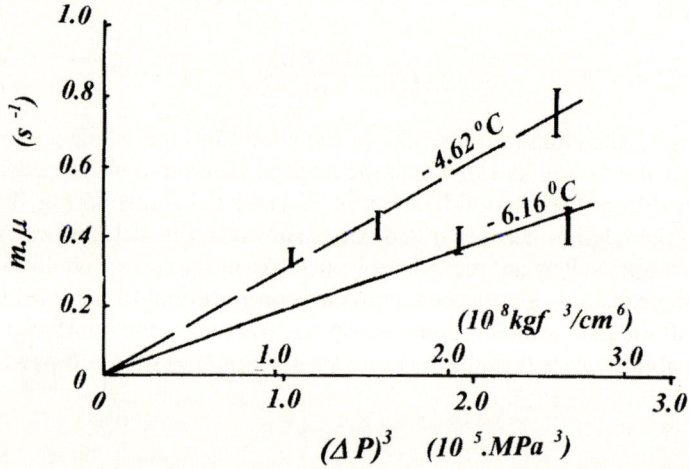

Fig. 4.16. Whole ice formation probability $m\mu$ in m mole water with respect to the supersaturation pressure ΔP

as shown in Fig. 4.16, from which we obtain $m\mu$ at an arbitrary ΔP also by an interpolation.

4.3.3 Activation Energies of Ice Formation

From the experimental results explained above the relation of $\ln m\mu$ and $\ln \tau$ to the reciprocal water temperature in the ice formation as shown in Fig. 4.17 can be obtained.

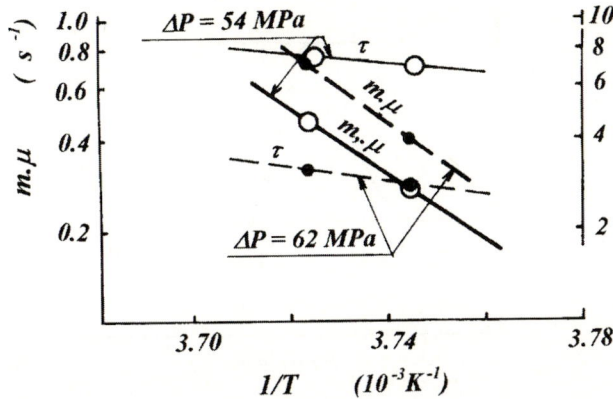

Fig. 4.17. Logarithms of the whole ice formation probability $m\mu$ in m mole water and nucleus growing period τ with respect to the reciprocal water temperature $1/T$

From the slopes of the straight lines in these diagrams and according to the equations

$$\mu = A \exp\left(-\frac{E_1 + E_2}{RT}\right), \tag{4.8}$$

$$\tau = B \exp\frac{E_2}{RT}, \tag{4.9}$$

the activation energies E_1 and E_2 are obtained, that is
$E_1 = 210 \pm 80$ kJ mol^{-1},
$E_2 = -5.0 \pm 20$ kJ mol^{-1} at $\Delta P = 550$ kgf cm^{-2} and
$E_2 = -12.5 \pm 20$ kJ mol^{-1} at $\Delta P = 630$ kgf cm^{-2}.
E_2 has a negative value and is almost equal to the latent heat of solidification 6.0 kJ mol^{-1}. It means that at the ice formation the heat of solidification should be taken off.

The energy process at the ice formation is schematically illustrated in Fig. 4.18. According to the classical nucleation theory, the activation energy must be $13-20 \times 10^4$ kJ mol^{-1}, with which any ice formation can scarcely take place.

As the whole probability of ice formation is proportional to the water quantity, the ice cannot be formed in a small quantity of water. With the temperature increase the probability of ice formation increases, but the growth of nucleus is delayed and the ice formation cannot proceed. Frequency factor A is proportional to the number of nucleus having the critical size, but at the saturated state the radius r_c of the critical size is infinitely large, consequently $A = 0$ and therefore no ice formation can be observed.

The stochastic theory can be applied to either the endothermic process or exothermic process.

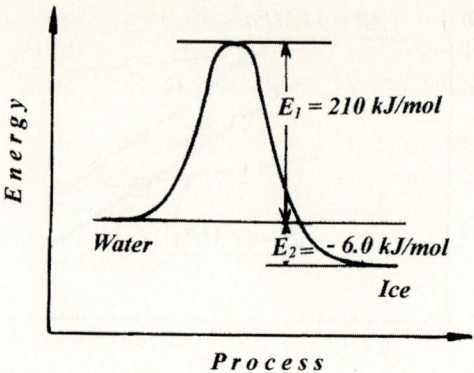

Fig. 4.18. Energy process at the ice formation in water

4.3.4 Ice Formation Under a Radiation of High Energy

As observed in the cloud chamber[39] or bubble chamber,[40, 41] a nucleation of liquid droplet or bubble in it by shooting some high energy particles into a supersaturated vapor or liquid, and the trace of the high energy particles is observed. Such high energy particles may promote the ice formation in water, too.

Using the same apparatus and method as described above, the ice formation in water is observed, putting a radioisotope (^{60}Co) near the apparatus and a

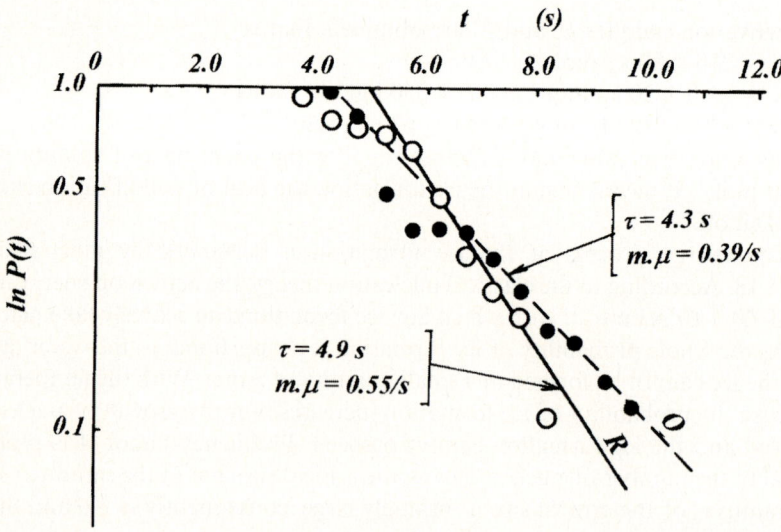

Fig. 4.19. Log $P(t)$ with respect to the induction period t of ice formation under a radiation from a radioisotope (*solid line*) and that without radiation (*broken line*)

4.3 Stochastic Theory for Ice Formation in Water

histogram of induction period of the ice formation is obtained. Though the energy of the radiation from the radioisotope is quantitatively unknown, we observed a clear difference between both the ice formation under and without radiation.

In Fig. 4.19, ln $P(t)$–t relation under the radiation from the radioisotope and that without any radiation is illustrated. The results suggest as follows:

	τ (s)	$m\mu$ (s^{-1})
R: under radiation	4.9	0.55
O: without radiation	4.3	0.39

Namely the nucleus growing period τ under the radiation from the radioisotope is longer than that without radiation, whereas the probability of ice formation μ under the radiation is larger than that without radiation.

The phenomenon is attributed as follows:

As an energy ε_r is supplied to the ice formation from the radiation, only an energy of $(E_1-\varepsilon_r)$ less than E_1, is necessary for the initiation of the ice formation and consequently the ice formation probability under the radiation is larger than that without radiation. On the other hand, during the nucleus growing period $(E_2 + \varepsilon_r)$ larger than E_2, must be taken off and therefore, the nucleus growing period under the radiation is longer than without radiation.

Applying Mach numbers $M_1 = u_1/a_1$ and $M_2 = u_2/a_2$, the following equations are obtained:

$$\frac{P_2}{P_1} = \frac{2\gamma M_1^2 - (\gamma-1)}{\gamma+1} \tag{5.13}$$

$$\frac{P_1}{P_2} = \frac{2\gamma M_2^2 - (\gamma-1)}{\gamma+1} \tag{5.13'}$$

$$\frac{\rho_2}{\rho_1} = \frac{u_1}{u_2} = \frac{\dfrac{P_2}{P_1} \cdot \dfrac{\gamma+1}{\gamma-1} + 1}{\dfrac{P_2}{P_1} + \dfrac{\gamma+1}{\gamma-1}} = \frac{(\gamma+1)M_1^2}{(\gamma-1)M_1^2 + 2} \tag{5.14}$$

$$\frac{T_2}{T_1} = \frac{P_2}{P_1} \frac{\rho_1}{\rho_2} = \frac{P_2}{P_1} \cdot \frac{\dfrac{P_2}{P_1} + \dfrac{\gamma+1}{\gamma-1}}{\dfrac{P_2}{P_1} \cdot \dfrac{\gamma+1}{\gamma-1} + 1} \tag{5.15}$$

$$= \left\{ \frac{2\gamma M_1^2 - (\gamma-1)}{\gamma+1} \right\} \cdot \left\{ \frac{(\gamma-1)M_1^2 + 2}{(\gamma+1)M_1^2} \right\}$$

According to these equations the states of the gas ahead of and behind the shock waves can be calculated, but we have to consider that the ratio of the specific heats is changed by the gas temperature. First one has to assume the value of T_2. Applying γ fitted the assumed T_2, T_2 is calculated. If T_2 is not equal to that first assumed, then again the temperature T_2 must be calculated using γ fitted to the calculated one. Such calculation must be repeated, till the correct value is obtained.

It is here to be noticed what the Hugoniot-curve means. The gas state does not change continuously from the state in front of the shock wave to that behind the shock wave, but the gas state changes discontinuously from that on front of the shock to that behind the shock, increasing its entropy. The Hugoniot-curve expresses only two states, those in front of and behind the shock waves. The tangent to Hugoniot-curve, therefore, means the line passing two states infinitely near each other without any entropy increase, i.e., isentropic change. The slope of the tangent is square of the sound velocity of the gas.

5.2 Simple Shock Tube

Shock tubes are apparatus to produce shock waves. Using a shock tube a hypersonic flow or high temperature gases can be produced within a few microseconds. Shock tubes are, therefore, applied to investigate the gasdynamics of flow having high velocity, or chemical reactions, especially ignition and combustion. We explain here a simple shock tube having a basic construction.

As shown in Fig. 5.4, a tube is divided by a diaphragm into two sections. In the first section a high pressure gas is filled as a shock driver gas, while the second section is filled by a test gas of low pressure. Breaking the diaphragm

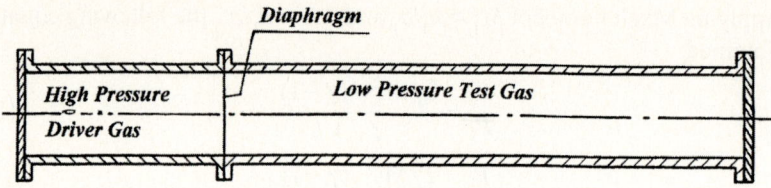

Fig. 5.4. Simple shock tube

suddenly, the high-pressure gas flows into the low-pressure gas, producing a shock wave in its front which propagates to right, while a rarefaction wave propagates to the left into the high-pressure gas with its sound velocity. The low-pressure gas is compressed by the shock wave, increasing its temperature and density, while the driver gas flowing into the second section expands, decreasing its temperature and density.

The relation between the initial pressure ratio P_0/P_1 of the driver gas to the low-pressure gas and that P_2/P_1 of the incident shock wave to the initial state of low-pressure gas is expressed by the following equation:

$$\frac{P_0}{P_1} = \frac{P_2}{P_1} \left\{ 1 - \frac{a_{11}(\gamma_0 - 1)(P_2/P_1 - 1)}{2a_0 \gamma_1 \sqrt{1 + \frac{\gamma_1 + 1}{2\gamma_1}(P_2/P_1 - 1)}} \right\}^{-2\gamma_0/(\gamma_0 - 1)} \quad (5.16)$$

where γ_0, a_0 are ratio of specific heats and sound velocity of the driver gas, γ_1, a_1 those of the low pressure test gas, respectively. Or, using the Mach number of the incident shock wave $M_1 = u_1/a_1$:

$$\frac{P_0}{P_1} = \frac{1 + \frac{2\gamma_1}{\gamma_1 + 1}(M_1^2 - 1)}{\left[1 - \frac{\gamma_0 - 1}{\gamma_1 + 1} \cdot \frac{a_1}{a_0} \cdot \frac{M_1^2 - 1}{M_1}\right]^{2\gamma_0/(\gamma_0 - 1)}} \cdot \quad (5.17)$$

The pressure P_2, temperature T_2, and density ρ_2 of the gas behind the incident shock wave in the shock tube are already described. In order to investigate chemical reactions, the observation of the gas behind the incident shock wave is not favorable, since the gas flows very rapidly and the flow may give some influences on the reactions. Therefore, the observation of the gas behind a shock wave reflected from the tube end is more favorable, as the gas scarcely flows and besides pressure as well as temperature is much higher than those behind the incident shock wave.

The pressure P_5, temperature T_5, and density ρ_5 behind the reflected shock wave can be calculated from the initial state of the gas and the Mach number M_1 of the incident shock wave according to the following equations:

$$\frac{P_5}{P_2} = \frac{\frac{P_2}{P_1}\frac{3\gamma - 1}{\gamma - 1} - 1}{\frac{P_2}{P_1} + \frac{\gamma + 1}{\gamma - 1}} \quad (5.16)$$

5.2 Simple Shock Tube

$$\frac{P_5}{P_1} = \frac{P_5}{P_2}\frac{P_2}{P_1} = \left[\frac{\frac{3\gamma-1}{\gamma-1}M_1^2 - 2}{M_1^2 + \frac{2}{\gamma-1}}\right]\left[\frac{\frac{2\gamma}{\gamma-1}M_1^2 - 1}{\frac{\gamma+1}{\gamma-1}}\right] \quad (5.17)$$

$$\frac{\rho_5}{\rho_1} = \frac{\rho_5}{\rho_2}\frac{\rho_2}{\rho_1} = \frac{\frac{P_2}{P_1}\frac{\gamma}{\gamma-1}}{\frac{P_2}{P_1}+\frac{1}{\gamma-1}} \cdot \frac{\frac{P_2}{P_1}\frac{\gamma+1}{\gamma-1}+1}{\frac{P_2}{P_1}+\frac{\gamma+1}{\gamma-1}}$$

$$= \frac{M_1^2 \frac{\gamma+1}{\gamma-1}\left(\frac{2\gamma}{\gamma-1}M_1^2 - 1\right)}{\left(M_1^2 + \frac{2}{\gamma-1}\right)\left(2M_1^2 + \frac{3-\gamma}{\gamma-1}\right)} \quad (5.18)$$

$$\frac{T_5}{T_1} = \frac{T_5}{T_2}\frac{T_2}{T_1} = \frac{\left(\frac{3\gamma-1}{\gamma-1}M_1^2 - 2\right)\left(2M_1^2 + \frac{3-\gamma}{\gamma-1}\right)}{\left(\frac{\gamma+1}{\gamma-1}\right)^2 M_1^2}. \quad (5.19)$$

According to these equations the pressure, temperature, and density of the gas behind the reflected shock wave can be calculated from the initial state of the gas and the Mach number of the incident shock wave, considering the ratio of specific heats γ corresponding to the gas temperature at each step.

The propagation of the incident and reflected shock waves, as well as that of the rarefaction waves and the discontinuous contact surface between the driver gas and test gas in a shock tube are illustrated in Cartesian coordinates, so-called shock diagram in which the horizontal axis express the position of each front of the waves and that of the contact surface, while the vertical axis shows the time after the diaphragm rupture.

Figure 5.5 shows an example of shock diagrams in a shock tube. 0_m means the position of the membrane in the shock tube. When the diaphragm bursts, the shock wave propagates along the line 0_m–S_1, while the front of the rarefaction wave travels in the opposite direction into the high-pressure driver gas along the line 0_m–R_1 with the sound velocity. After a certain expansion period the tail of the rarefaction wave propagates along the line 0_m–R_2. The contact surface propagates along the line 0_m–S_2 with the same velocity as the flow velocity of w_2. The reflected shock propagates along the line S_1–S_2 which is separated into two waves S_2–S_3 and S_2–R_1S_1 after collision with the contact surface at S_2. The state of the gas behind the reflected shock waves having the conditions expressed by (5.16)–(5.19) is kept only in the region 5 in this diagram. Its duration period and region are namely limited.

In order to keep the state behind the reflected shock waves much longer, tailored shock waves can be applied, in which the contact surface after the collision of the incident shock with the reflected one stands still, namely, S_2–S_4 line is

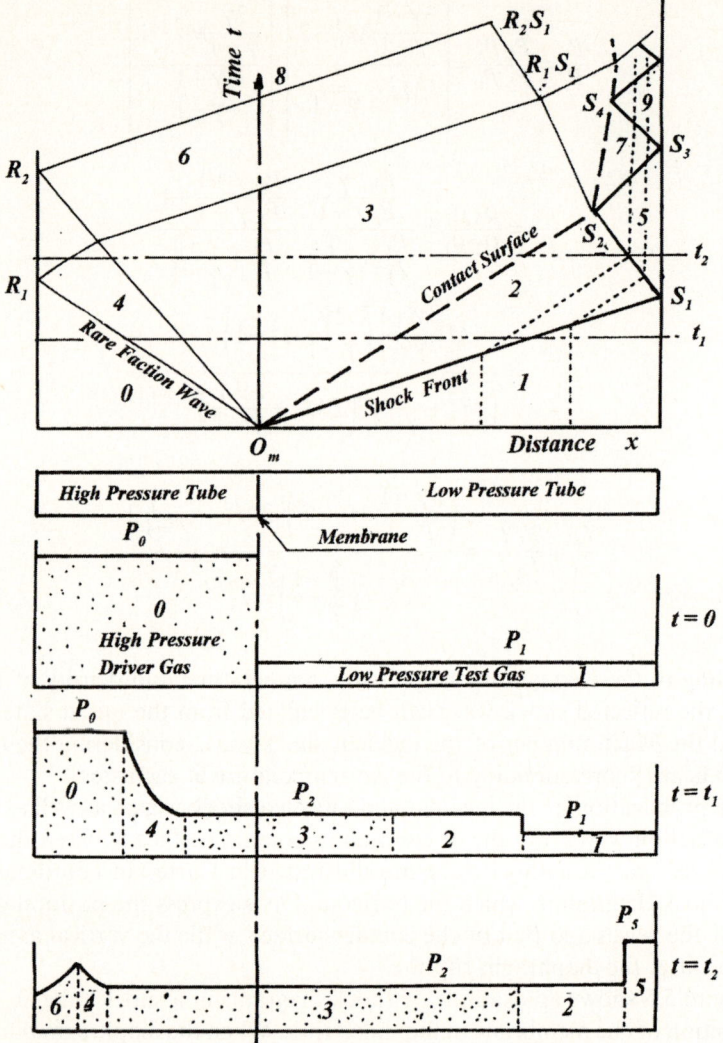

Fig. 5.5. Shock diagram and pressure process in a simple shock tube

parallel to the time-axis. Then the state of region 5 is kept till the rarefaction wave R_1–R_1S_1 reaches.

In Fig. 5.6a, b, and c the ratios of the pressure P_2, density ρ_2, and temperature T_2 behind incident shock waves to those P_1, ρ_1, T_1 in front of the shock wave with respect to the Mach number M_1 of the incident shock wave for gases having different ratios γ of specific heats are illustrated, while in Fig. 5.7a, b, and c those P_5, ρ_5, T_5 behind the reflected shock wave to the initial gas state are shown.

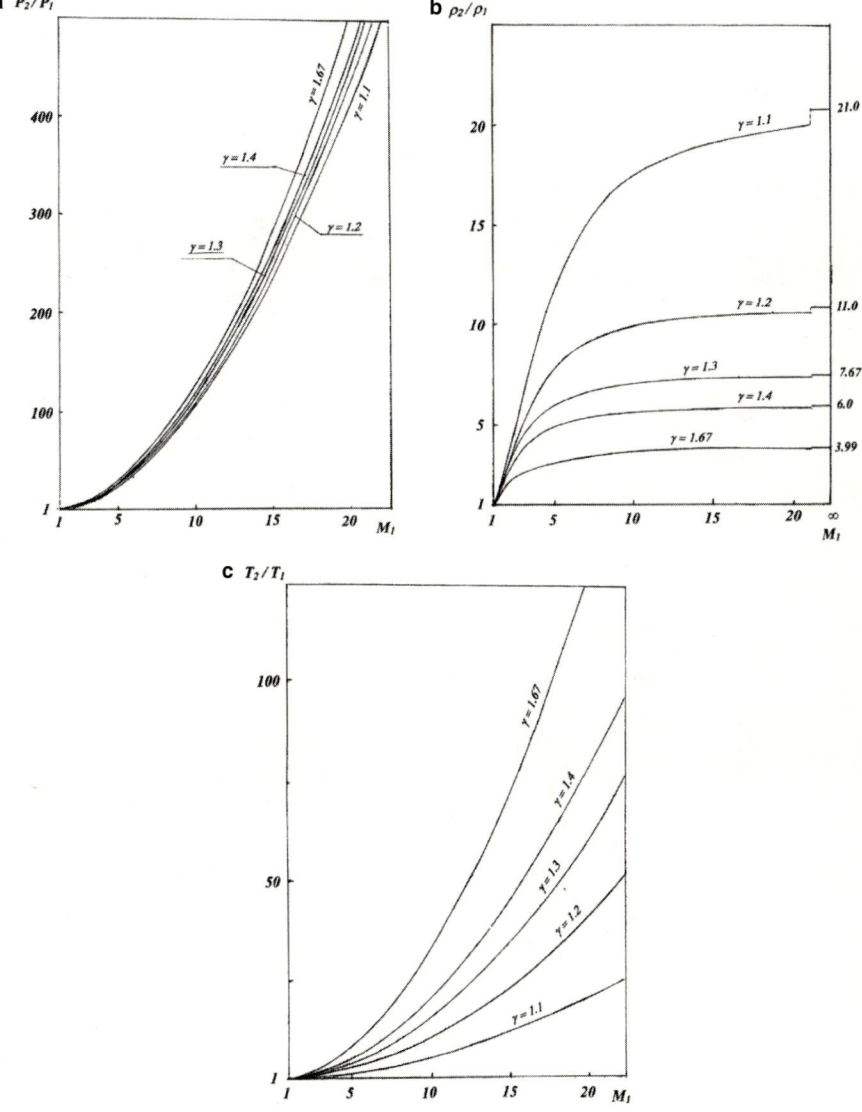

Fig. 5.6. (a) P_2/P_1 in relation to M_1, (b) ρ_2/ρ_1 in relation to M_1, (c) T_2/T_1 in relation to M_1

In Fig. 5.8a, b, and c the Mach number M_1 of incident shock wave in relation to the ratio of initial pressure P_0 of the driver gas to that P_1 of test gas in the case of monatomic gas ($\gamma = 1.67$) and diatomic gas ($\gamma = 1.4$) are shown, together with the ratios of sound velocities a_0/a_1 between both the driver and test gases as parameters.

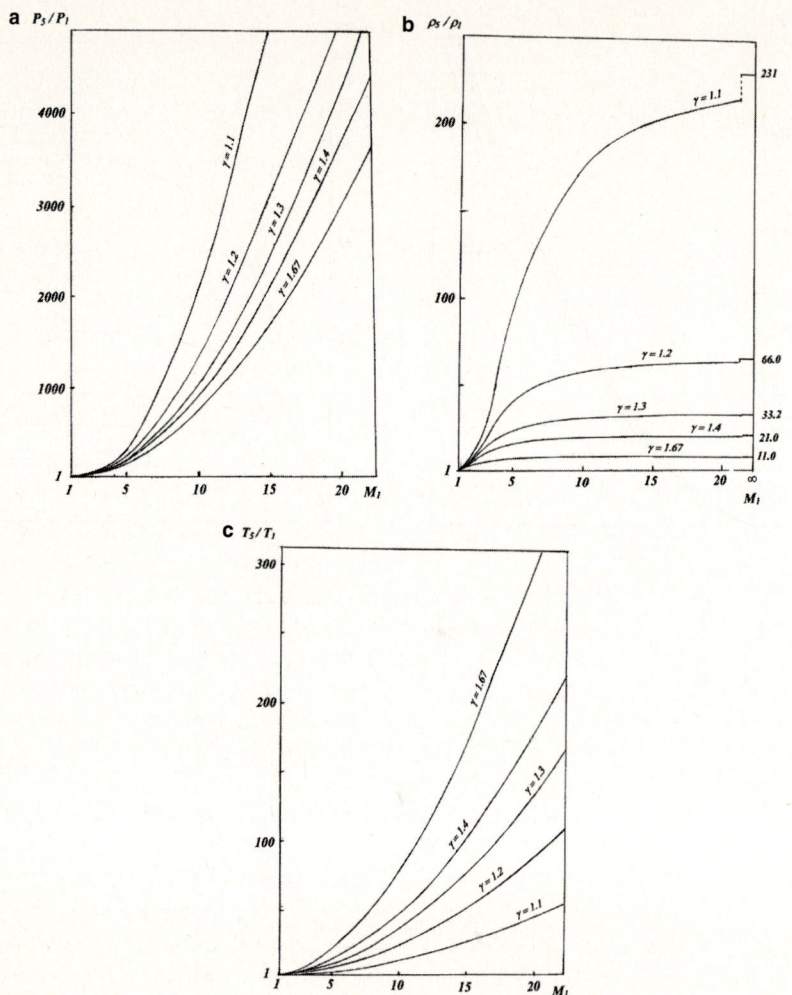

Fig. 5.7. (a) P_5/P_1 in relation to M_1, (b) ρ_5/ρ_1 in relation to M_1, (c) T_5/T_1 in relation to M_1

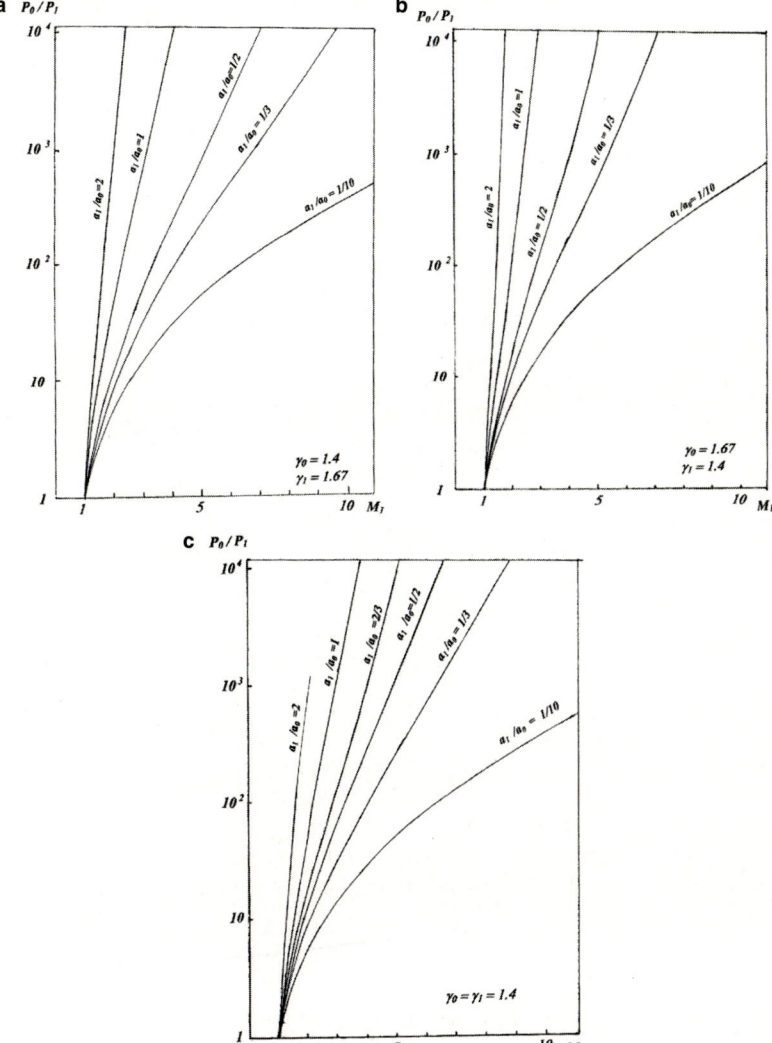

Fig. 5.8. (a) P_0/P_1 in relation to M_1 at $\gamma_0 = 1.4$ and $\gamma_1 = 1.67$, (b) ρ_5/ρ_1 in relation to M_1 at $\gamma_0 = 1.67$ and $\gamma_1 = 1.4$, (c) T_5/T_1 in relation to M_1 at $\gamma_0 = \gamma_1 = 1.4$

6

Stochastic Ignition Theory

The ignition is also a distinctly irreversible process and has many stochastic phenomena in it. In the classical theories it is assumed that the spontaneous ignition in a homogeneous mixture proceeds homogeneously in the whole mixture space, but practically we never observe such phenomenon in which the ignition takes place everywhere in the mixture at the same time. In reality the ignition starts from a few points in the mixture.

A spontaneous ignition in a stoichiometric isooctane-air mixture in a cylindrical vessel schematically shown in Fig. 6.1 under a rapid adiabatic compression by a piston which was driven by a high-pressure air was observed through a narrow slit. Its picture taken on a film rolled around a rotating cylinder is shown in Fig. 6.2 together with its pressure variation. As this photograph shows, the ignition takes place at a few points, from which flames propagate in the whole space. The induction period of ignition is thus different corresponding to the part of the mixture. The ignition induction period, therefore, fluctuates with a probability, which can be calculated from the fluctuation of the ignition induction period.[44]

6.1 Probability of Ignition Behind Shock Waves[45]

In order to investigate a spontaneous ignition in a combustible mixture, it is first necessary to bring the mixture suddenly into a state in which an ignition can take place, namely, to raise the temperature as quickly as possible. For this purpose it is convenient to apply a shock tube and ignite the mixture behind the reflected shock wave in the shock tube, where the mixture is heated to a temperature between several hundred to a few thousand centigrade in a few microseconds without any influential flow.

In this method

1. The incident shock waves should not be so strong that a spontaneous ignition can take place in the mixture behind them

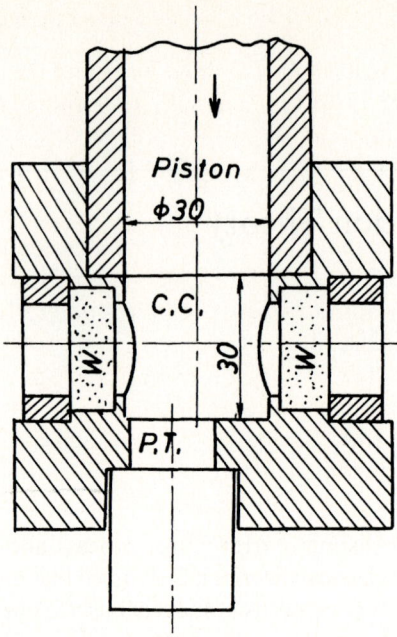

Fig. 6.1. Experimental apparatus of adiabatic compression for spontaneous ignition. Compression ratio: 9.5 (by Martinengo)

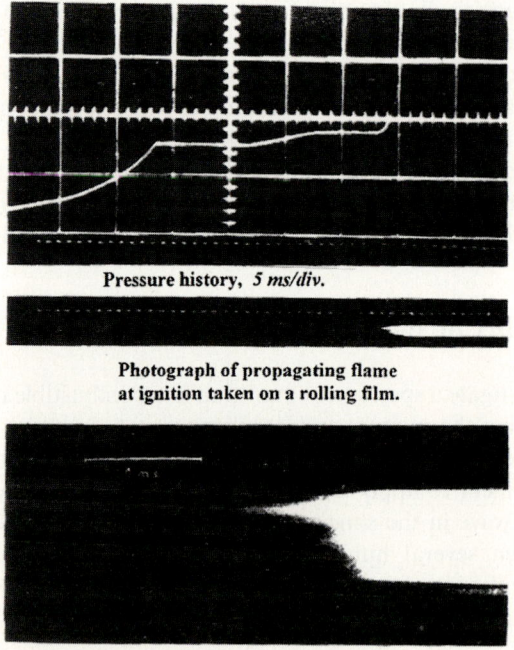

Pressure history, 5 ms/div.

Photograph of propagating flame at ignition taken on a rolling film.

Enlarged photograph of the picture above.

Fig. 6.2. Ignition in a stoichiometric isooctane–air mixture under an adiabatic compression. Initial mixture temperature is 80°C and compression pressure 16 bar (the phenomena progress from left to right)

6.1 Probability of Ignition Behind Shock Waves

2. The spontaneous ignition in the mixture behind the reflected shock waves should occur, before the contact surface or rarefaction waves give some influences on the ignition.

In this case, therefore, only the ignition in the region 5 in the shock diagram illustrated in Fig. 5.5 should come into question. Besides, as it is unknown where the ignition takes place, the whole space of the mixture must be observed. The induction period of ignition, therefore, is to be the period from the instant when the incident shock waves reach the tube end wall and begin to reflect to the ignition instant.

In a shock tube as shown in Fig. 6.3 a combustible mixture is filled in the low pressure tube, in which shock waves driven by the high pressure gas propagate and reflect at the tube end. The luminescence of ignition in the mixture behind the reflected shock waves is observed by a photomultiplier aligned along the tube axis behind a plexiglas window set at the tube end. We can take the period from the shock reflection at the tube end to the beginning of the ignition luminescence as the induction period of ignition.

As already mentioned, the ignition having an ignition induction period shorter than t_r, should be taken into the experimental results. Repeating the experiment and measurement under the same condition a histogram of ignition induction period t can be obtained. By normalizing such a histogram, a probability density $q(t)$, then $P(t) = \int_t^\infty q(t)dt$ can be obtained. As already in Chap. 3 explained, the probability of ignition μ (mol^{-1} s^{-1}) is expressed as follows:

$$m\mu = -\frac{d \ln P(t)}{dt}, \tag{3.13}$$

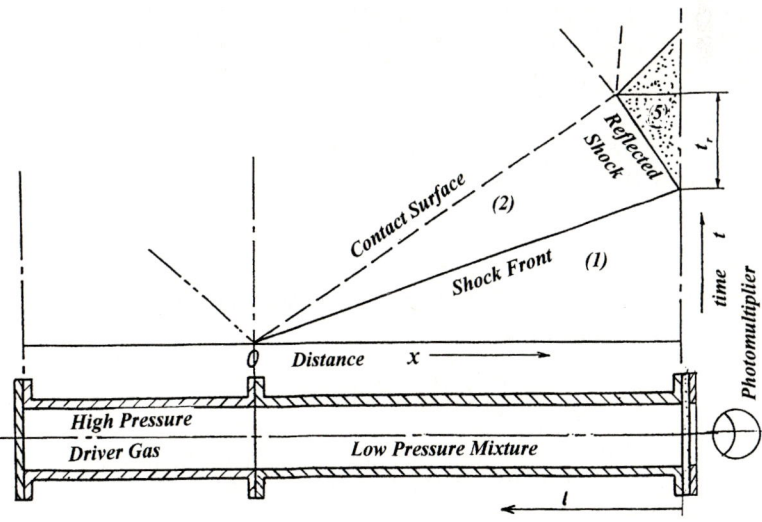

Fig. 6.3. Shock tube and shock diagram applied for the investigation of ignition in a mixture

in which the molar number of the mixture m heated behind the reflected shock waves, however, increases with the propagation of the reflected shock waves, namely

$$m = D \frac{F}{V_m} t, \qquad (6.1)$$

where D is the propagation velocity of the reflected shock wave, F the cross-section area of the shock tube, V_m the molar volume of the mixture corresponding to the shock heated state, and t time from the reflection of the incident shock from the end wall. Substituting this relation into (3.13),

$$D \frac{F}{V_m} \mu(t) t = -\frac{d \ln P(t)}{dt}. \qquad (6.2)$$

If $\mu(t)$ is constant in time and expressed by μ_t, we obtain

$$\mu_t = 2 V_m \frac{\ln P(0) - \ln P(t)}{F D t^2}. \qquad (6.3)$$

μ_t can be obtained from the slope of the line in the diagram of $\ln P(t)$ vs. t^2. In this case, however, the growing period τ of the ignition nuclei must be considered. Therefore $t = t_i - \tau$, where t_i is the observed induction period of ignition.

In an irreversible phenomenon the fluctuations are observed not only in time but also in space. In the spontaneous ignition not only the induction period but also the position of ignition fluctuates. The ignition probability can be obtained from the fluctuation of ignition position, too. If l is distance from the end of the shock tube to the ignition position and t the ignition induction period at the spontaneous ignition behind reflected shock waves in a shock tube,

$$l = Dt \qquad (6.4)$$

Substituting this relation into (6.3), the ignition probability μ_l is expressed by the following equation:

$$\mu_l = 2 V_m D \frac{\ln P(0) - \ln P(l)}{F l^2}, \qquad (6.5)$$

where l is also the length of the heated mixture in which the reflected shock waves propagate in time t and

$$P(l) = \int_l^\infty q(l) dl, \qquad (6.6)$$

in which $q(l)$ is the probability density corresponding to the position. Thus, the ignition probability μ_l of 1 mol mixture per unit time can be calculated from $P(l)$–l^2 diagram. Under the same condition $\mu_t = \mu_l$.

6.2 Spontaneous Ignition in a Hydrogen–Oxygen Mixture Behind Shock Waves

As already described, chemical kinetics has been applied to explain the phenomena of the ignition and combustion, especially the explosion limits, i.e., spontaneous ignition limits of the hydrogen–oxygen mixture. Namely, the inverse S-form explosion limits, well known as "explosion peninsula" among combustion scientists, as shown in Fig. 2.4, has been explained by the thermal explosion theory and the chain-branching kinetics with a consideration of the wall effect. These theories, however, are valid under the assumption of a homogeneous and steady process, that is, valid merely for a reversible process. Both the classical theories, therefore, cannot be applied to such a distinctly irreversible process like ignition which fluctuates in time and space. Instead of these, the stochastic theory is to be applied to investigate the ignition phenomena in a hydrogen–oxygen mixture.

The spontaneous ignition in a hydrogen–oxygen mixture behind reflected shock waves in a shock tube has been investigated by many scientists. As an example, Fig. 6.4 shows a diagram of logarithm of the ignition induction period τ_m in relation to the reciprocal mixture temperature $1/T$ reported by Solouhkin.[46] In this diagram we recognize some fluctuations in the ignition induction period.

Such fluctuations of ignition induction period are observed in every ignition phenomenon and have been attributed to heterogeneity of the mixture or impurities. Especially in a shock tube having small diameter a larger fluctuation is

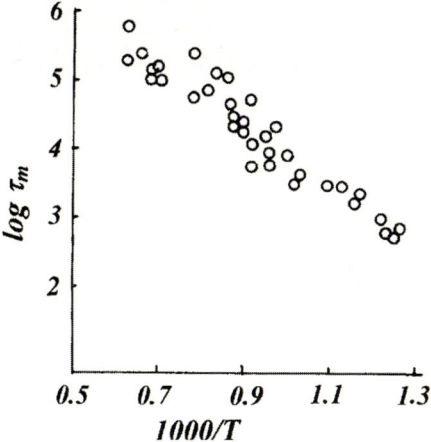

Fig. 6.4. Ignition induction period τ_m in a stoichiometric hydrogen–oxygen mixture with respect to the reciprocal mixture temperature $1/T$

observed and the reason for it is attributed to the boundary layer in the flow behind shock waves and one says shock tube having a larger diameter than 100 mm should be applied to the experiment of ignition.

Usually the so-called effective activation energy E_e has been calculated according to an equation in Arrhenius' formula $\ln \tau_m = \text{Const.} \exp(-E_e / RT)$ from the diagram of τ_m with respect to $1/T$. From the experimental results in a shock tube having a smaller diameter, a larger activation energy has been obtained. This is also a reason for applying a larger shock tube in the experiment of ignition.

In order to examine if the heterogeneity, impurities, or boundary layer play some important roles on the ignition, it is necessary to use shock tubes having different diameters in the experiments, applying the stochastic ignition theory.[45]

6.2.1 Experimental Method

The experiments are carried out by using two shock tubes of stainless steel having different inner-diameter, 49 mm and 26 mm, respectively. Both the shock tubes consist of 1 m long driver tube and a 3 m long low-pressure tube, as schematically shown in Fig. 6.5a. Hydrogen–oxygen mixture ($H_2 + 9O_2$) is filled in the low-pressure tube under an arbitrary pressure from 20 to 80 kPa and room temperature of 293 K, while nitrogen or helium gas is filled in the driver tube under an arbitrary pressure from 250 to 750 kPa. Breaking aluminum

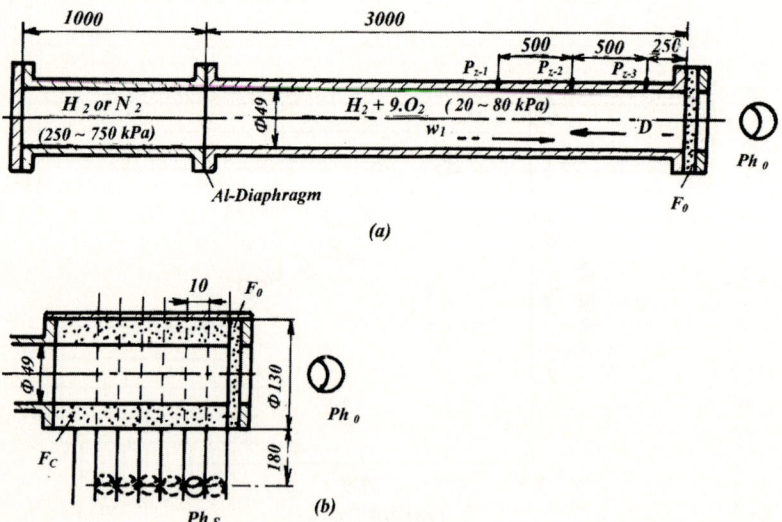

Fig. 6.5. Measurement system of ignition in a hydrogen–oxygen mixture behind reflected shock waves

diaphragms of 0.05 mm or 0.1 mm thickness inserted between the driver tube and low-pressure tube, shock waves having a Mach number from 2.2 to 2.5 are driven into a hydrogen–oxygen mixture.

The propagation velocity w_1 of the incident shock waves is measured by the piezoelectric pressure transducers P_{z1}, P_{z2}, and P_{z3} set in different positions on the shock tube. The mixture is compressed behind the shock waves reflected from the end to a density between 0.2 and 0.9 kg m^{-3} and a temperature between 800 and 1,100 K. The state of the mixture behind reflected shock waves can be calculated from the shock propagation velocity w_1 and the initial state. The spontaneous ignition in the mixture behind the reflected shock waves is investigated.

The time interval between the arrival of the incident shock wave at the plexiglas window of the tube end and appearance of the ignition luminescence observed by a photomultiplier Ph$_0$ aligned along the tube axis behind the plexiglas window is taken as the induction period of ignition t_i.

The ignition position is measured using an apparatus shown in Fig. 6.5b. At the end of the shock tube, a plexiglas tube F_c having a length of 10 cm and the same diameter as the shock tube is attached, dividing its inner space into several segments of 10 mm width side by side along the tube axis. In this direction a second photomultiplier Ph$_s$ is set so that the luminescence of the ignition in each segment can be separately observed through an oscilloscope, moving Ph$_s$ successively to the position of each segment.

The ignition in the whole space is observed by the first photomultiplier Ph$_0$ through the same oscilloscope at the same time. Comparing both the oscillograms of ignition luminescence observed by Ph$_0$ and Ph$_s$, we can recognize, in which segment the ignition takes place.

The experiments are carried out at different temperatures and densities of the mixture behind the reflected shock waves. In order to obtain the histograms of the induction period of ignition as well as those of the ignition position which fluctuates over a range much wider than the experimental error, the experiment under the same condition is repeated many times and about 50 fitted results whose density and temperature of the mixture behind the reflected shock waves fluctuate within 1.5% are selected for the calculation. An example of such histograms of ignition induction period is shown in Fig. 6.6.

6.2.2 Experimental Results

From the histograms of the ignition induction period t_i obtained in our experiments, as shown in Fig. 6.6, we obtain ln $P(t)$–t_i diagrams according to (3.11), from which the ignition nucleus growing period τ can be estimated, as shown in Fig. 6.7a. Then we obtain a straight line of ln $P(t)$ in relation to $t^2 = (t_i - \tau)^2$ from each diagram, as shown in Fig. 6.7b. From the slope of the straight line the probability of ignition μ_t (mol^{-1} ms^{-1}) which is constant in time can be calculated according to (6.3).

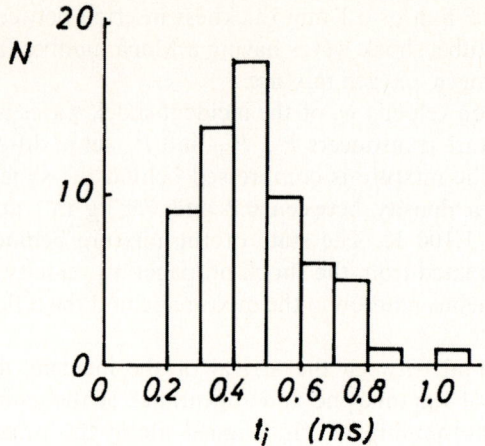

Fig. 6.6. Histogram of the induction period t_i of ignition in a $(H_2 + 9O_2)$ mixture behind reflected shock waves in a shock tube

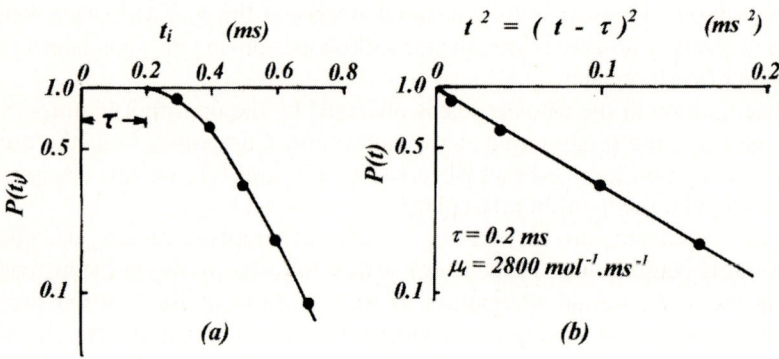

Fig. 6.7. (a) Log $P(t)$ with respect to ignition induction period t_i and (b) log $P(t)$ with respect to t in a $(H_2 + 9O_2)$ mixture having a temperature of 819 K and density of 0.40 kg m^{-3} $t = t_i - \tau$, where τ is the ignition nucleus growing period

From a histogram of ignition positions obtained in our experiments, the probability density $q(l)$ in relation to the distance l from the shock tube end shown in Fig. 6.8a is obtained. From the probability density we obtain the relation between $\ln P(l)$ and l^2, as shown in Fig. 6.8b, then further the probability of ignition μ_1 (=4,600 mol^{-1} ms^{-1}) can be calculated according to (6.5). In Fig. 6.8b $\ln P(t)$ in relation to $t^2 = (t_i - \tau)^2$ obtained in the experiment under the same conditions is also illustrated, from which the ignition probability μ_t is calculated to be 4,600 mol^{-1} ms^{-1}. This proves $\mu_t = \mu_1$ under the same conditions, as the

6.2 Spontaneous Ignition in a Hydrogen–Oxygen Mixture

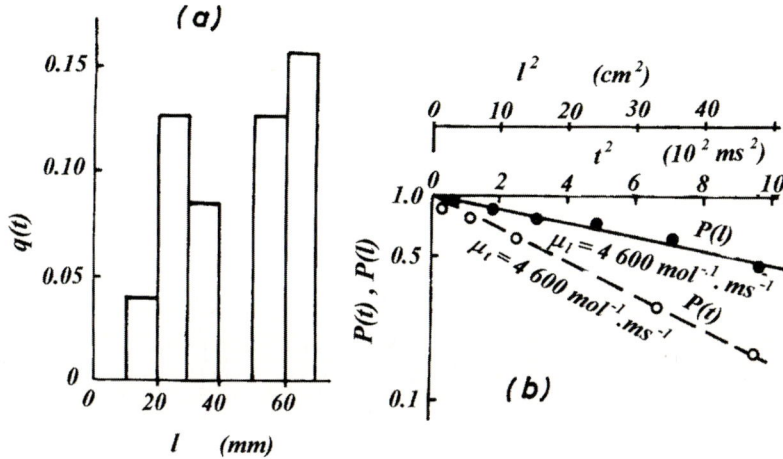

Fig. 6.8. (a) Histogram of the ignition position l, where l is distance from the shock tube end and (b) log $P(t)$ with respect to t^2, and log $P(l)$ to l^2, respectively, in a $(H_2 + 9O_2)$ mixture having a temperature of 856 K and a density of 0.36 kg m^{-3}.

stochastic ignition theory suggests. Table 6.1 represents all the experimentally obtained results. The experimental error of the ignition probability is estimated to be with in ±15%.

In (3.15) and (3.17), the frequency factor A, constant B, and energy W released by the initiation reaction and kept in the reaction system depend on the collision number of the particles which participate in the ignition reaction. As the mixture temperature in the experiments is varied only from 800 to 1,100 K and the velocity of the particles is proportional to $\sqrt{T}$, the influence of the temperature difference on the collision number and the frequency factor A is estimated to be less than the experimental error. According to (3.15) and (3.17), therefore, the logarithm of the ignition probability μ as well as that of ignition nucleus growing period τ is expressed by a straight line as a function of the reciprocal mixture temperature $1/T$.

Figure 6.9 represents such relations among ln μ, ln τ, and $1/T$ for different mixture densities.

According to (3.15) and (3.17), (E_1+E_2-W), (E_2-W), E_1, A, and B can be calculated from the diagram of ln μ and ln τ vs. $1/T$ in Fig. 6.9, which are listed in Table 6.2.

The activation energy E_1 for the initiation reaction is almost independent of the mixture density and always about 42 kJ mol^{-1}, while (E_2-W) depends on the mixture density, the frequency factor A as well as the energy W released by the initiation reaction and supplied to the following development of the reaction after dispersion by collision depends on the mixture density ρ.

Table 6.1. Summarized experimental results

number of experiment	A-11	A-12	A-13	A-14	A-21	A-22	B-11	B-12	B-13	C-11	C-12
shock tube diameter (mm)	49	49	49	49	26	26	49	49	49	49	49
M_1	2.21	2.27	2.34	2.59	2.35	2.50	2.20	2.28	2.46	2.33	2.47
T (K)	819	856	878	1,032	895	974	807	848	941	872	949
ρ (kg m^{-3})	0.40	0.36	0.43	0.41	0.44	0.43	0.74	0.76	0.82	0.19	0.21
D (cm ms^{-1})	34.1	34.5	34.9	35.7	35.1	35.9	34.4	34.6	35.5	34.8	35.6
τ (ms)	0.20	0.13	0.12	0.05	0.13	0.07	0.03	0.02	0.01	0.18	0.17
μ_t (10^3 mol^{-1} ms^{-1})	2.8	4.6	7.0	34.3	7.1	2.3	2.7	5.6	20.0	8.8	1.5
μ_l (10^3 mol^{-1} ms^{-1})		4.6									
$t_m{}^2$ (ms)	0.45	0.32	0.25	0.11	0.42	0.22	0.21	0.17	0.09	0.39	0.32

M_1 is Mach number of the incident shock waves, T and ρ temperature and density of the mixture behind the reflected shock waves, respectively, and D propagation velocity of the reflected shock waves

6.2 Spontaneous Ignition in a Hydrogen–Oxygen Mixture

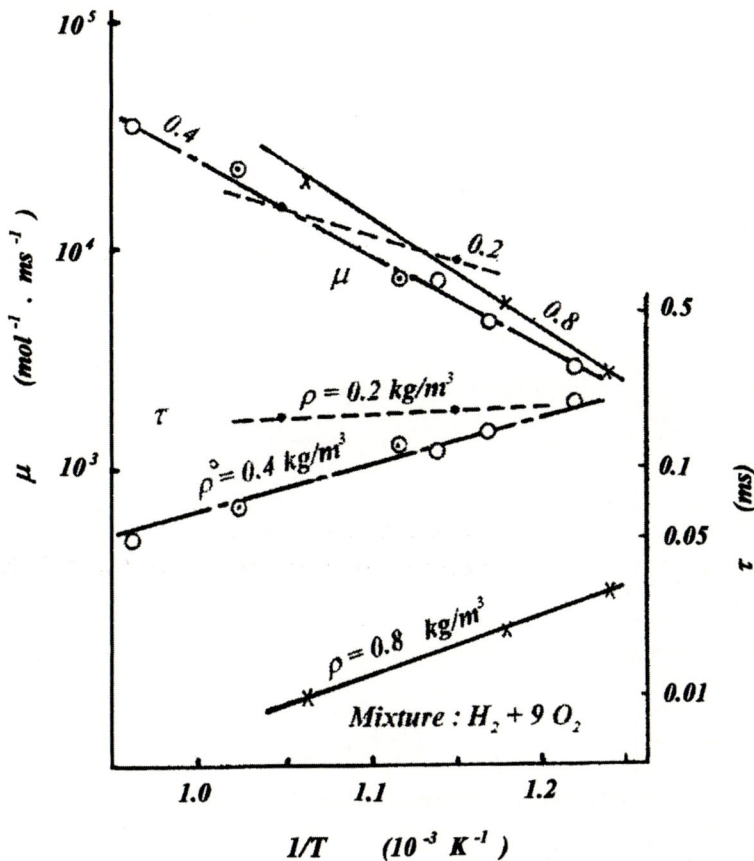

Fig. 6.9. Logarithms of ignition probability μ (mol^{-1} ms^{-1}) and ignition nucleus growing period τ (ms) in a (H$_2$ + 9O$_2$) mixture with respect to the reciprocal mixture temperature 1/T

Table 6.2. Results calculated from experimental results

ρ (kg m^{-3})	0.2	0.4	0.8
A (mol^{-1} ms^{-1})	5.5×10^6	7.8×10^7	4.2×10^9
B (ms)	7.7×10^{-2}	5.5×10^{-4}	9.6×10^{-5}
$E_1 + E_2 - W$ (kJ mol^{-1})	48.1	79.4	96.1
$E_2 - W$ (kJ mol^{-1})	6.3	37.6	54.3
E_1 (kJ mol^{-1})	42.0	42.0	42.0

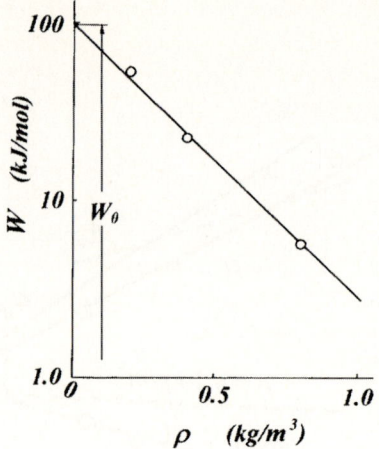

Fig. 6.10. Relation between the energy W supplied from the initiation reaction to the development one and mixture density ρ

From the values A, (E_1+E_2-W) and (E_2-W) experimentally obtained under several different conditions, we estimate that A is approximately proportional to ρ^5, that is

$$A = b\rho^5, \tag{6.7}$$

where b is a proportionality constant, while $W = W_0 (1 - \alpha_c)^n$, where W_0 is the energy released by the initiation reaction and estimated to be 105 kJ mol^{-1} from the experimental results, α_c is a coefficient of energy loss due to collision between particles, and n the collision number during the reaction and proportional to ρ, as shown in the diagram of Fig. 6.10, namely

$$W = W_0(1 - \alpha_c)^{a\rho}, \tag{6.8}$$

where a is a proportionality constant. We can, thus, further estimate the ignition probability μ of a mixture at an arbitrary temperature T and density ρ of the mixture.

From the experimental results, the energy loss coefficient α_c is estimated to have an order of 10^{-4}. This suggests that the energy transport at the ignition reaction is carried out not by collisions among the molecules, radicals, or atoms, but by that between electrons and molecules, radicals, or atoms, considering particle mass.

6.2.3 Explosion Limits

As the ignition is a stochastic phenomenon, there is no ignition limit which strictly separates the ignition region from the non-ignition region, but a combustible mixture has always some ignition probability, i.e., an ability of ignition corresponding to its state. The empirically observed spontaneous ignition limits, i.e., the

6.2 Spontaneous Ignition in a Hydrogen–Oxygen Mixture

explosion limits, have been decided, observing a combustible mixture in a state for a certain definite period, and if an explosion occurs during the period, the state of the mixture belongs to the explosion region, while it belongs to non-explosion region, if no explosion is observed. The so-called explosion limits, therefore, should be the state where the mixture has a certain constant ignition probability.

As already explained, E_1 and E_2 which are independent of the mixture density, A and B at an arbitrary mixture density can be estimated from the experimental results, (6.7) and (6.8). Applying these values to (3.15), the probability μ of spontaneous ignition in the mixture ($H_2 + 9O_2$) having an arbitrary density ρ at an arbitrary mixture temperature T can be estimated and expressed in a $\ln \mu$ against $1/T$ diagram, as illustrated in Fig. 6.11. The higher the mixture density, the larger the frequency factor A, and the steeper are the slopes of the straight lines in the diagram of $\ln \mu$ against $1/T$. That is, the straight lines of $\ln \mu$ cross with each other somewhere in the diagram of $\ln \mu$ vs. $1/T$.

As the mixture should have a constant value of the overall ignition probability $m\mu$ at the explosion limit, we can deduce the temperature, density, and pressure of the mixture at the explosion limits according to (3.15) or the diagram of $\ln \mu$ against $1/T$ for the mixture ($H_2 + 9O_2$). For example, the state of the 1 mol mixture having $m\mu = 100$ ms^{-1} is shown in Fig. 6.11 as a straight line parallel to $1/T$-axis or in a pressure-temperature diagram in Fig. 6.12, which has an inverse S-form just like the so-called explosion peninsula well known in hydrogen–oxygen

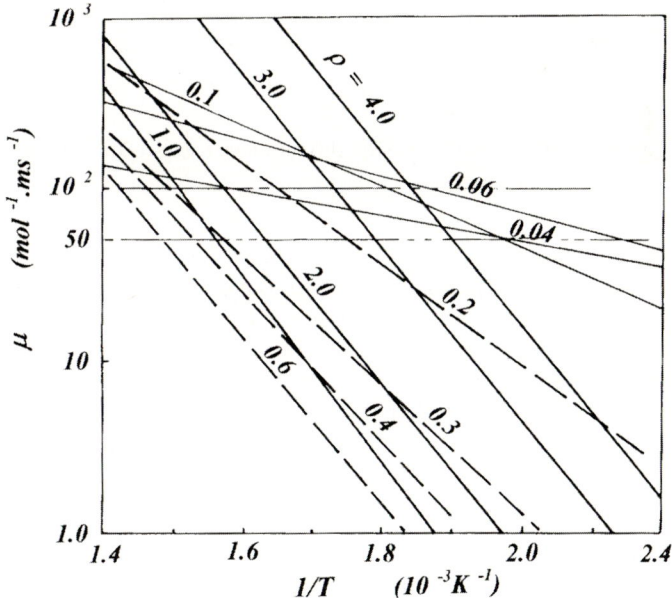

Fig. 6.11. Theoretically estimated ignition probability μ with respect to the reciprocal temperature $1/T$ of mixture ($H_2 + 9O_2$) having different density ρ (kg m^{-3})

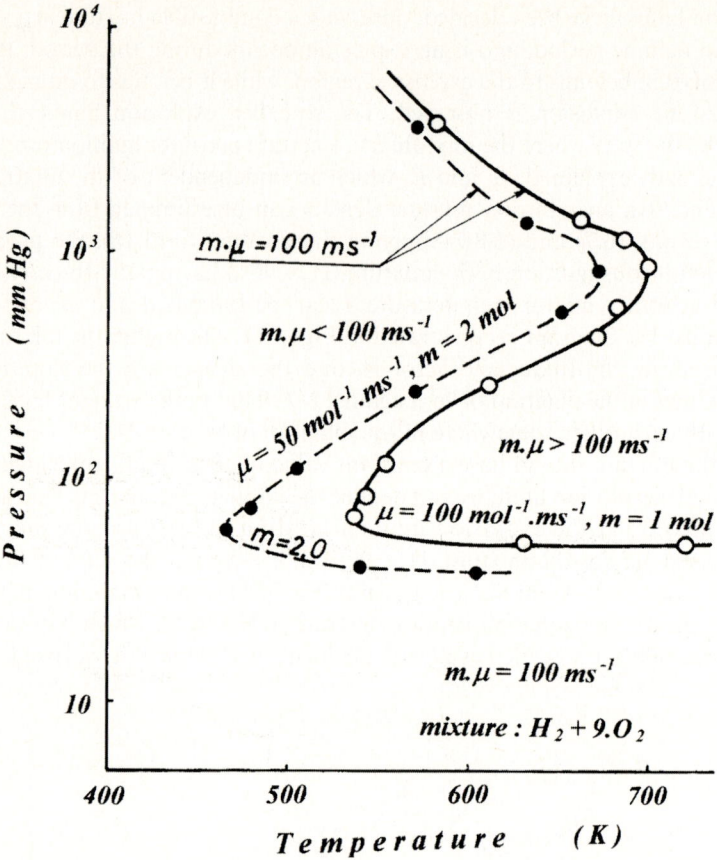

Fig. 6.12. Explosion limits of ($H_2 + 9O_2$) mixture according to the stochastic ignition theory

mixtures. Only the pressure of the mixture on these explosion limits is somewhat higher than the empirically observed ones in the stoichiometric mixture, shown in Fig. 2.7. As the reason for this discrepancy, the differences of the mixture ratio, experimental method, and others should be considered.

As the stochastic ignition theory can be applied to the ignition in other combustible mixtures, it is expected that such explosion limits of peninsula form may appear at the ignition in other mixtures, for instance, in $CO + O_2$, $P + O_2$, $(C_2H_5)_3P + O_2$, etc.

The fact that the probability of spontaneous ignition is independent of time, as shown in Fig. 6.7b, means that the intermediate products during the ignition which should govern the ignition itself by the chain-branching kinetics do not play any role in ignition. Such intermediates are produced only as a result of the ignition reaction. Important for ignition is, therefore, not its chemical reaction process, but its energy process.

6.2.4 Quantity Effect on the Ignition

The overall ignition probability per unit time of m mole mixture is expressed by $m\mu$ and proportional to the quantity of the mixture. The more mixture there is, the more easily it ignites. At an ignition limit of $m\mu$ = constant, μ decreases with increase of mixture quantity. Therefore, the ignition limit shifts corresponding to the mixture quantity, as shown also in Fig. 6.12 in which the explosion limits of 2 mol hydrogen–oxygen mixture having the same overall ignition probability $m\mu = 100$ ms^{-1} is shown by the curve of $\mu = 50$ mol^{-1} ms^{-1}. As well known in the classical theory, the second explosion limit of hydrogen–oxygen mixture shifts toward the lower temperature region upon enlarging the vessel of the mixture, i.e., upon increasing the mixture quantity.

In the investigation of the ignition, the average induction period t_m for ignition is often measured and discussed. According to (3.18), the average induction period is $t_m = \tau + 1/(m\mu)$ and depends on the ignition nucleus growing period τ and the overall ignition probability $m\mu$. In the case of large $m\mu$, i.e., in the case of large mixture quantity m or of large ignition probability μ, the average induction period t_m is close to τ. In this case, the effective activation energy E_e estimated from the relation between the logarithm of the average induction period and the reciprocal mixture temperature according to the Arrhenius' formula is close to $(E_2 - W)$, while in the case of small overall ignition probability $m\mu$, namely in the case of small mixture quantity m or of small μ, the effective activation energy is close to $(E_1 + E_2 - W)$.

The experimental results illustrated in Fig. 6.13 which represents the logarithms of the average induction period t_m in the hydrogen–oxygen mixture ($H_2 + 9O_2$)

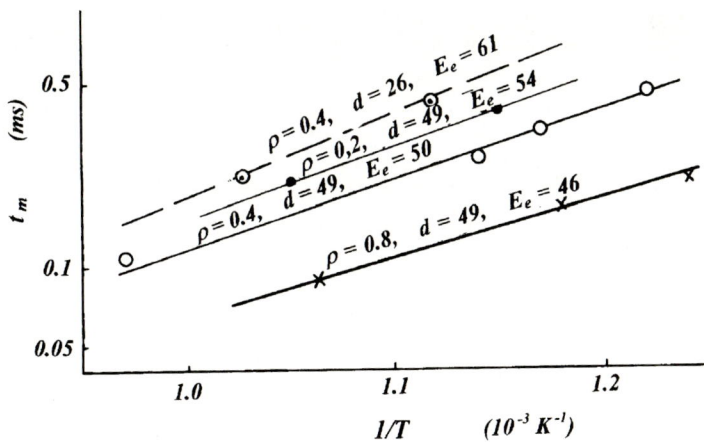

Fig. 6.13. Logarithm of average induction period t_m for spontaneous ignition behind reflected shock waves in shock tubes having different inner-diameter d (mm) with respect to the reciprocal temperature $1/T$ of the mixture having different density ρ (kg m^{-3})

behind reflected shock waves with respect to the reciprocal mixture temperature $1/T$, support the tendency mentioned above. The effective activation energy E_e obtained from t_m measured in the shock tube of 26 mm inner-diameter is about 60 kJ mol^{-1}, while that obtained from the measured results in the shock tube of 49 mm inner-diameter is about 50 kJ mol^{-1}.

On the other hand, as shown in Table 6.1 or Fig. 6.9, all values of $\ln \mu$ and $\ln \tau$ of the same mixture having the same density, obtained from the experimental results by two shock tubes having different inner diameter with respect to the reciprocal mixture temperature fall, respectively, on the same straight line. This suggests that the ignition reaction process is independent of the shock tube size, and affected neither by the boundary layer nor by the tube wall which has been used to explain the effect of the vessel size on the explosion limits according to the chain-branching kinetics.

6.2.5 Reaction Mechanism

As the explosion limits of the hydrogen–oxygen mixture, the so-called explosion peninsula, can be explained as the state where the mixture has the same ignition probability, the same reaction proceeds within the three explosion limits and such three explosion limits appear because of the difference of the collision number among the reacting particles, although the detail of the ignition mechanism is not yet clear.

From the results listed in Table 6.2, the relation between the energy levels and the reaction process is schematically illustrated in Fig. 6.14. Considering the variation of energy level in the process, the reactions following the initiation reaction, i.e., those in the development period agree with those in the classical theory. Considering that the probability of the spontaneous ignition is independent of the time during the ignition process, the intermediate products play no role in the ignition, while the chain-branching kinetics assume that the

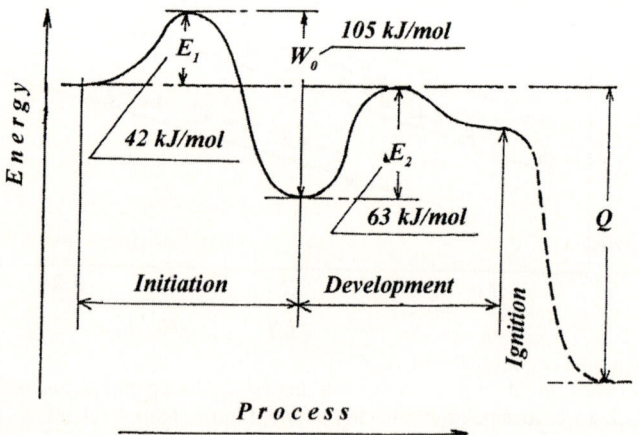

Fig. 6.14. Energy process of the ignition reaction in a hydrogen–oxygen mixture

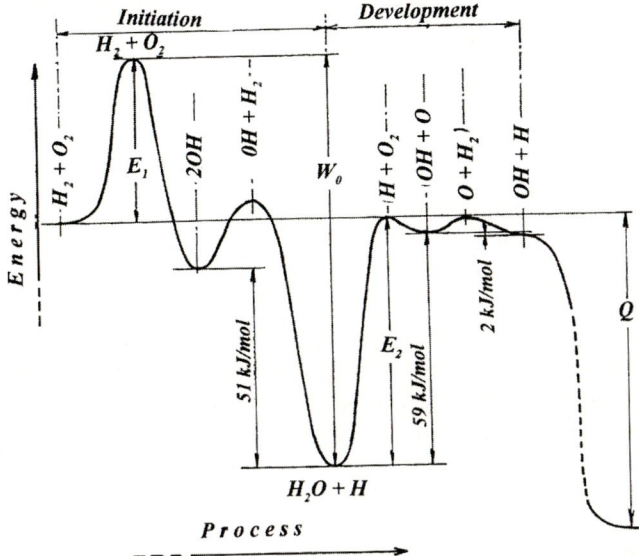

Fig. 6.15. A reaction model of ignition in a hydrogen–oxygen mixture proposed according to the stochastic ignition theory

intermediates produced during the ignition control the reactions. According to the stochastic ignition theory, the intermediates appear only as the results of the reaction, but never control the ignition.

In order to satisfy these conditions, the initiation reaction must take place in some parts of the mixture having a temperature much higher than the average one, producing active radicals much enough to push the following development reactions forward. The initial state of the mixture, thus, governs the ignition. As an example, the following reaction mechanism may be proposed, as schematically illustrated in Fig. 6.15.

1. $H_2 + O_2 \rightarrow 2OH$ initiation reaction
2. $OH + H_2 \rightarrow H_2O + H$ development reactions
3. $H + O_2 \rightarrow OH + O$
4. $O + H_2 \rightarrow OH + H$
5. $H + OH + M \rightarrow + M$

During the ignition nucleus growing period, the same reactions as those from 2 to 4 take place around the ignition nucleus and further propagate, driven by the heat of reaction 5.

According to the stochastic ignition theory, there is no explosion limit and so long as a combustible mixture exists, there is always some ignition probability and danger of explosion. In order to avoid any explosion,

1. Never make a combustible mixture
2. The mixture should be divided into many small parts to minimize the ignition probability.

6.3 Spontaneous Ignition in Hydrocarbon–Air Mixtures

As oils of hydrocarbons are often used as fuels for engines, the combustion and ignition in mixtures of air, and different hydrocarbons have long time been investigated by many scientists and engineers.

In hydrocarbon–air or hydrocarbon–oxygen mixtures we usually observe two kinds of spontaneous ignition; the initiation of a cool flame having blue color and hot ignition. In a pressure–temperature diagram the hot ignition is observed in a region having a higher temperature and pressure than those of the cool flame, as shown in Fig. 6.16. The classical ignition theory divides the ignition region into two parts, namely, the cool flame region and the hot ignition region, and explains the cool flame appears as a prereaction of the hot ignition. In order to explain the ignition mechanism by the chemical kinetics, enormously many elementary reactions have been proposed, but the ignition mechanism is not yet clear.

In this book the experimental results of spontaneous ignition mainly in mixtures of paraffin hydrocarbons, i.e., methane, ethane, ethylene, n-hexane, and air behind reflected shock waves in a shock tube as well as the influence of lead on the ignition are explained according to the stochastic ignition theory.[47, 48]

6.3.1 Ignition Probability and Mechanism in Paraffin–Fuel–Air Mixtures

The experiments of spontaneous ignition in mixtures of methane, ethane, n-hexane, or other paraffin hydrocarbons, and air having different concentrations are carried out using a shock tube just like that in hydrogen–oxygen mixtures explained in Chap. 5.

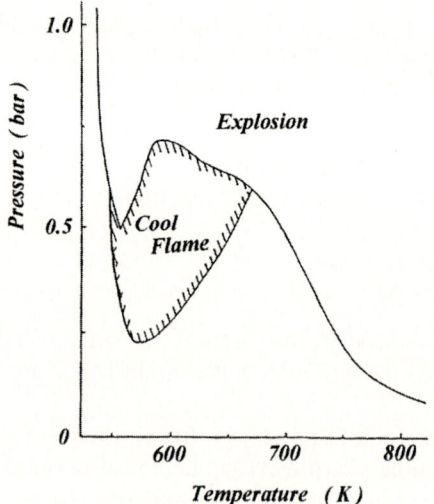

Fig. 6.16. Explosion limits of propane–oxygen mixture having a cool flame zone in a pressure-temperature (P–T) diagram

Probability of Spontaneous Ignition

The shock tube used in the experiments is composed of a driver tube having a length of 1 m and a low pressure tube of steel having a length of 3 m and a square cross-section of 3 cm × 3 cm. A mixture of a fuel excess ratio or equivalence ratio of fuel to air from 1.0 to 0.1 is introduced into the low pressure tube at room temperature of 20°C and an arbitrary pressure, while high pressure air or nitrogen gas is filled in the driver tube at an arbitrary pressure, so that shock waves are driven in the mixture after rupture of aluminum diaphragm set between both the tubes to keep a state in the mixture behind the shock waves reflected from the tube end and having a temperature from 700 to 900 K, a pressure of about 300 kPa and a specific volume from 18×10^{-3} to 24×10^{-3} m^3 mol^{-1}. The ignition is observed by a photomultiplier aligned along the tube axis behind a plexiglas window at the shock tube end just like in the experiments of the ignition in hydrogen–oxygen mixtures explained in Chap. 5. The time interval from the arrival of the incident shock at the plexiglas window at the tube end to the appearance of the ignition luminescence is taken as the induction period of ignition t_i.

By repeating each experiment under the same conditions, the histogram of induction period t_i of spontaneous ignition in the mixture behind the reflected shock waves, consequently the probability density $q(t_i)$ is obtained, from which further the diagram of ln $P(t)$–$(t_i - \tau)^2$, then ignition probability μ and ignition nucleus growing period τ are calculated according to (6.3) or obtained from the diagram. In Figs. 6.17–6.19 logarithms of ignition probability ln μ and that of ignition nucleus growing period ln τ of the mixtures of n-hexane–air, ethane–air, methane–air, and ethylene–air are illustrated with respect to the reciprocal mixture temperature $1/T$.

In Table 6.3 the values of activation energies $(E_1 + E_2 - W)$, $(E_2 - W)$, E_1, and frequency factor A and B calculated according to the equations

$$\mu = A \exp\left[-\frac{E_1 + E_2 - W}{RT}\right], \qquad (3.15)$$

$$\tau = B \exp\left[\frac{E_2 - W}{RT}\right]. \qquad (3.17)$$

are listed. As shown in Figs. 6.17–6.19, the ignition probability μ of lean mixture (having smaller equivalence ratio Φ) is larger than that of rich mixture (having larger equivalence ratio Φ) in the higher mixture temperature, while those in lower mixture temperature show an inverse tendency.

The average induction period t_m of ignition is expressed as follows:

$$t_m = \tau + \frac{1}{m\mu}. \qquad (3.18)$$

As shown in Fig. 6.20, the equivalence ratio having the minimum average induction period of ignition in n-hexane–air mixtures is different according to the

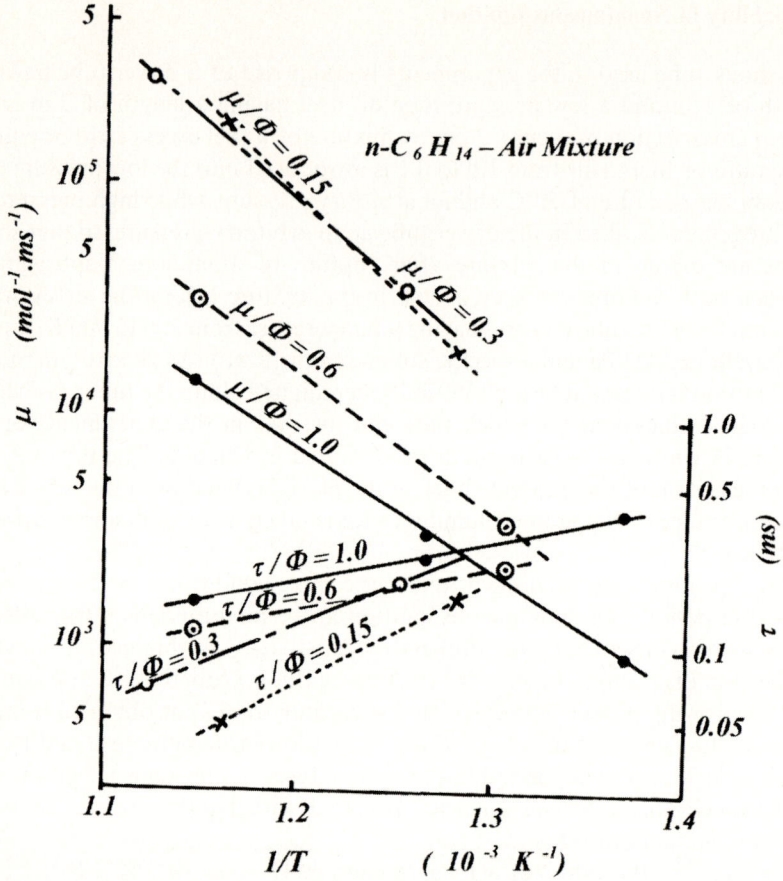

Fig. 6.17. Logarithm of ignition probability μ and that of ignition nucleus growing period τ in n-hexane–air mixtures having different equivalence ratios Φ with respect to the reciprocal mixture temperature $1/T$

mixture temperature. The higher the temperature, the leaner the mixture having the minimum average ignition induction period t_m.

In the classical theories the effective activation energy is estimated from the mean induction period of ignition. In Fig. 6.21 some mean induction periods of ignition in n-hexane–air mixtures having equivalence ratios $\Phi = 1.0$ and 0.375 in a closed vessel under adiabatic compression experimentally obtained by other scientists are illustrated with respect to the reciprocal mixture temperature $1/T$, comparing with those estimated from the experimental results using a shock tube according to (3.18), considering the mixture quantities m (mole number), which are shown in the table:

As the ignition instant by other scientists is taken at that of sudden pressure increase and the ignition nucleus is much larger than that in our experiments in which the ignition is taken at the beginning of the ignition luminescence, the

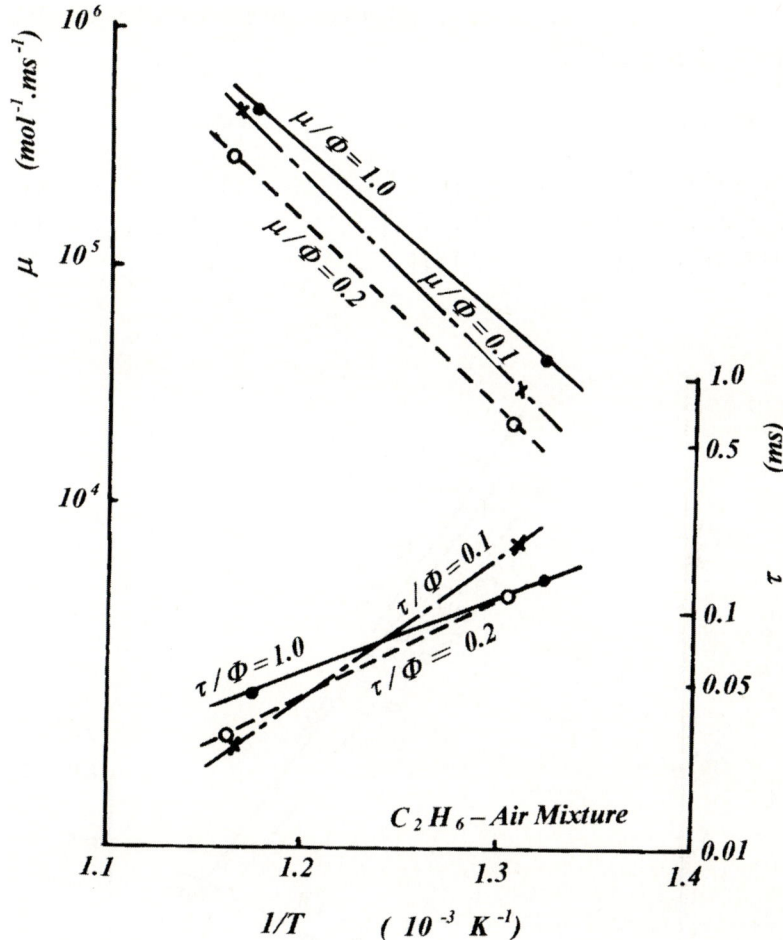

Fig. 6.18. Logarithm of ignition probability μ and that of ignition nucleus growing period τ in ethane–air mixtures having different equivalence ratios Φ with respect to the reciprocal mixture temperature $1/T$

induction period in the former experiments is much longer than those of the latter, but both the effective activation energies agree with each other well.

Activation Energies

As shown in Table 6.3, the activation energy E_1 for the initiation reaction has an almost constant value from 60 to 80 kJ mol^{-1} in the mixture of n-hexane-air and ethane-air, but that in methane-air and ethylene-air mixture has also an almost constant but double value from 120 to 135 kJ mol^{-1}. In Fig. 6.22 the activation energy E_1 in relation to the carbon atom number in paraffin hydrocarbons is illustrated.

6 Stochastic Ignition Theory

On the other hand the molecular structure of fuels are expressed as follows.

Molecular Structure

```
   n-hexane              ethane        methane     ethylene

   H H H H H H           H H             H          H   H
   | | | | | |           | |             |          |   |
H-C-C-C-C-C-C-H         H-C-C-H        H-C-H        C=C
   | | | | | |           | |             |          |   |
   H H H H H H           H H             H          H   H
```

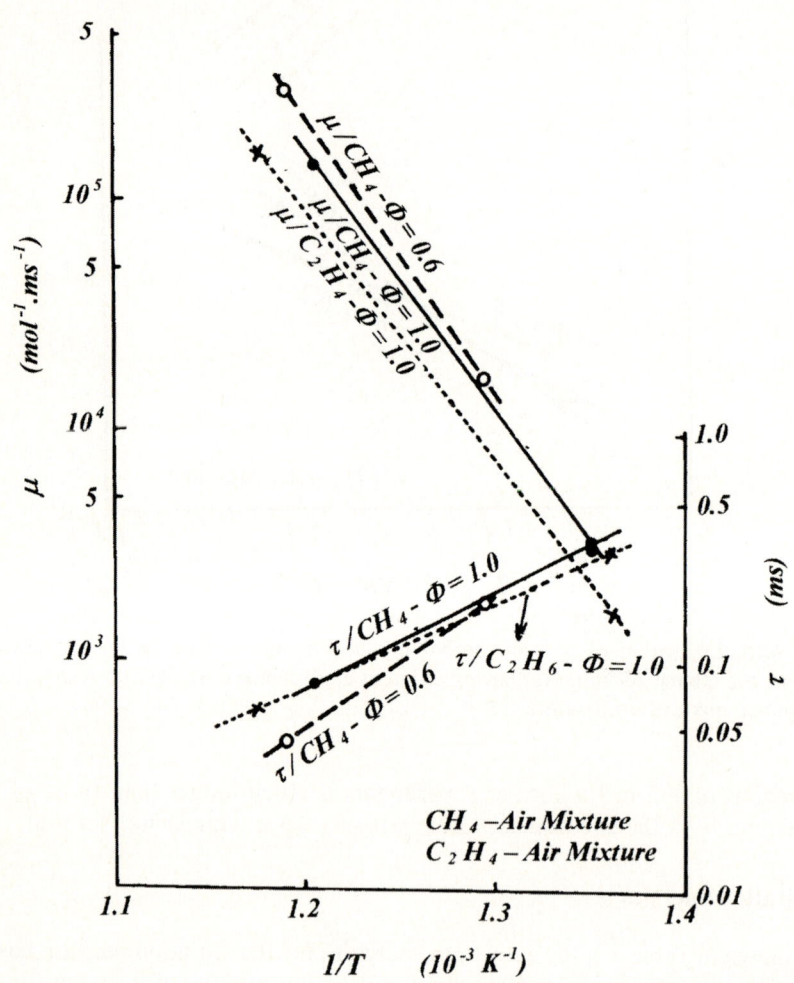

Fig. 6.19. Logarithm of ignition probability μ and that of ignition nucleus growing period τ in mixtures of methane–air and ethylene–air having different equivalence ratio Φ with respect to the reciprocal mixture temperature $1/T$

6.3 Spontaneous Ignition in Hydrocarbon–Air Mixtures

Table 6.3. Summarized experimental results

mixture	Φ	$E_1 + E_2 - W$ (kJ mol^{-1})	$E_2 - W$ (kJ mol^{-1})	E_1 (kJ mol^{-1})	A (mol^{-1} ms^{-1})	B (ms)
n-hexane–air						
n-C$_6$H$_1$ + air	1.0	102	35	67	1.76×10^{10}	1.28×10^{-3}
	0.6	115	36	79	2.38×10^{11}	1.1×10^{-3}
	0.3	135	68	67	2.32×10^{13}	7.3×10^{-6}
	0.15	155	86	69	3.80×10^{14}	3.3×10^{-7}
ethane–air						
C$_2$H$_6$ + air	1.0	134	64	70	7.35×10^{13}	5.6×10^{-6}
	0.2	148	79	69	2.69×10^{14}	4.8×10^{-7}
	0.1	155	81	74	1.13×10^{15}	3.3×10^{-7}
methane–air						
CH$_4$ + air	1.0	220	82	138	8.37×10^{18}	6.3×10^{-7}
	0.6	234	112	123	9.2×10^{19}	8.2×10^{-13}
ethylene–air						
C$_2$H$_4$ + air	1.0	200	73	237		

According to the Table 6.4 showing the binding energies between two atoms, the binding energy of $\equiv$C–C$\equiv$ in *n*-hexane and ethane is the weakest, while there is only $\equiv$C–H in methane and $\equiv$C–H is the weakest in ethylene. Therefore, in the ignition reaction of *n*-hexane–air and ethane–air mixtures the $\equiv$C–C$\equiv$ bond must be broken first and reacts with oxygen molecules, while in that of methane–air and ethylene–air mixtures the $\equiv$C–H bond is broken and reacts with oxygen. Considering that the bond energy of $\equiv$C–C$\equiv$ is much smaller that of $\equiv$C–H, the activation energy E_1 of the initiation reaction in *n*-hexane–air as well as in ethane–air mixture must be much smaller than that in methane–air and ethylene–air mixtures.

The reactions following the initiation one must proceed similarly, first the weakest bonds are broken and then the next weakest ones follow it.

Frequency Factor

The values of A listed in Table 6.3 suggest that the frequency factor A depending on the collision between fuel molecules and oxygen molecules must have a certain relation between the concentrations of fuel and oxygen. The frequency factor A of *n*-hexane–air mixture is nearly proportional to sixth power of the inverse equivalence ratio $1/\Phi$, namely

$A \propto (1/\Phi)^6$ in *n*-hexane–air mixture
$A \propto (1/\Phi)^4$ in ethylene–air mixture
$A \propto (1/\Phi)^4$ in methane–air mixture

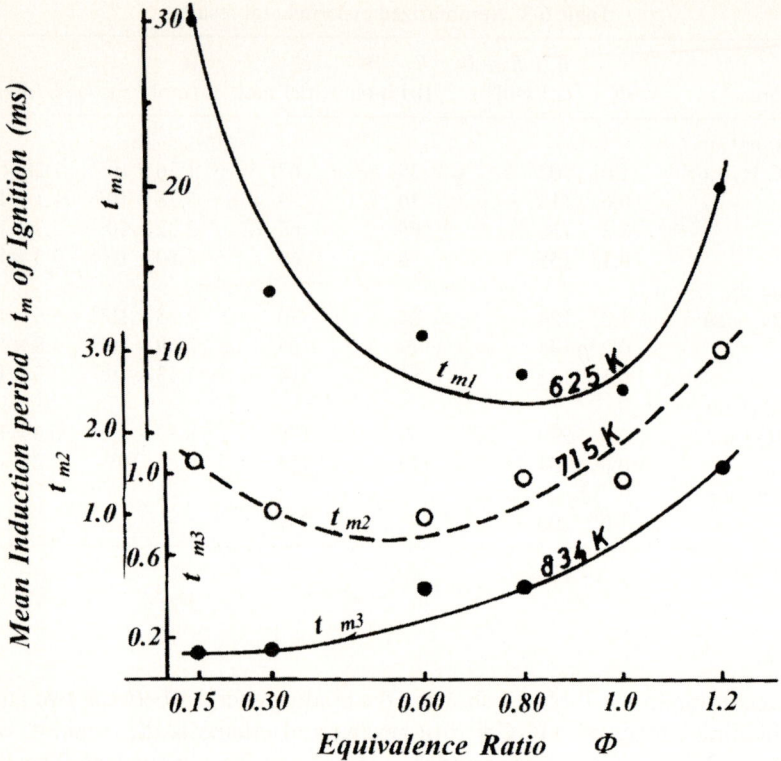

Fig. 6.20. Mean induction period t_m of ignition in *n*-hexane–air mixtures having different initial temperatures with respect to the equivalence ratio Φ

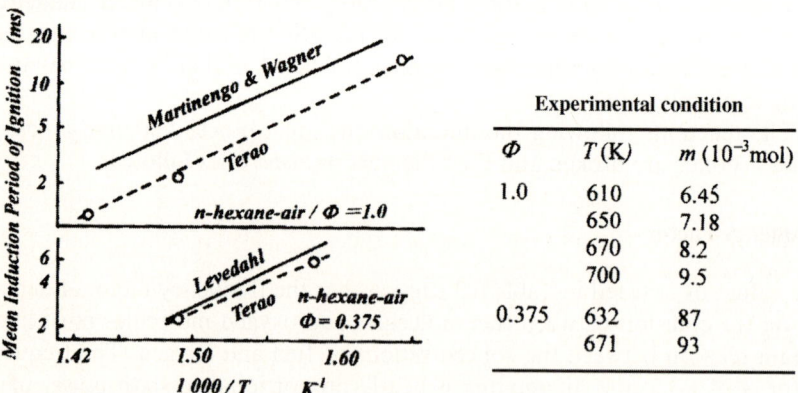

Fig. 6.21. Mean induction period of ignition in *n*-hexane–air mixtures having different equivalence ratios Φ in a closed vessel under adiabatic compression with respect to the reciprocal mixture temperature $1/T$ in comparison with those estimated according to the stochastic ignition theory[44, 49]

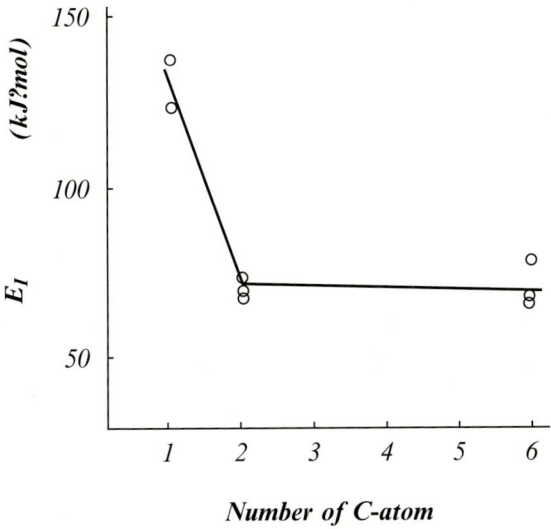

Fig. 6.22. Activation energy E_1 for the initiation reaction in relation to the number of C-atom in different paraffin

Table 6.4.[50] Binding energy of chemical bond

chemical bond	binding energy (kJ mol^{-1})	chemical bond	binding energy (kJ mol^{-1})
H–H	432.2	≡C–C≡	244.9
–O–O–	145.9	≡C–H	364.9
–O–H	460.6	=C=C=	418.0
≡C–O–	292.6	=C≡C=	514.1
		=C=O	593.6–635.4

On the other hand Φ is proportional to the ratio of fuel concentration C_f to oxygen concentration C_o, i.e., $1/\Phi \propto C_o/C_f$ and is proportional to the probability that an oxygen molecule collides with a certain fuel molecule.

Though the combustion does not finish, the ignition can be recognized when the ignition reaction proceeds so far that the ignition nucleus can develop by itself. The ignition is thus recognized when 5 to 7 oxygen molecules in the n-hexane–air mixtures, 3 to 5 ones in ethane–air and methane–air mixtures take part in the reaction with the fuel molecules.

Rigorously said, the translation velocity of each molecule is proportional to square root of its temperature and should be considered at their collision, but the influence of temperature on A may be so small to be neglected in such a narrow temperature range as from 700 to 900 K.

Ignition Mechanism

Concluding the results described above, the ignition mechanism of hydrocarbon-air mixtures can be explained as follows:

1. In n-hexane–air mixtures, the initiation reaction starts first, breaking one of the chemical bonds of C–C and taking part in the reaction with one or two O_2 molecules, then 3 to $6O_2$ take part in the reaction with other C–C bonds. After all C–C are exhausted, $\equiv$C–H bonds come into the reaction with O_2 and the ignition takes place
2. In ethane–air mixtures, the C–C bond is first broken and takes part in the reaction with one or two O_2, then C–H bonds are broken one after another, taking part in the reaction with 2 to $4O_2$ molecules and the ignition occurs
3. In methane–air mixtures, one of C–H bonds is first broken and takes part in the reaction with an O_2 molecule, then the other C–H bonds are broken one after another, taking part in the reaction with 2 or $3O_2$ molecules
4. In ethylene–air mixtures, one of C–H bonds is first broken, taking part in the reaction with one O_2 molecule, then the other C–H bonds are broken one after another, taking part in the reaction with O_2 molecules and the ignition is observed.

In the reaction mechanism described above, not only a single fuel molecule takes part in the reaction with O_2 molecules, but one of fuel groups comes into reaction with O_2 molecules under the highest probability for the reaction explained above, that is, other reactions, for example, one of C–H bonds in n-hexane-air mixture can first be broken even under a very small probability.

Thus the ignition reaction proceeds stepwise, taking part in the reaction with O_2 molecules one after another.

Assuming

Z, the collision frequency between fuel and oxygen molecules in 1 mol mixture during the unit time

a, an arbitrary constant

C_f, C_o, the fuel and oxygen concentrations in the mixture, respectively

l, the number of O_2 molecule necessary for the ignition

ε_{CC}, the activation energy for the reaction of C–C bond with O_2 molecule

ε_{CH}, the activation energy for the reaction of C–H bond with O_2 molecule

p_j, q_j, the numbers of C–C and C–H bonds existing in the reacting fuel molecule at jth step reaction, respectively

w_{j-1}, the energy supplied from the forgoing reaction to jth reaction

R, the gas constant, and

T, the mixture temperature

6.3 Spontaneous Ignition in Hydrocarbon–Air Mixtures

Then the ignition probability μ is expressed by the following equation:

$$\mu = Z\left(a\frac{C_o}{C_f}\right)^1 \left\{p_1\exp\left(-\frac{\varepsilon_{CC}}{RT}\right) + q_1\exp\left(-\frac{\varepsilon_{CH}}{RT}\right)\right\}$$
$$\left\{p_2\exp\left(-\frac{\varepsilon_{CC}-w_1}{RT}\right) + q_2\exp\left(-\frac{\varepsilon_{CH}-w_1}{RT}\right)\right\}\cdots$$
$$\left\{p_j\exp\left(-\frac{\varepsilon_{CC}-w_{j-1}}{RT}\right) + q_j\exp\left(-\frac{\varepsilon_{CH}-w_{j-1}}{RT}\right)\right\}\cdots$$
$$\left\{p_l\exp\left(-\frac{\varepsilon_{CC}-w_{l-1}}{RT}\right) + q_l\exp\left(-\frac{\varepsilon_{CH}-w_{l-1}}{RT}\right)\right\}. \quad (6.9)$$

In general $\varepsilon_{CC} < \varepsilon_{CH}$, so $\exp\{-(\varepsilon_{CC})/(RT)\} \gg \exp\{-(\varepsilon_{CH})/(RT)\}$. Therefore (6.9) should practically be expressed as follows:

$$\mu = Z\left(a\frac{C_o}{C_f}\right)^1 p_1\exp\left(-\frac{\varepsilon_{CC}}{RT}\right) p_2 p_3 p_4 \cdots p_{s+1}$$
$$\exp\left(-\frac{s\varepsilon_{CC}-sw}{RT}\right) q_{s+2} q_{s+3} q_{s+4} \cdots q_{s+t+1} \exp\left(-\frac{t\varepsilon_{CC}-tw}{RT}\right), \quad (6.10)$$

where s is the number of C–C bond coming into the reaction during the ignition nucleus growing period, while t that of C–H bond, and $s + t + 1 = l$. As $E_1 = \varepsilon_{CC}$ and $E_2 = s\varepsilon_{CC} + t\varepsilon_{CH}$, $W = sw + tw$, we obtain our basic formula

$$\mu = A\exp\left[-\frac{E_1+E_2-W}{RT}\right]. \quad (4.15)$$

In the ignition nucleus growing period τ are the same relations as in the ignition probability. Considering that τ is inversely proportional to the reaction velocity, we obtain the following formula described already, as $E_2 = s\varepsilon_{CC} + t\varepsilon_{CH}$ and $W = sw + tw$,

$$\tau = B\exp\left[\frac{E_2-W}{RT}\right]. \quad (4.17)$$

As explained already, the intermediates are produced from the ignition reaction, but never give any influence on the ignition reaction. The break of chemical bonds and reactions with O_2 molecules are defined corresponding to the initial mixture conditions.

6.3.2 Ignition Limits of Hydrocarbon Fuel-Air Mixtures

According to the stochastic ignition theory, there is no ignition limit in any hydrocarbon fuel, too. The so-called ignition limits of hydrocarbon fuels, therefore, should be explained according to the stochastic ignition theory. First we have to investigate the phenomenon of the initiation of the cool flame having blue color and that of the hot ignition.[51]

At the measurement of the induction period of spontaneous ignition in a hydrocarbon–air mixture behind reflected shock waves using a shock tube, the cool flame appearance and hot ignition can be separately observed applying a filter, as the wave length of the light emitted from the cool flame and that from the hot ignition are different.

As illustrated in Fig. 6. 23, two photomultipliers are set outside of the plexiglas window at the tube end so far that both the photomultipliers can observe the whole inside space of the shock tube. One of the photomultipliers has a filter on its front which cannot pass the light having a wave length shorter than $520\ m\mu$, i.e. blue light. By this method we can recognize the hot ignition, if both the photomultipliers observe the light emitted at the same instant, and only the initiation of the cool flame, if only the photomultiplier without filter catches the emission.

The same shock tube used for the experiment of ignition in n-hexane-air mixtures having a square cross section of $3\ cm \times 3\ cm$ in Fig. 6.23, is applied to the experiments of spontaneous ignition in a stoichiometric n-heptane-air mixture behind reflected shock waves.

In Fig. 6.24 an example of the results observed by an oscilloscope is shown, in which a pressure variation P composed of those measured at two different

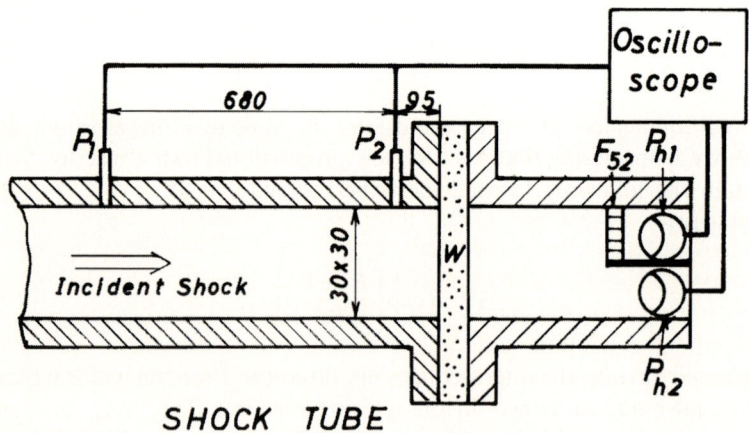

Fig. 6.23. Arrangement of measurement apparatus at the end part of shock tube. P_{z1}, P_{z2}: pressure transducer, W: plexiglas window, F_{52}: filter, P_{h1}, P_{h2}: photomultiplier

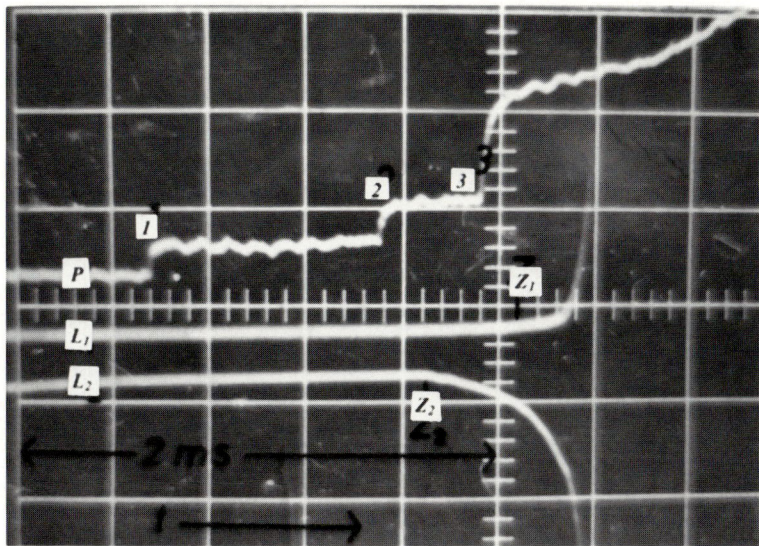

Fig. 6.24. Pressure variation P, luminescence L_1 of hot ignition observed by P_{h1} and that L_2 of cool flame observed by P_{h2} in a stoichiometric n-hexane–air mixture. 1 is passage instant of incident shock front at P_{z1}, 2 that at P_{z2}, Z_1 hot ignition, Z_2 initiation of cool flame *0.4 ms/div*

positions in the low pressure tube, L_1 the light emission from the ignition observed by the photomultiplier with the filter, and L_2 that without the filter are registered. In this example, the emission is observed in L_2 earlier than L_1. This means, that first a cool ignition takes place and then a little later a hot ignition follows.

Probabilities of Cool Flame Initiation and Hot Ignition

Repeating an experiment many times under the same condition, a histogram of the induction period of the cool flame ignition and that of the hot ignition are obtained, from which both the ignition probabilities can be calculated.

The experiments for observing the spontaneous ignition in a stoichiometric n-heptane–air mixture are carried out under the mixture pressure of 3.4 and 5.0 bars at several different temperatures from 650 to 800 K.

Concluding the experimental results, some diagrams of ignition probability of cool flame μ_k, that of hot ignition μ_g, ignition nucleus growing period τ_k of cool flame and that τ_g of hot ignition in relation to the reciprocal mixture temperature $1/T$ are illustrated in Fig. 6.25.

These results suggest that the frequency factors as well as the activation energies of the cool flame initiation are quite different than those in the hot ignition. This means that the reaction processes are different with each other.

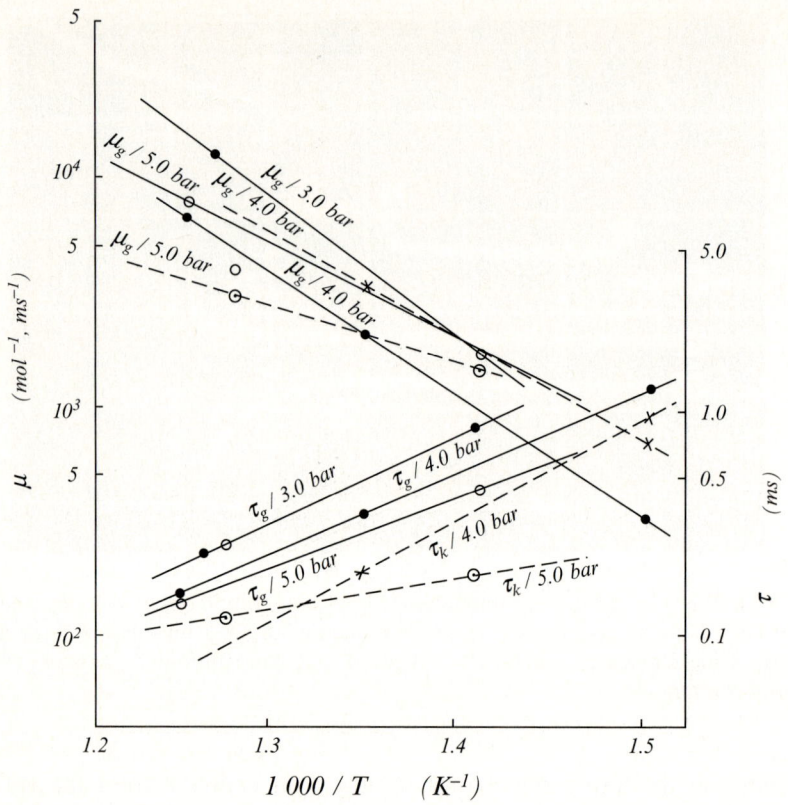

Fig. 6.25. Logarithms of probability μ_k and nucleus growing period τ_k of cool flame initiation as well as those of μ_g and τ_g of hot ignition in a stoichiometric *n*-heptane–air mixture under different pressure with respect to the reciprocal mixture temperature $1/T$

Ignition Limits

The ignition limits should be the state where the mixture has a certain constant ignition probability, as already explained in Chap. 5. In this case, however, there are two ignition limits, i.e., the limit of cool flame initiation and that of hot ignition, as shown on the right side in Fig. 6.26 by a broken line *KC* of $\mu_k = 10^2$ (mol^{-1} ms^{-1}) and a solid line *CG* of $\mu_g = 10^2$ (mol^{-1} ms^{-1}). Besides, there is another region. This limit for which μ_k is equal to μ_g is shown by a solid line *ABC* in the right diagram of Fig. 6.26.[51]

In the region surrounded by the curves *ABC* and *KC* the probability μ_k of the cool flame initiation is always higher than the probability μ_g of the hot ignition, while in the region surrounded by the curves *ABC* and *CG* μ_g is always higher than μ_k. The former region should be the cool flame region, while the latter is the hot ignition region. These limits of cool flame initiation and hot ignition suggested by the stochastic ignition theory agree qualitatively rather well with the empirically observed ones shown in the left diagram of Fig. 6.26.

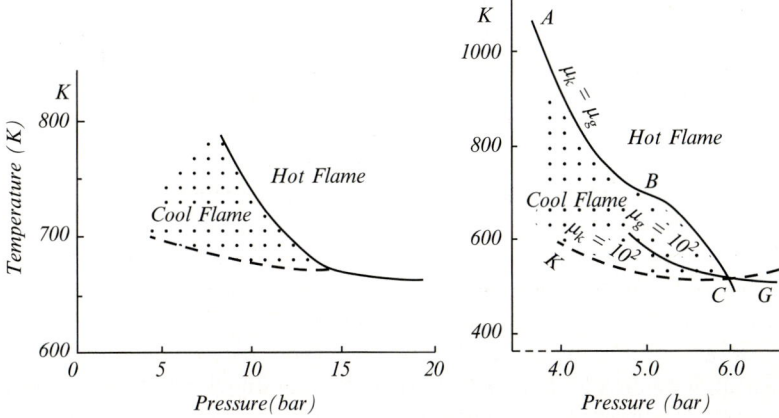

Fig. 6.26. Limits of cool flame and hot ignition in a hydrocarbon–air mixture. *Left*: empirically observed ones,[50, 52] *right*: those according to the stochastic ignition theory

6.3.3 Influences of Tetraethyl Lead on the Ignition

In the spark-ignition engine, the higher the compression ratio, the higher is the thermal efficiency of the engine. In the combustion chamber of a spark-ignition engine having a high compression ratio, however, a so-called knock phenomenon[52] takes place causing an abnormal combustion accompanied by surging pressure waves and decreasing the thermal efficiency. The knock phenomenon is attributed to a spontaneous ignition occurring in some corners of combustion chamber before arrival of the flame propagating from the spark plug. In order to suppress the knock phenomenon, therefore, a small quantity of tetraethyl lead is added to the fuel, as the lead in tetraethyl lead hinders the ignition. By adding such an antiknock additive to the fuel, the compression ratio is increased and a higher thermal efficiency in the spark-ignition engine can be expected.

The effect of lead in tetraethyl lead on the suppression of ignition is also an important reason for supporting the chain-branching kinetics, namely the lead suppresses the ignition by deactivating the active radicals or atoms during the ignition process.

Before going to explain the antiknock effect of lead according to the stochastic ignition theory, the remarkable experimental works of the antiknock effect of tetraethyl lead reported by Sturgis should be introduced.

Sturgis' Works[53]

In spark-ignition engines so-called knocking phenomenon[52] is often observed in the case of high compression ratio or at the drive by an ignition occurring too early, accompanied by a high frequency noise like hammering a hard metal. One of the explanations has been proposed as follows:

A flame initiated by a spark ignition in a fuel–air mixture compressed in a combustion chamber of engine propagates compressing further the gas in the combustion chamber. The unburned mixture near the wall of the combustion chamber is also compressed, increasing the pressure as well as the temperature so high that a spontaneous ignition takes place in the mixture. From the ignition point a flame propagates so rapidly accompanied by a shock wave, which propagates in the combustion chamber, repeating reflections on the wall of the combustion chamber and producing a metallic noise.

Otherwise several explanations have been proposed, but there is not yet any decided theory. In any case, however, the knocking phenomenon can be suppressed in a mixture having less ability of ignition. Therefore, one has tried to find some additives decreasing the ignition ability of fuel, so-called antiknock additives. Finally the tetraethyl lead having the strongest antiknock ability was found and has long time been applied to increase the antiknock ability of fuel. Today, however, the tetraethyl lead is no longer used, as the lead injures human health. This antiknock effect of tetraethyl lead, however, has supported the chain-branching kinetics.

Based on the chain-branching kinetics and with the idea in which the prereaction must play the most role for the ignition, Sturgis investigated the reaction mechanism in a mixture of *n*-heptane–air and that contained a small quantity of tetraethyl lead during the spontaneous ignition in a CFR-engine.

CFR-engine is a small spark-ignition engine developed for investigating the antiknock ability of fuel by varying its compression ratio over a wide range. At the experiments of Sturgis a stoichiometric *n*-heptane–air mixture was introduced into the CFR-engine driven by an electrical motor at 900 rpm, varying the compression ratio so that the peak temperature of each cycle takes an arbitrary value between 600 and 800 K. In the preignition state at the peak temperature the mixture was sampled and analyzed. The *n*-heptane–air mixture adding tetraethyl lead of 0.8 cm^3 in the mixture of 1,000 cm^3 was also examined in the same way, and both the results were compared.

The measured results in the *n*-heptane–air mixture are shown in Fig. 6.27, while those in the mixture containing tetraethyl lead in Fig. 6.28. The results in Fig. 6.27 suggest that higher aldehydes, ketones, and unsaturates are formed before the cool flame appearance, but hydrogen peroxide and formaldehyde are observed almost at the same time of the cool flame appearance.

The results in Fig. 6.28 suggest that the formation of hydrogen peroxides, HOOH, decreases much by adding tetraethyl lead to the fuel in comparison with that in the fuel without tetraethyl lead.

In Fig. 6.29 the ignition limits of the both mixtures having or without tetraethyl lead are illustrated in a pressure–temperature diagram. This suggests that the cool flame limit is not affected by adding tetraethyl lead, but that hot ignition is moved to a region having higher pressure and temperature adding tetraethyl lead.

HOOH is formed by a binding of 2OH, which are active chain carriers and promote the reaction to the ignition. The decrease of HOOH by addition of

6.3 Spontaneous Ignition in Hydrocarbon–Air Mixtures

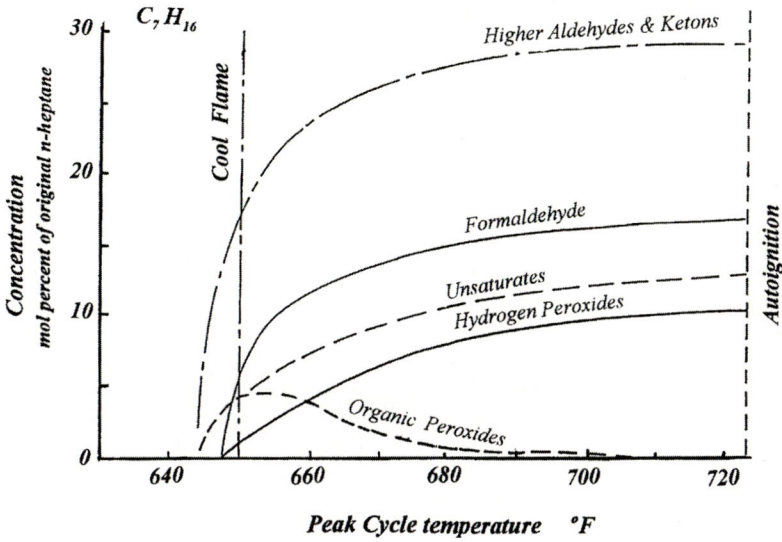

Fig. 6.27. Exhaust gas composition of a stoichiometric *n*-heptane–air mixture

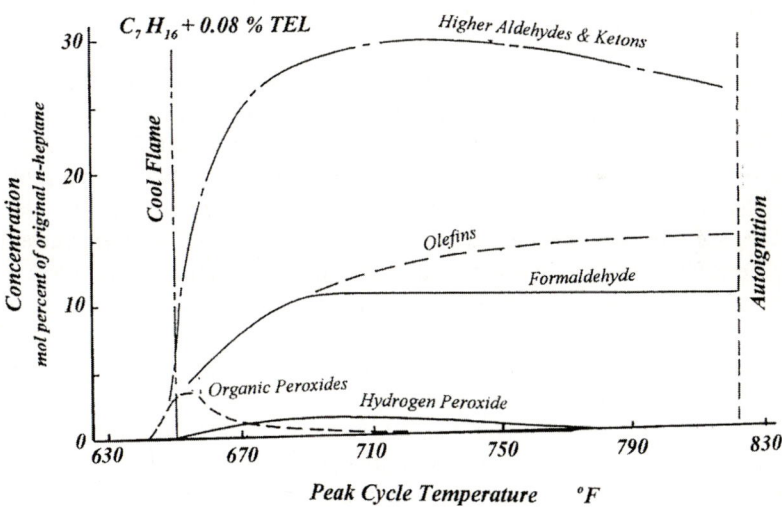

Fig. 6.28. Exhaust gas composition of a stoichiometric *n*-heptane–air mixture containing 0.08% TEL. TEL: tetraethyl lead

tetraethyl lead means that the lead in the additives takes away the activity of reaction intermediates and suppresses the formation of OH, as the chain-branching kinetics suggests.

In the following sections we try to explain the knocking phenomena described above according to the stochastic ignition theory.

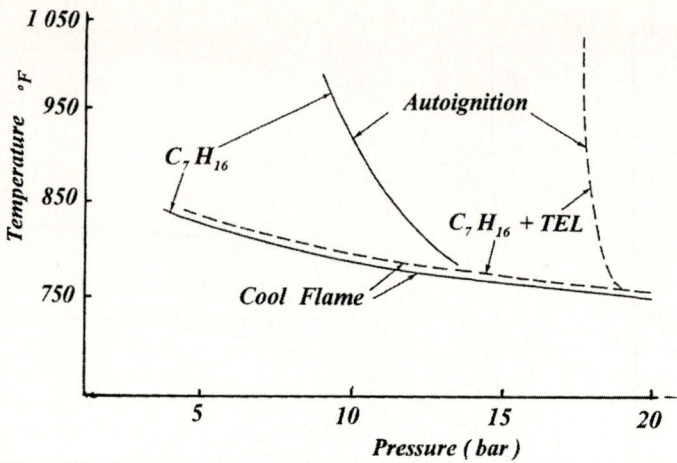

Fig. 6.29. Effect of TEL on ignition limits of *n*-heptane–air mixture

Application of the Stochastic Ignition Theory[54]

Just like the investigation of the ignition limits in hydrocarbon–air mixtures, the experiments for investigating the influence of tetraethyl lead on the ignition in *n*-heptane–air mixtures having an equivalence ratio $\Phi = 0.9$ and 0.6 with and without tetraethyl lead of 0.08 mol in 100 mol *n*-heptane are carried out, using the same shock tube and measurement apparatus as in Sect. 6.3.2. The luminescence of ignition is observed by a photomultiplier with a filter cutting off the light having shorter wave length than 520 mμ and that without any filter to measure the induction period of hot ignition and cool flame appearance at the same time. The pressure of the mixture behind the reflected shock waves where the ignition takes place can take an arbitrary value between 370 and 450 kPa with a temperature between 700 and 850 K. Under the same condition an experiment is carried out about 100 times, so that the histograms of hot ignition and cool flame appearance are obtained, from which the probabilities μ_g, μ_k of hot ignition and cool flame appearance as well as the ignition and cool flame nucleus growing periods τ_g, τ_k are estimated according to (6.3).

Several experiments are carried out under different conditions. The logarithms of the probabilities μ_g and μ_k and those of nucleus growing periods τ_g and τ_k in relation to the reciprocal mixture temperature $1/T$ are illustrated in Figs. 6.30 and 6.31. Figure 6.30 shows those in the mixture of $\Phi = 0.6$ and Fig. 6.31 those in the mixture of $\Phi = 0.9$. Solid lines mean the results in the mixture of *n*-heptane–air without any additives and the broken lines that containing tetraethyl lead having a mole concentration of 0.08% in *n*-heptane.[54]

Let us assume that ε_b is an activation energy of lead in a reaction with particles, for example, radicals, intermediates, particles of fuel or oxygen and b the concentration of lead, then the ignition probability μ_b of the mixture containing lead is expressed by the following equation:

6.3 Spontaneous Ignition in Hydrocarbon–Air Mixtures

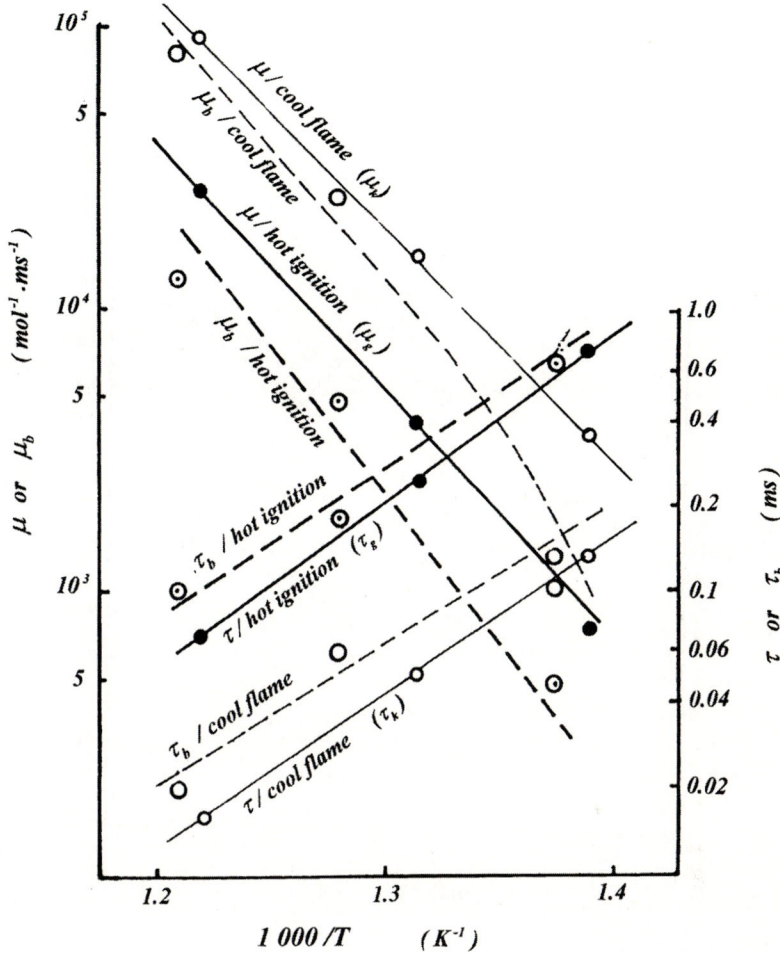

Fig. 6.30. Logarithms of μ, μ_b, τ, and τ_b with respect to the reciprocal mixture temperature $1/T$. μ is the probability of cool flame initiation or hot ignition and τ the growing period of nucleus at the cool flame initiation or hot ignition in an n-heptane–air mixture having an equivalence ratio $\Phi = 0.6$, while μ_b and τ_b the probability and nucleus growing period of cool flame initiation or hot ignition, respectively, in the same mixture containing TEL of 0.08%

$$\mu_b = A \left\{ \exp\left(-\frac{\varepsilon_{CC}}{RT}\right) - b \exp\left(-\frac{\varepsilon_b}{RT}\right) \right\}$$

$$\left\{ \exp\left(-\frac{\varepsilon_{CC} - w}{RT}\right) - b \exp\left(-\frac{\varepsilon_b}{RT}\right) \right\}^s \quad (6.11)$$

$$\left\{ \exp\left(-\frac{\varepsilon_{CH} - w}{RT}\right) - b \exp\left(-\frac{\varepsilon_b}{RT}\right) \right\}^t$$

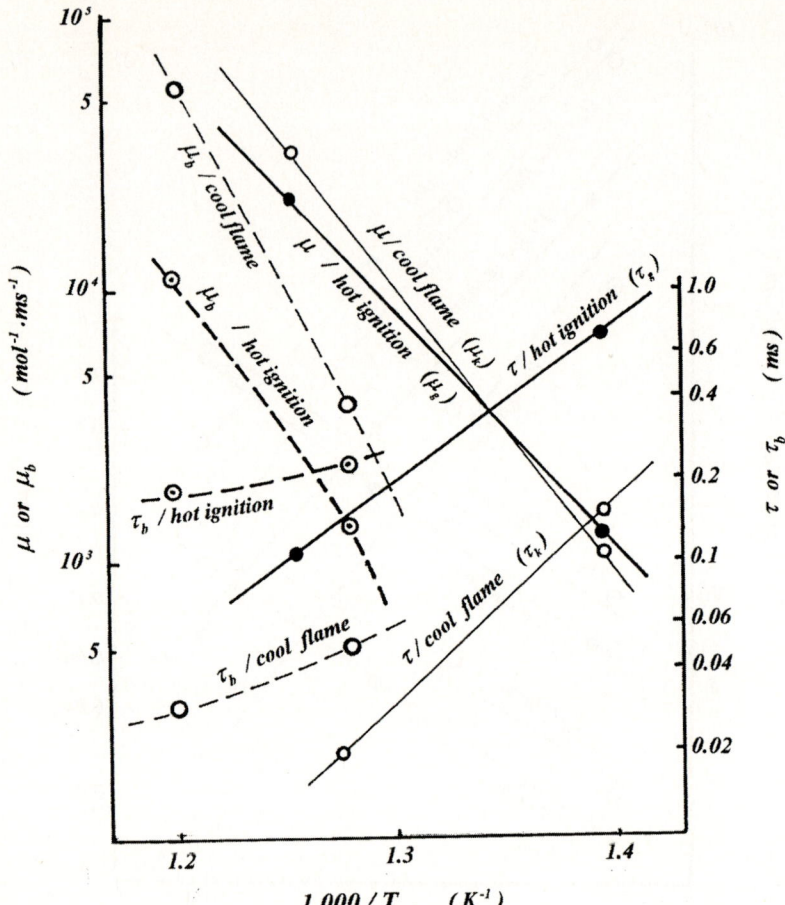

Fig. 6.31. Logarithms of μ, μ_b, τ, and τ_b with respect to the reciprocal mixture temperature $1/T$. μ is the probability of cool flame initiation or hot ignition and τ the growing period of nucleus at the cool flame initiation or hot ignition in an n-heptane–air mixture having an equivalence ratio $\Phi = 0.9$, while μ_b and τ_b the probability and nucleus growing period of the cool flame or hot ignition, respectively, in the same mixture containing 0.08% TEL

Developing the equation into a Maclaurin's series under an assumption of $\varepsilon_{CC} - w \approx \varepsilon_{CH} - w$ and ignoring the higher order terms, we obtain the following relation:

$$\mu_0 \approx A\left\{\exp\left(-\frac{E_1 + E_2 - W}{RT}\right) - b\exp\left(-\frac{\varepsilon_b + E_2 - W}{RT}\right)\right\}. \quad (6.12)$$

In the same way we obtain the following relation for the ignition nucleus growing period τ_b:

$$\tau \approx B\left\{\exp\left(\frac{E_2 - W}{RT}\right) + (l-1)b\exp\left(-\frac{\varepsilon_b - (\varepsilon_{CH} - w) - (E_2 - W)}{RT}\right)\right\}. \quad (6.13)$$

6.3 Spontaneous Ignition in Hydrocarbon–Air Mixtures

On the other hand, as

$$\mu = A \exp\left(-\frac{E_1+E_2-W}{RT}\right),$$

$$\tau = B \exp\left(\frac{E_2-W}{RT}\right),$$

$$\mu - \mu_b \approx Ab \exp\left(-\frac{\varepsilon_b+E_2-W}{RT}\right), \tag{6.14}$$

$$\tau_0 - \tau \approx B(l-1)b \exp\left\{-\frac{\varepsilon_b-(\varepsilon_{CH}-w)-(E_2-W)}{RT}\right\}. \tag{6.15}$$

Diagrams of ln $(\mu-\mu_b)$ and ln $(\tau_b-\tau)$ with respect to the reciprocal mixture temperature $1/T$ are illustrated in Fig. 6.32 as straight lines. From the slope of the straight lines the values of $(\varepsilon_b + E_2-W)$ and $[\varepsilon_b-(\varepsilon_{CH}-w)-(E_2-W)]$ can be

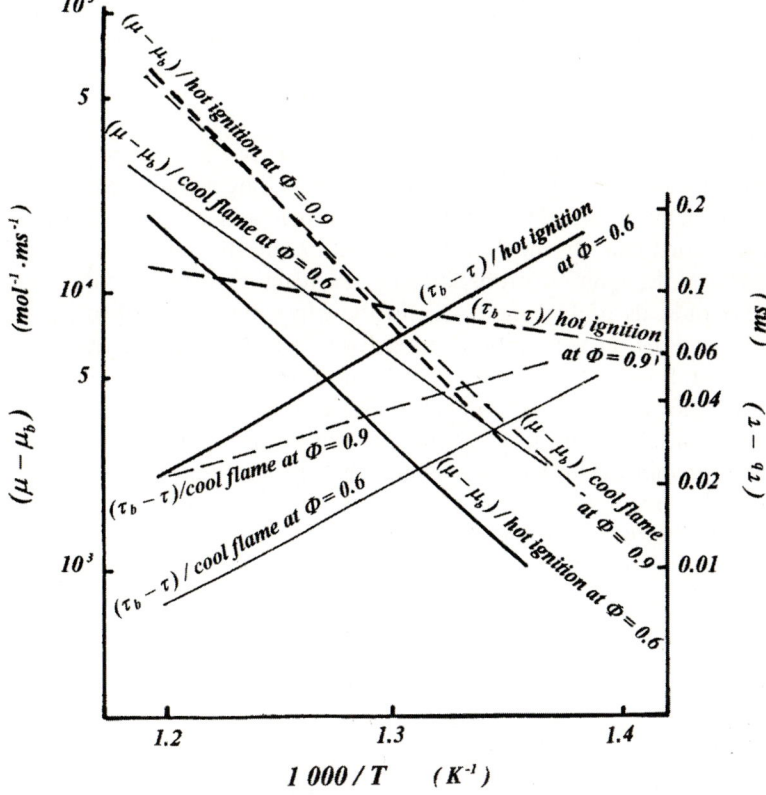

Fig. 6.32. Logarithms of $(\mu-\mu_b)$ and $(\tau_b-\tau)$ in n-heptane–air mixtures having different equivalence ratios Φ with respect to the reciprocal mixture temperature $1/T$

estimated. These values are independent of the fuel concentration and always almost constant. That is

$$\varepsilon_b + E_2 - W \approx 146 \text{ kJ mol}^{-1}, \quad (6.16)$$

$$\varepsilon_b - (\varepsilon_{CH} - w) - (E_2 - W) \approx -92 \text{ kJ mol}^{-1}. \quad (6.17)$$

In the investigation of the ignition limits in the n-heptane–air mixture described in Sect. 6.3.2, $E_2 - W = 104$ kJ mol^{-1} is reported. From (6.16), therefore, ε_b is estimated to be 42 kJ mol^{-1}. Also from the same investigation, the ignition reaction in the n-heptane–air has a reaction order of 4 to 6. $(\varepsilon_{CH} - w)$ is then estimated to have a value of 21 to 33 kJ mol^{-1}. Considering this value in (6.17), ε_b should have a value of 33 to 46 kJ mol^{-1}, which agrees with that estimated from (6.16) quite well. Thus, the activation energy ε_b of the reaction between lead and other elements or radicals is to be about 42 kJ mol^{-1}.

Effect of Lead

As described above, the activation energy of lead in the reaction with intermediates, radicals, fuel, or oxygen is estimated to be 42 kJ mol^{-1}, which is somewhat less than that (65–70 kJ mol^{-1}) in the reaction of –C–C– bond with oxygen. This means:

1. Any reaction of lead with some active chain carriers does not need such a high activation energy. On the other hand, as long as the mixture is leaner than stoichiometric, the quantities of intermediates or active radicals must be proportional to the fuel concentration in the mixture. If the lead decreases the ignition probability by the reaction with the intermediates, active radicals of fuel molecules in the experiments described above, the effect of lead on the ignition probability must be independent of the fuel concentration in the mixture, as the concentration of the lead in the fuel is always the same. The experimental results, however, suggest that the larger the rate of lead to the air, the larger the effect of lead for decreasing the ignition probability. This lets us conclude, that the ignition suppression effect of lead takes place not by a destruction of the active chain carriers through the reaction, but by the oxidation of lead.
2. In this case, as lead must react only on activated oxygen molecules, the activation energy of 42 kJ mol^{-1} of lead is a reasonable value for the reaction. If the lead has less activation energy, it reacts on the oxygen molecules having a lower temperature, while the oxygen molecules having a higher temperature come into the reaction with the fuel. Lead, therefore, cannot suppress the combustion reaction. On the other hand, if lead has a higher activation energy, the activated oxygen molecules come into the reaction with the fuel and the lead cannot play any role for the reaction. In order to suppress the combustion reaction, therefore, only the activated oxygen

molecules in the nonequilibrium state should be taken off before the combustion reaction. Lead has the most reasonable value of activation energy for it and only a very small quantity lead, therefore, can play so large a role.

The explanations described above and experimental results carried out according to the stochastic ignition theory can be applied to explain the phenomena and experimental results observed by Sturgis as follows:

(1) *Delay of hot ignition by addition of tetraethyl lead.* As in Figs. 6.29 and 6.30 observed, the ignition probabilities of cool flame initiation and hot ignition are decreased, while the nucleus growing periods of cool flame and hot ignition are elongated by addition of a small quantity of tetraethyl lead to *n*-heptane. Therefore, the mean induction period $t_m = \tau + 1/(m\mu)$ is also elongated in the same rate. The induction period of hot ignition is ten times longer than that of the cool flame, but both the induction periods are delayed in the same rate by adding tetraethyl lead. For example, at the ignition in an *n*-heptane–air mixture having an equivalence ratio $\Phi = 0.9$ in which the *n*-heptane contains 0.08% tetraethyl lead, the induction period of cool flame initiation is elongated only about 0.04 ms, while that of hot ignition is elongated about 0.25 ms at 800 K ($1,000/T = 1.25$), as shown in Fig. 6.33. The initiation of cool flame, therefore, seems to be not delayed by the addition of tetraethyl lead, while the delay of hot ignition is clearly recognized.

(2) *Decrease of OH concentration.* As the lead takes part in the reaction with the activated oxygen molecules and takes them off from the mixture, the intermediates produced by higher order reaction with oxygen is affected by the lead more than those produced by lower reaction order. The reactions

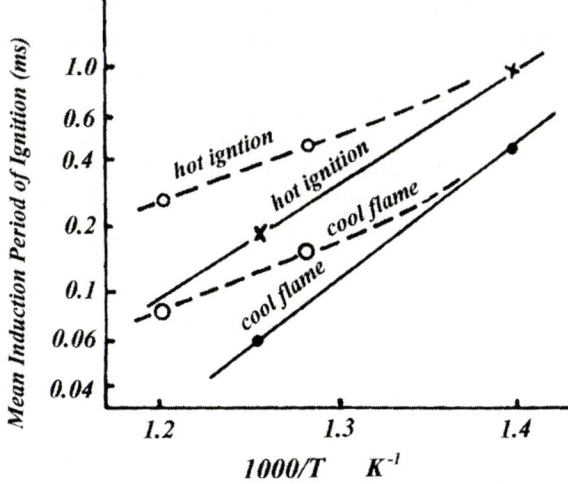

Fig. 6.33. Mean induction periods of cool flame initiation and hot ignition in an *n*-heptane–air mixture of $\Phi = 0.9$. *Solid lines*: without *TEL*; *broken lines*: with 0.08% TEL

producing OH are, thus, affected most by the lead and the production of OH is suppressed very much, while the radicals like aldehyde are affected much less than OH.

(3) *Antiknock effect for engines having ultra high compression ratio.* It is also known that Mn (manganese) has a much higher antiknock effect than lead for spark-ignition engines having ultra high compression ratio (higher than 12). In the mixture in such a high compression ratio engine the mixture temperature becomes so high and the activated oxygen molecules having higher energy are produced so much that the lead reacts on the oxygen molecules having relatively low energy, remaining those having higher energy to react on the fuel. In order to suppress the ignition in such a high compression ratio engine, one needs other additives having higher activation energy like Mn.

(4) *Very small addition of tetraethyl lead shows a large antiknock effect.* As the lead reacts only on activated oxygen molecules, very small addition of tetraethyl lead to the fuel can play a large antiknock effect.

(5) *Suppression of combustion velocity.* By addition of tetraethyl lead to the fuel the flame propagation velocity in the mixture of the fuel and air is decreased a few percents, as the lead reacts with the activated oxygen molecules and take them off from the mixture. The effect, however, is not too much, as the temperature of the mixture at the flame front is higher than 1,000 K, in which the activated oxygen molecules are produced much more than lead.

Ignition Suppression Effect of Reducing Agents

As explained above, the ignition is suppressed by the oxidation of lead. Every reducing agent, therefore, must suppress the ignition. In order to confirm it, we carried out an experiment of spontaneous ignition in an *n*-heptane–air mixture adding a reducing agent, sodium sulfate $NaSO_4$ behind reflected shock waves using the same shock tube as in the experiments described above.

As $NaSO_4$ cannot be directly mixed into the *n*-heptane, a water solution of sodium sulfate is prepared, then the vapor of the solution is introduced into the *n*-heptane–air mixture, which contains 7.5% water vapor and 0.15% $NaSO_4$ vapor. In order to observe the effect of the reducing agent sodium sulfate on the ignition, the experiments of spontaneous ignition are carried out in an *n*-heptane–air mixture having an equivalence ratio of 0.9, 7.5% water vapor and 0.15% $NaSO_4$ and in the same mixture having the same equivalence ratio and water vapor concentration but without $NaSO_4$.

The mean induction periods of ignition in the both mixtures in different temperatures are measured and the results are illustrated in the diagram of Fig. 6.34. In the diagram the logarithms of the mean induction periods in the mixture with the reducer $NaSO_4$ by broken lines.

The results suggest that both the induction periods of cool flame and hot ignition are elongated by addition of the reducer. The rate of both the delays of

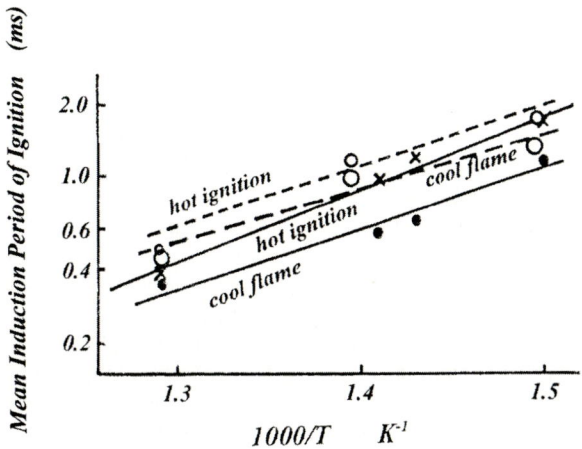

Fig. 6.34. Mean induction periods of cool flame initiation and hot ignition in an n-heptane–air mixture of $\Phi = 0.9$. *Solid lines*: without any reducer; *broken lines*: with 0.15% $NaSO_4$

induction periods are not as much as by tetraethyl lead, but the same tendency is observed. The reducing agent has really an effect of ignition suppression and this also proves that the effect of ignition suppression of tetraethyl lead is caused by its reaction with activated oxygen molecules.

6.4 The Chain-Branching Kinetics and Stochastic Ignition Theory

Concluding the theoretical and experimental results, and discussions described in the foregoing chapters, the classical ignition theory, i.e., the chain-branching kinetics and the stochastic ignition theory are compared here as follows:

the chain branching kinetics	the stochastic ignition theory
(1) The ignition reaction is mainly governed by the concentration of the intermediates produced during the reaction	(1) The intermediates produced during the ignition reaction are observed as the results of the reaction, but do not play any role on the ignition
(2) A combustible mixture has explosion or ignition limits separating its pressure-temperature diagram into two regions, explosion or ignition and nonignition region	(2) Every combustible mixture has never any explosion limits. As long as a combustible mixture exists, there is always a probability of explosion or ignition
(3) In a hydrogen–oxygen mixture an explosion limit, so-called explosion peninsula having an inverse S-form. It is divided into three parts: on the	(3) The empirically observed border lines in the pressure–temperature diagram of a mixture are those on which the mixture has the same ignition

(*Continued*)

the chain branching kinetics	the stochastic ignition theory
first lowest and second middle ones the chain carriers should be deactivated on the vessel surface or in the gas phase, while the third upper one shows a thermal explosion limit	probability. In a hydrogen–oxygen mixture the limit seems to consist of three limits, but on each limit the reaction proceeds with the same mechanism
(4) Because of the deactivation effect of chain carriers on the vessel surface, the explosion limits expands to the lower pressure and temperature side with enlarging the vessel, while the quantity effect on the ignition is not considered	(4) The ignition probability is proportional to the mixture quantity. With increasing the mixture quantity, the whole ignition probability increases, increases too, and the explosion limits seems to expand to the lower pressure and temperature side
(5) Because of the deactivation effect of metal molecules, the spontaneous ignition in fuel–air mixture is suppressed and the antiknock ability of the mixture is raised by adding tetraethyl lead	(5) The lead in tetraethyl lead reacts with activated oxygen molecules, decreases their concentration and suppresses the ignition
(6) The so-called two-step ignition in hydrocarbon–air or oxygen mixtures accompanying cool flame and hot ignition is explained as process in chain-branching reactions	(6) The ignition limits of hydrocarbon–air or oxygen mixture consist of the curves on which both the probabilities of cool flame initiation and hot ignition have the same value, and those on which each probability has a certain constant value

As described above, all of the phenomena which have been explained only by the chain branching kinetics are explained by the stochastic ignition theory. As already explained both the classical theories, i.e., thermal explosion theory and chain-branching kinetics can only be applied to reversible phenomena, but the explosion or ignition is a distinctly irreversible phenomenon to which the both classical theories can never be applied.

7
Ignition in a Fuel Spray

As described in Chap. 6, the ignition is a stochastic phenomenon in which the ignition induction period as well as the ignition position, even in a homogeneous combustible mixture, fluctuates over a wide range. From the histogram of the induction period or distribution of the ignition position, we can obtain the ignition probability which governs the fluctuating phenomena and gives us much information about the ignition.

In heterogeneous mixtures, especially in fuel sprays which are often practically used in industry, the ignition has also a stochastic character and its probability has spatially and temporally different values having some fluctuations. In this chapter the theoretical and experimental methods to obtain the distribution of the ignition probability in an n-octane spray injected into high-temperature air behind reflected shock waves in a shock tube and its significance are explained.

Before going into the main theme, however, it is necessary to know the mixture state having the highest ignition ability. Therefore, in Sect. 7.1, the most inflammable state of the mixture is explained.

7.1 The Most Inflammable State of a Fuel–Air Mixture[55]

Although the knowledge of spontaneous ignition in a combustible mixture much leaner than the stoichiometric concentration is very important for the analysis of the combustion in a fuel spray as well as the combustion on the surface of liquid and solid fuels, we still have little information on it, except the oxidation in mixtures diluted with inert gas.

Applying the stochastic ignition theory, therefore, the spontaneous ignition in an n-octane–air mixture, as an example, having an equivalence ratio Φ, the ratio of the fuel concentration to the stoichiometric one, less than 0.4, behind reflected shock waves in a shock tube is investigated.

7.1.1 Ignition Probability in Lean *n*-Octane–Air Mixtures

The shock tube of stainless steel having an inner-diameter of 50 mm used in the experiments is schematically illustrated in Fig. 7.1. A shock wave driven by He gas propagates through an *n*-octane–air mixture having Φ of 0.2 or 0.4 in the low-pressure tube and is reflected from the tube end wall. The mixture is spontaneously ignited behind the reflected shock wave. The state of the mixture before the ignition behind the reflected shock wave can be examined from the initial mixture state and the propagation velocity of the incident shock waves measured by three piezoelectric pressure transducers set at different positions in the tube.

Observing the ignition with a photomultiplier Ph set on the tube axis outside the end of the shock tube through a filter F through which only light having a wavelength longer than 480 mμ can pass, the induction period of hot ignition in the mixture is measured, as described already in the previous chapters. The period from the arrival and reflection of the incident shock at the tube end wall W (plexiglas window) to the detection of light emission of ignition by the photomultiplier is taken as the induction period t_i.

Repeating the same experiment many times, a histogram of the induction period t_i corresponding to each condition can be obtained. Normalizing the histogram, the probability density $q(t)$, then further the probability $P(t) = \int_t^\infty q(t) dt$ can be obtained, where $t = t_i - \tau$ and τ is the nucleus growing period in which the ignition nucleus grows to a flame having a measurable size and practically obtained from the histogram of the induction period of ignition as the minimum one.

According to the following equation, as already explained in the Chap. 6 as (6.3):

$$\mu = 2V_m \frac{\ln P(0) - \ln P(t)}{F.D.t^2}. \tag{7.1}$$

The ignition probability μ (mol.$^{-1}$ ms^{-1}) can be calculated, where V_m is the molar volume of the mixture behind the reflected shock wave, F the cross-section area of the tube and D the propagation velocity of the reflected shock wave.

The ignition probability μ as well as the ignition nucleus growing period τ obtained from the experiments with respect to the mixture temperature T behind the reflected shock wave is illustrated in Fig. 7.2.

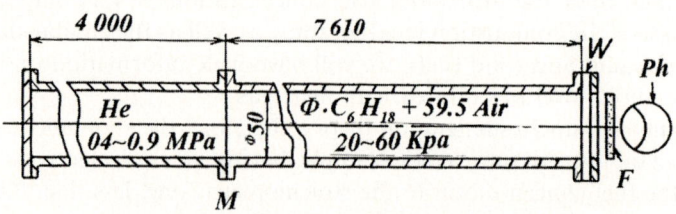

Fig. 7.1. Shock tube applied to the experiments

7.1 The Most Inflammable State of a Fuel–Air Mixture

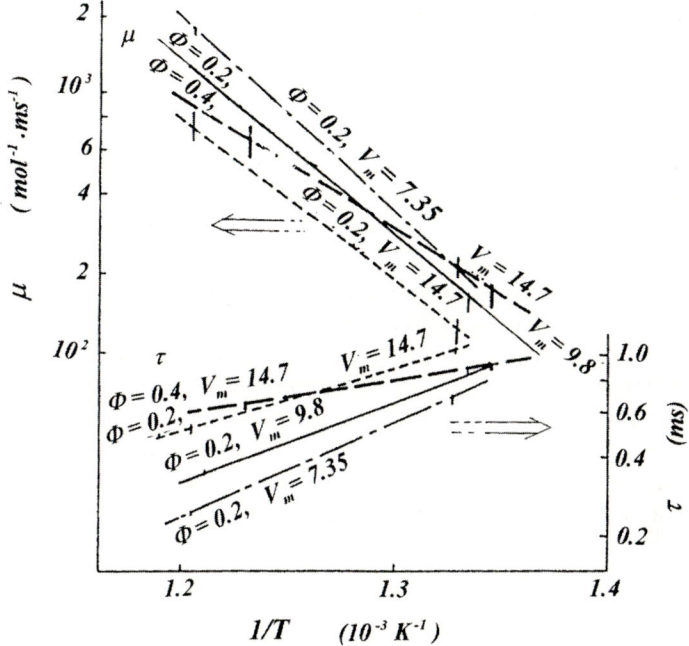

Fig. 7.2. Ignition probability μ and nucleus growing period τ with respect to the reciprocal mixture temperature $1/T$. Φ is the equivalence ratio and V_m (10^{-3} m^3 mol^{-1}) the molar volume of the mixture

Combining the relationships of μ and τ with the fuel concentration and mixture density already described in the previous chapters, we obtain the following equations:

$$\mu = A'(a\Phi.V_m)^n . \exp\left\{-\frac{E_1+E_2-W_0(1-\alpha_c)^{C/\Phi V_m}}{RT}\right\} \quad (7.2)$$

and

$$\tau = B'(a\Phi.V_m)^m . \exp\left\{\frac{E_2+W_0(1-\alpha_c)^{C/\Phi V_m}}{RT}\right\}, \quad (7.3)$$

where A' and B' are arbitrary constants, n and m the number of reaction steps of fuel molecule with oxygen, E_1 and E_2 the activation energies of the initiation and development of the ignition reaction, respectively, W_0 is the energy released in the initial reaction, a and C are proportionality constants, α_c is a coefficient of the energy loss due to the collision between reacting particles, R the gas constant, and T the mixture temperature.

From the experimental results shown in Fig. 7.2., we obtain $E_1 = 63$ kJ mol^{-1}, $E_2 = 192$ kJ mol^{-1}, $W_0 = 178$ kJ mol^{-1}, $\alpha_c = 5.5 \times 10^{-4}$, $a = 1.0$ m^3 mol^{-1}, $C = 1.0$ m^3 mol^{-1}, $A' = 2.3 \times 10^{-3}$ mol^{-1} ms^{-1}, $B' = 6.2 \times 10^6$ ms, $n = -5$, and $m = 4$.

7.1.2 The Most Inflammable State of the Mixture

Substituting the values of E_1, E_2, W_0, α_c, a, C, A', B', n, and m into (7.2) and (7.3), we obtain the relationships of μ and τ with Φ for different temperatures and molar volumes of the mixture as shown in Fig. 7.3.

Each curve of $\ln \mu$ has a maximum value at a certain equivalence ratio Φ, while that of $\ln \tau$ has a minimum value. The mixture should show the most inflammable state in the region having the maximum ignition probability μ_{max} and the minimum growing period τ_{min}, where the mixture has the shortest mean induction period of ignition corresponding to the quantity M_m(mol) of ignitable mixture, as the mean ignition induction period t_m is expressed by the following equation: $t_m = \tau + 1/(M_m \mu)$.

According to (7.2) and (7.3), the most inflammable state of the mixture is expressed in a diagram of the mixture temperature T with respect to the equivalence ratio Φ for different molar volumes V_m as shown by hatched areas in Fig. 7.4. We can thus draw the following conclusions: the higher the mixture temperature is, the leaner is the most inflammable mixture, while the higher the mixture density is (the lower the molar volume of the mixture), the richer is the most inflammable mixture.

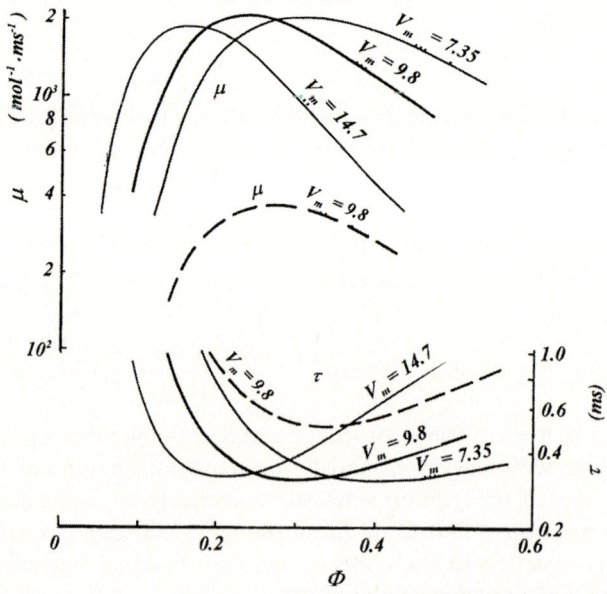

Fig. 7.3. Ignition probability μ and ignition nucleus growing period τ in n-octane–air mixture of 830 K (*solid lines*) and 750 K (*broken lines*) with respect to the equivalence ratio Φ. Molar volume V_m in 10^{-3} m^3 mol^{-1}

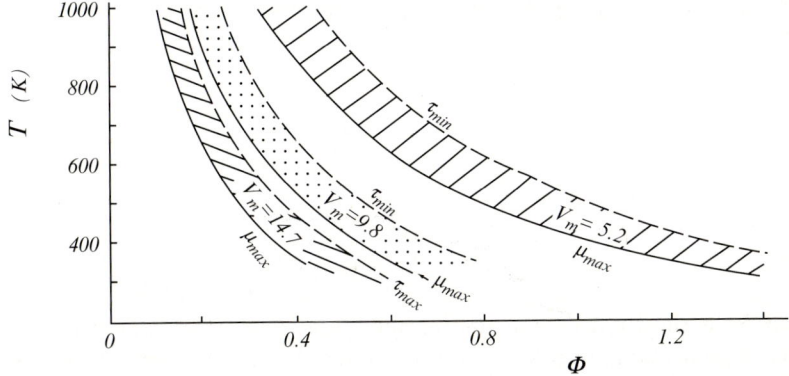

Fig. 7.4. The most inflammable state of *n*-octane–air mixture (hatched area) in the diagram of the mixture temperature T vs. the equivalence ratio Φ. The molar volume V_m 10^{-3} m^3 mol^{-1}

7.2 Ignition Probability in a Fuel Spray[55–57]

In heterogeneous mixture, especially in fuel sprays practically used in industry, the ignition probability has spatially and temporally different values. In this section, the theoretical and experimental methods for obtaining the distribution of the ignition probability in an *n*-octane spray injected into high-temperature air behind reflected shock waves in a shock tube and its significance are explained.

7.2.1 Partial Ignition Probability

If $q(t)$ is the probability density obtained from the histogram of the ignition induction period and $P(t) = \int_t^\infty q(t)\mathrm{d}t$, the ignition probability $w(t)$ in a mixture can be obtained according to the following equations, as already explained in the previous chapters:

$$w(t) \cdot P(t)\, \mathrm{d}t = -\mathrm{d}P, \tag{7.4}$$

$$w(t) = -\frac{\mathrm{d}\ln P(t)}{\mathrm{d}t}. \tag{7.5}$$

These relationships are retained in either the homogeneous or heterogeneous mixture.

To obtain the distribution of ignition probability in a heterogeneous mixture, the partial ignition probability in each section should be obtained after dividing the total space into many sections 1, 2, 3,. . ., *j*,. . ., *n*. With many repetition of the same experiment and measurement of the ignition induction period under the same conditions, the histogram of the ignition induction period in each section, as well as the whole space, can be obtained.

The overall ignition probability $w_0(t)$ in the entire mixture space is deduced from the histogram according to (7.5). On the other hand, we obtain from (7.4) the following relations:

$$w_0(t) \cdot N_0 \cdot P(t) \cdot \Delta t = -\Delta N \tag{7.6}$$

and

$$w_j(t) \cdot N_0 \cdot P(t) \cdot \Delta t = -\Delta N_j, \tag{7.7}$$

where N_0 is the total number of repeated experiments on which the ignition under the same conditions is observed. ΔN and ΔN_j are the numbers of the ignitions which take place during the period from t to $t + \Delta t$ in the space and in jth section, respectively, as shown in Fig. 7.5, and $w_j(t)$ is the partial ignition probability in jth section. From (7.6) and (7.7), we obtain the following equation:

$$w_j(t) = w_0(t) \frac{\Delta N_j}{\Delta N}. \tag{7.8}$$

According to this equation, the partial ignition probability $w_j(t)$ in each section can be obtained from the value of $w_0(t)$ and the histogram of the ignition induction period in each section and in the whole space.

7.2.2 Experiments Using a Shock Tube

The shock tube of stainless steel for the experiments is schematically illustrated in Fig. 7.6.

The low-pressure tube has an inner-diameter of 50 mm and a length of 7.6 m, while the high-pressure tube has an inner-diameter of 62.5 mm and a length of 1.0 m. The high pressure tube is filled with He gas at an arbitrary pressure between 2.0 and 3.0 MPa, while the low-pressure tube is filled with dry air at an arbitrary pressure between 30 and 40 kPa and at room temperature of 20°C.

On breaking the polyester film between both the tubes with a cutter, a shock wave with a Mach number between 2.0 and 2.5 propagates through the

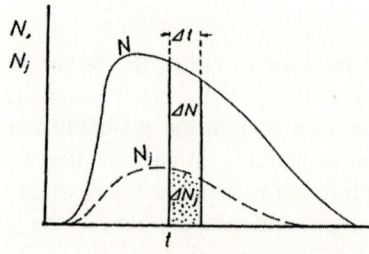

Fig. 7.5. Frequencies N in the whole space and N_j in jth section having an ignition period t

7.2 Ignition Probability in a Fuel Spray

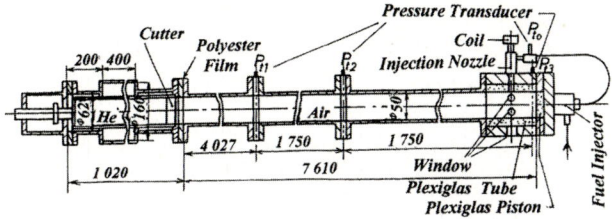

Fig. 7.6. Shock tube applied to the experiments

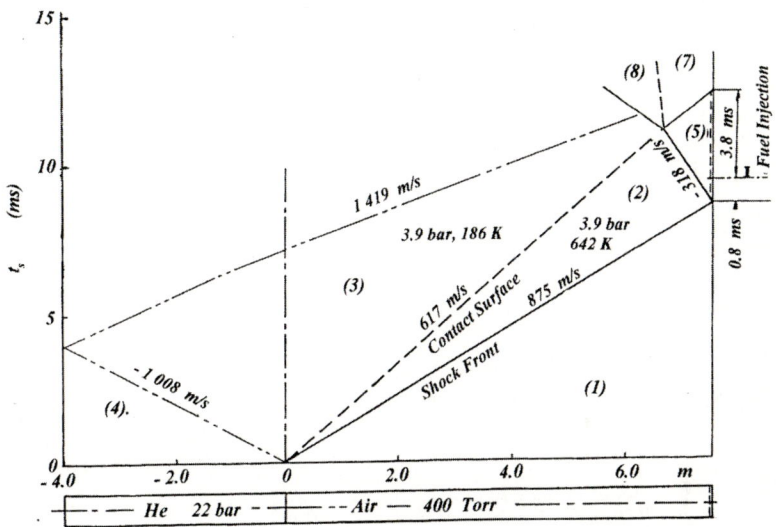

Fig. 7.7. Shock diagram

low-pressure air. Behind the shock wave reflected from the tube end, the air is compressed to a pressure between 1.5 and 2.2 MPa and heated to a temperature between 1,000 and 1,300 K. This state is maintained at more than 4.0 ms, as shown in the shock diagram in Fig. 7.7.

The pressure, temperature, and density of the air behind the reflected shock waves are deduced from the initial state of the air and the propagation velocity of the incident shock measured by observing the passage instants of the shock front at three piezoelectric pressure transducers set at different positions on the low pressure tube. As each experiment under the respective conditions must be repeated many times, fluctuations of the propagation velocity of the incident shock is kept within 1.0%.

In order to inject a liquid fuel of *n*-octane into the shock heated air, an apparatus schematically shown in Fig. 7.8. is attached at the end of the

7 Ignition in a Fuel Spray

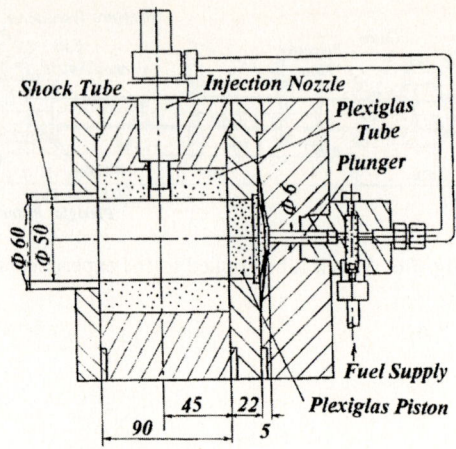

Fig. 7.8. Apparatus of fuel injection

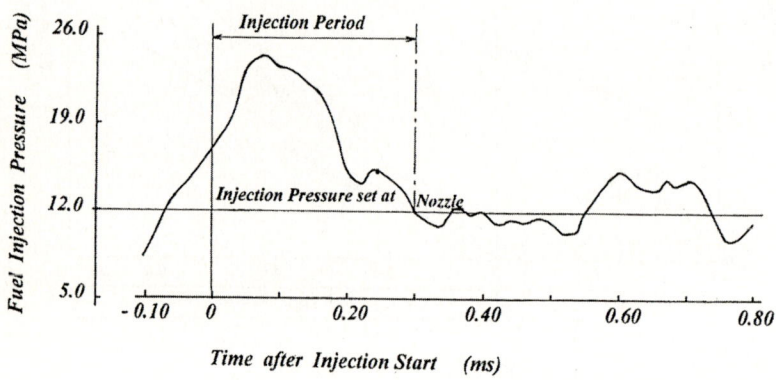

Fig. 7.9. Injection pressure history observed in the plunger room

shock tube. A piston of plexiglas having the same diameter as the inner-diameter of the shock tube and a thickness of 15 mm is set on the end plate of the shock tube. A stainless steel plunger of 6 mm in diameter is connected to the back side of the plexiglas piston. Every time when the incident shock is reflected from the end of the low pressure tube, the plexiglas piston is pushed a few mm against the plunger, which within 0.4 ms compresses the liquid n-octane fuel in the plunger room and the injection pipe to a pressure higher than 17 MPa. The fuel is then injected into the air behind the reflected shock waves through an injection nozzle mounted at the end of the shock tube. In Fig. 7.9 an example of the injection pressure histories of the fuel measured in the plunger room is illustrated.

7.2 Ignition Probability in a Fuel Spray   107

7.2.3 Induction Period of Ignition in the Fuel Spray

Measurement of the ignition induction period is carried out in an *n*-octane spray injected into air with a temperature of 1,270 K and a pressure of 1.5 MPa behind the reflected shock wave in the shock tube. Figure 7.10 shows some examples of shadow photographs of the *n*-octane spray injected into air having the same density (1.8 kg m^{-3}) at room temperature as that described earlier behind the reflected shock wave.

Since we have free space around the shock tube, we can observe the ignition in the fuel spray injected into the shock tube from all direction outsides the tube. We divide the space optically into 25 sections, as shown in Fig. 7.11. Using

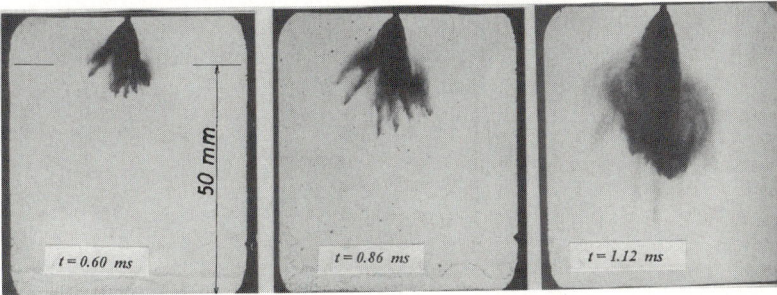

Fig. 7.10. Shadow photograph of an *n*-octane spray injected into air having a temperature of 295 K and density of 1.83 kg m^{-3} taken at time *t* ms after the injection

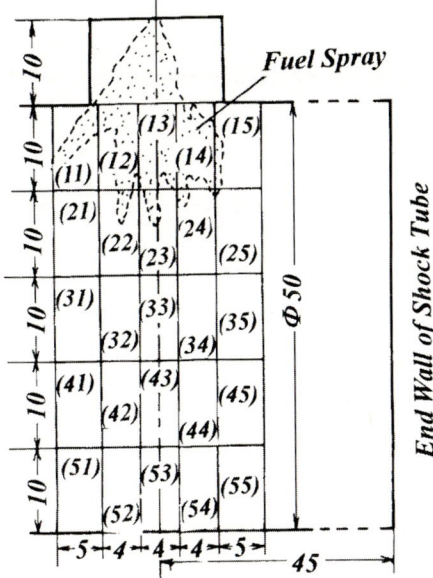

Fig. 7.11. Optical division of the fuel spray space (dimension in mm)

several photomultipliers at the same time, we observe the ignition in each section separately through an oscilloscope by which the motion of the injection valve is also recorded. The period from the beginning of fuel injection to the first luminescence of ignition is taken to be the induction period of ignition.

The same experiment is repeated many times under the assumption that the spray keeps in every time the same form and process, so that a histogram of the ignition induction period in each section, as well as that in the whole space, can be obtained. An example of such histograms in a section and that in the whole space are shown in Fig. 7.12.

7.2.4 Distribution of the Ignition Probability

From the histogram of the ignition induction period observed in the whole space of the fuel spray, shown in Fig. 7.12 (right), we obtain the relation between $\ln P(t)$ and t, as shown in Fig. 7.13, then according to (7.5) the variation of the ignition probability $w_0(t)$, as shown in Fig. 7.14.

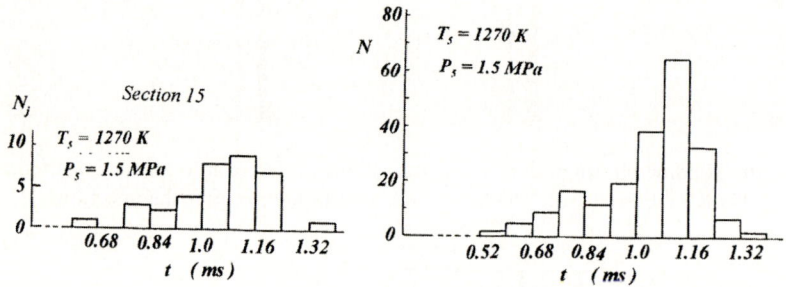

Fig. 7.12. Histogram of ignition induction period t in section 15 (left) and in the whole space of fuel spray

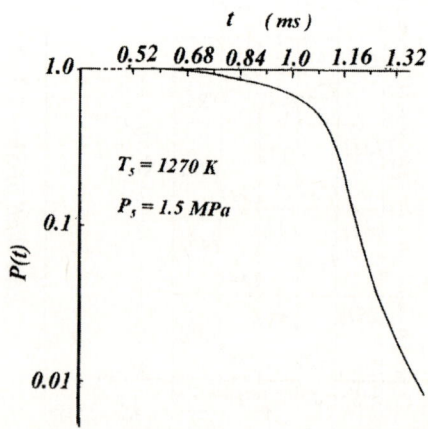

Fig. 7.13. $P(t)$ with respect to the time t after injection start

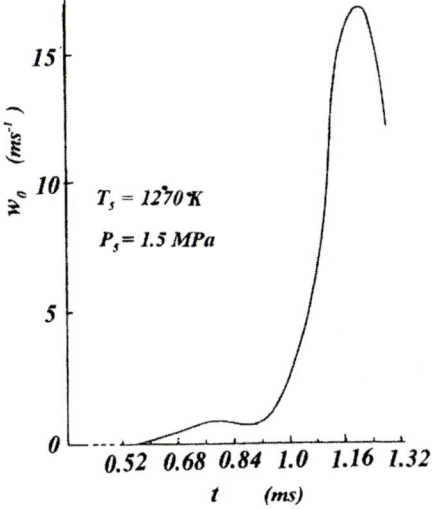

Fig. 7.14. Ignition probability $w_0(t)$ in the whole spray space with respect to time t after injection start

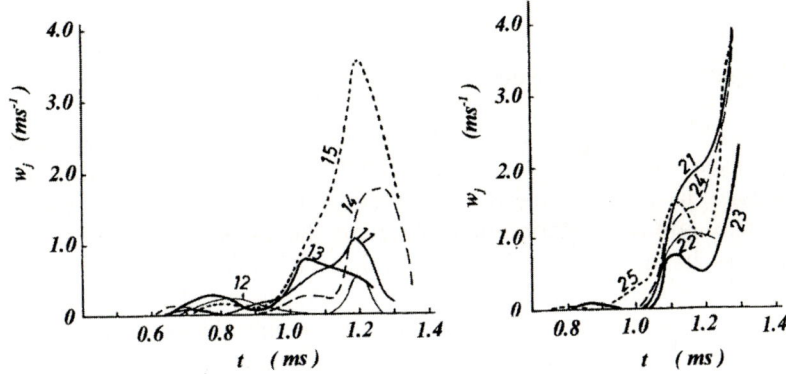

Fig. 7.15. Partial ignition probability $w_j(t)$ in two sections with respect to the time t after injection start. Number on each curve means the section number of the fuel spray space

From the value of $w_0(t)$ and the histogram of the ignition induction period in each section and in the whole space, we can obtain the partial ignition probability $w_j(t)$ in each section according to (7.8). In Fig. 7.15 two examples of the variations of the partial ignition probability in some sections are illustrated.

Observing the temporal variations of the partial ignition probability in all sections, we can obtain the spatial distribution of the ignition probability $w_j(t)$ in the fuel spray at an arbitrary moment. Figure 7.16 illustrates some distributions of the ignition probability in the *n*-octane spray, different moments after the beginning of the fuel injection.

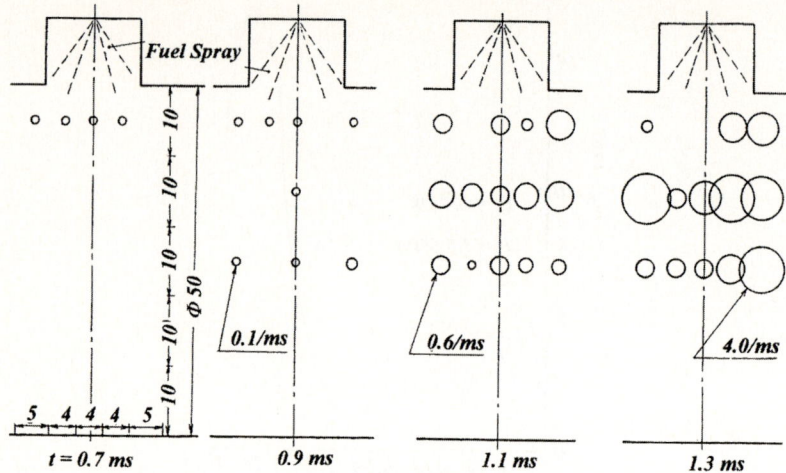

Fig. 7.16. Spatial distribution of the ignition probability at different time t after the injection start. The circular areas represent the values of the partial ignition probability (dimensions in mm)

7.2.5 Ignition and Combustion in a Fuel Spray

The spatial and temporal distribution of ignition probability in an n-octane spray injected into air at 1,270 K and 1.5 MPa compressed behind the reflected shock wave in a shock tube is thus experimentally obtained, by dividing the spray space into many sections and observing the ignition in each section.

The results shown in Fig. 7.16 suggest that the ignition first takes place in the upper region near the injection nozzle during a short period in the early stage after the onset of injection, but afterward the most inflammable regions move downward, increasing the ignition probability. The process is similar, with some delay, to that of fuel spreading into the air.

The distribution of the ignition probability, therefore, should mainly depend on the vaporization, diffusion, and mixing of the fuel with the high-temperature air. Considering these phenomena quantitatively and applying the values of ignition probability in an n-octane–air mixtures with different mixture ratios, densities and temperatures experimentally obtained in the previous section, we tried to compute the partial ignition probability w_j in an n-octane spray injected into air in the same state as that in the experiments described earlier, i.e., 1,270 K and 1.5 MPa. The fuel is injected under the same conditions in the experiment explained earlier, but we assume that the spray has a symmetrical form, as shown in Fig. 7.17 and a fuel particle velocity at the injection nozzle of $U_p = -1.46 \cdot 10^3 \, t + 133$ m s^{-1}, estimated from the measured fuel pressure at injection, where t is the time (s) after the onset of injection. The computed results are illustrated in Fig. 7.18.

7.2 Ignition Probability in a Fuel Spray   111

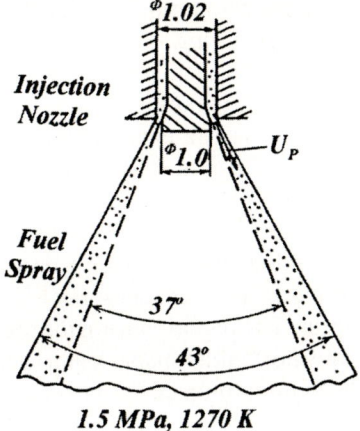

Fuel particle velocity $U_p = -1.46 \cdot 10^5 + 133.0$ m/s
Fuel droplet diameter $d_p = 33.15 \cdot U_p^{-0.39}$ μm
Injection period : 0. 32 ms
Injection quantity : 3.0 mg

Fig. 7.17. *n*-octane spray model for computing the partial ignition probability

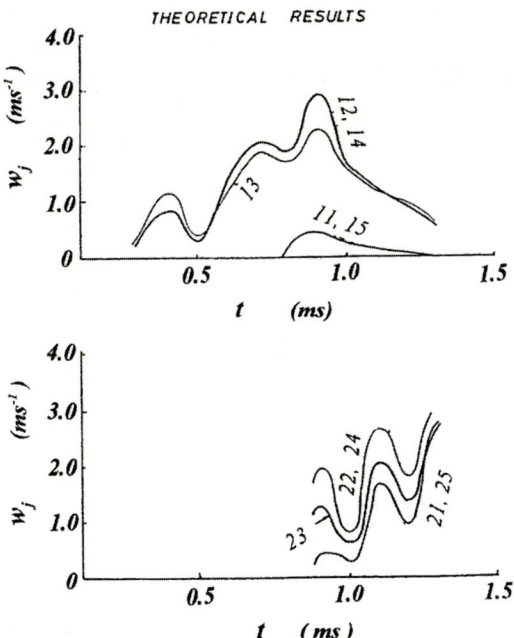

Fig. 7.18. Computed partial ignition probability w_j in each section of the *n*-octane spray with respect to time t after the injection start. Number on each curve means the section number

The tendency of the computed ignition probability w_j agrees well with that experimentally obtained, considering the difference in the configuration between the theoretically assumed state and the real state in the experiment. Using the spatial and temporal distribution of the ignition probability in a fuel spray experimentally obtained, we can examine the assumption for estimating the process of vaporization, diffusion, and mixing of the fuel with the high temperature air.

On the other hand, the period from ignition to flame propagation in the whole fuel spray is usually a few hundred microseconds. Considering that the maximum ignition probability in the whole space of fuel spray is 15 ms^{-1}, the ignition starts from a few points in the spray. The regions having the highest ignition probability, therefore, mean not only the most inflammable regions, but also the regions where the flame produced by the ignition propagates most quickly.

7.2.6 Conclusions

Applying the stochastic ignition theory to a fuel spray injected into high-temperature air, the spatial and temporal distribution of ignition probability in the fuel spray can be obtained by observing the ignition in different regions in the spray separately, but at the same time.

It is possible, by this method, to know where and when the ignition of a fuel spray takes place most easily and through which lane the flame propagates most quickly.

The stochastic ignition theory, thus, can be applied not only to a homogeneous mixture but also to a heterogeneous mixture changing its state, as long as the same process can be repeated.

8
Ignition by Electric Sparks

As ignitions of a combustible mixture by some external energy are also distinctly irreversible phenomena in which some nonequilibrium and heterogeneous states appear, the stochastic ignition theory developed for the spontaneous ignition can be applied to investigate them. One of the most popular methods of ignition by an external energy is the ignition by electric sparks, which is applied not only to internal and external engines, but also to many combustion systems in the industry, housekeeping and others.

Giving an electric potential of several thousand volts to a pair of electrodes made of metallic bars having a diameter from 0.5 to 2 mm set in an explosive mixture, a breakdown, subsequently an electric discharge accompanied by a spark takes place. Then the mixture between the electrodes is heated to a temperature higher than 10,000 K, ignited during a period from several microseconds to several milliseconds and a flame spreads from the ignition point. The spark ignition has been investigated by many scientists and engineers, especially since the internal combustion engines appeared, and many different electric discharge system as well as numerous spark plugs have been developed and sent to the market.

Several different ignition mechanisms by electric sparks have also been proposed by the researchers. The following two ideas on the ignition are well known:

1. The ignition reaction is stimulated by ions produced by the electric discharge and the mixture is ignited by the ions.
2. The ignition takes place in the mixture heated by the Joule's heat of the current between the electrodes.

Both the hypotheses have been long time discussed, but after many experimental investigations the second one is now in general accepted.

On the other hand, dependence of the ignition on the characteristics of the electric discharge and the distance of the gap between the electrodes have come into question and been investigated.

In this book, first such classical theories and then an application of stochastic ignition theory to the spark ignition are explained.

8.1 Igniter Using Induction Coils[58, 59]

The electric discharge in gases itself has been an interesting, an important physical problem and investigated by many scientists. First we short explain the electric discharge apparatus most used for ignition in the internal combustion engines, other industrial apparatus and machines.

In Fig. 8.1 a typical discharge circuit for producing the discharge spark composed of a battery B having low voltage (6–24 V), primary and secondary coil L_1 and L_2, condensers C_1 and C_2, resistances R_1 and R_2, switch S, and spark plug G is schematically illustrated. Switching the current i supplied to the primary coil L_1 by S off, a high voltage proportional to the current variation di/dt is produced in the secondary coil L_2 and an electric discharge accompanied by a spark takes place between both the electrodes of the spark plug G. As shown in Fig. 8.2 the electric potential between the electrodes first rises very quickly to a high voltage of several thousand volts, at which a breakdown takes place, discharging the current through the gas between the electrodes, then rapidly drops to a lower voltage of a few hundred volts kept for a certain period almost constant. After the period the voltage decreases, damping its oscillation corresponding to the characteristics of the circuit.

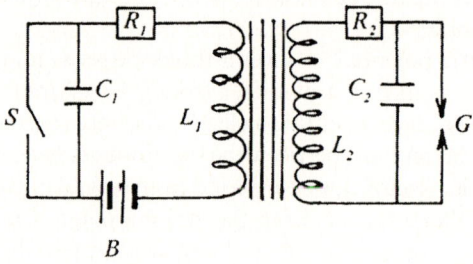

Fig. 8.1. Discharge circuit for spark ignition, B: battery, L_1, L_2: primary and secondary coils, S: switch, G: discharge electrodes, C_1, C_2: condensers, and R_1, R_2: resistances, respectively

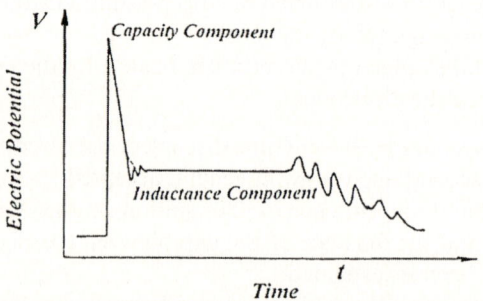

Fig. 8.2. Electric potential variation between discharge electrodes through induction coils

The first part of the discharge by the high potential of several thousand volts during a very short period at the beginning of the breakdown is called "capacity component" and the next part under an almost constant low potential kept longer period "inductance component." The voltage, current, and duration of each component can be regulated to some extent by the condensers C_1, C_2 and resistances R_1 and R_2.

8.1.1 Spark Ignition and Characteristics of Electric Discharge[59]

As the inductance and capacity components appear during the period of discharge using a couple of inductance coils, which of these components can more contribute to the ignition has come into question. Many experiments of ignition by each component have been carried out by many different scientists, separating both the components, but no evident difference between both the components has been found and the results suggest that each component can contribute to ignition much the same, so long it has energy much enough.

8.1.2 Ignition and Gap Distance between the Electrodes

In all apparatus of spark ignition, like spark plugs of internal combustion engines as well as burners of turbines and others, the most effective distance of the gap between the electrodes has come into question and been investigated by many scientists and engineers. The most favorable gap distance of spark plugs of automobile engines, for example, is about 1.0 mm, but neither longer nor shorter.

Lewis and others carried out many experiments of electric spark ignition in explosive mixtures, changing the gap distance and measuring the energy supplied to the discharge circuit at the ignition limit. In a stoichiometric mixture of a natural gas (83% CH_4 + 17% C_2H_6) and air contained in a vessel under atmospheric pressure and room temperature, a pair of stainless steel needles having a diameter of 1/16 in. is set, connecting to a condenser having a capacity of C. The electric potential between both the electrodes can be changed from 0 to −30 kV. Varying the gap distance between the electrodes from 0.01 to 0.15 in., a discharge takes place under a voltage V between the electrodes corresponding to the gap distance and ignites the mixture. The relation between the energy $CV^2/2$ given to the circuit at the ignition limit and the electric gap distance is illustrated in Fig. 8.3.

As the diagram suggests, in the region of the gap distance longer than 0.09 in., the minimum energy for the ignition is almost constant, but in the region having shorter gap distance, the minimum ignition energy given to the circuit at the ignition limit increases rapidly with the shortening of the gap. The reason for it is attributed to the cooling effect of the electrodes namely, first, the mixture between the electrodes is heated by the spark and ignited, but the ignition nucleus is reduced in size by cooling effect of the metallic electrode. If the reaction nevertheless proceeds and releases heat, then the ignition nucleus produced by the spark develops further to the ignition, but if not, it vanishes by the cooling.

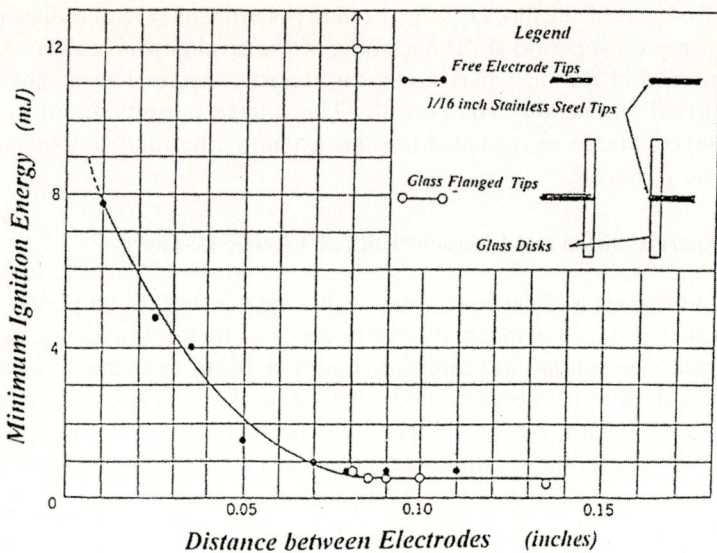

Fig. 8.3. Minimum ignition energy of discharge spark in relation to the electrode gap, by Lewis

The state where the heat released by the reaction and lost by cooling are balanced, is the limit of ignition. The less the gap distance, the more is the heat loss by cooling and the more is the ignition energy to be supplied from the spark.

In order to confirm the theory, an experiment of spark ignition using a pair of electrodes flanged by glass disks, as shown above right in Fig. 8.3, was carried out. In this case, the minimum ignition energy increases at a rather long gap distance of 0.08 in. rapidly to almost infinitely large value, that is, it cannot ignite more. The ignition nucleus may be cooled by the glass disks and disappears. In order to avoid the cooling effect, some spark plugs having thinner electrodes have been produced. The results of many experiments in combustible mixtures changing the mixture ratio, or mixture pressure have been reported, but those changing the mixture temperature are curiously much seldom.

8.2 Application of the Stochastic Ignition Theory to Spark Ignition

Generalizing the basic equations of the ignition probability expressed by an Arrhenius' formula, the stochastic ignition theory can also be applied to the ignition by external energies. Spark ignition is one of such ignitions by external energies, a distinctly irreversible process and has a stochastic character. In this case, however, the electric spark itself is an irreversible process, accompanied by

8.2 Application of the Stochastic Ignition Theory to Spark Ignition

some fluctuating phenomena. The fluctuation observed at the spark ignition is, therefore, a product of both the probabilities caused by the discharge and ignition. Considering it, we can separately obtain each probability from the results of experiments repeated many times, but if the fluctuation of the spark is much less than that of ignition, we can easily obtain the ignition probability and analyze the phenomena according to the stochastic ignition theory.[60, 61]

Considering the energy supplied from the external source to the ignition, the relation among the capacity and inductance components, gap distance between the electrodes and ignition, the mechanism of the spark ignition is investigated and discussed in the next chapters.

8.2.1 Stochastic Theory of Ignition by External Energies[60, 61]

Ignition by the external energy is also a stochastic phenomenon and its induction period fluctuates over a wide range. From a histogram of the ignition induction period t we can obtain a probability density $q(t)$ of ignition in relation to the induction period t and the following equations just like at the spontaneous ignition:

$$P(t) = \int_t^\infty q(t)dt, \tag{3.11}$$

$$m\mu\, P(t)\, dt = -dP, \tag{3.12}$$

$$m\mu(t) = -\frac{d\ln P(t)}{dt}, \tag{3.13}$$

where m is the mole number of the mixture in the reaction state excited by the external energy and $\mu(t)$ the probability of ignition in one mole mixture per unit time.

The ignition probability $\mu(t)$ as well as the period τ in which the ignition nucleus grows to a measurable size and no ignition is observed is expressed by the following equations as already described:

$$\mu(t) = A\exp\left(-\frac{E_1 + E_2 - W}{RT}\right), \tag{3.15}$$

$$\tau(t) = B\exp\left(\frac{E_2 - W}{RT}\right), \tag{3.17}$$

where A is the frequency factor, E_1 and E_2 are the activation energies for the initiation reaction and development or growing reaction, respectively, W the energy released by the initiation and supplied to the development or growing reaction, R the gas constant, T the mixture temperature and B a constant.

The ignition probability μ is, however, composed of two probabilities, probability α for the initiation and that β for the development, namely, $\mu = \alpha\beta$, where α is proportional to $\exp(-E_1/RT)$ and β to $\exp[-(E_2-W)]$, as already explained. Assuming the energy supplied from the external energy source to the initiation reaction is e_1 and that to the development reaction e_2, the following relations are obtained:

$$\alpha \propto \exp\left(-\frac{E_1 - e_1}{RT}\right), \tag{8.1}$$

$$\beta \propto \exp\left(-\frac{E_2 - W - e_2}{RT}\right). \tag{8.2}$$

Thus, we obtain the following equations generalized for ignition by external energy

$$\mu(t) = A \exp\left(-\frac{E_1 + E_2 - W - e_1 - e_2}{RT}\right), \tag{8.3}$$

$$\tau = B \exp\left(\frac{E_2 - W - e_2}{RT}\right). \tag{8.4}$$

These generalized formulae can be applied to the ignition by an electric discharge spark in which some energy is supplied from the spark to the initiation reaction.

8.2.2 Experiments

The experiments of spark ignition are carried out using an equipment composed of a combustion chamber, fuel supply, an electric discharge circuit for spark and measurement system schematically illustrated in Fig. 8.4.

A stoichiometric *n*-hexane–air mixture is introduced into a steel vessel having a cubic space of 30 mm edge length and ignited by an electrical discharge spark generated by conventional induction coils for automobile engines between

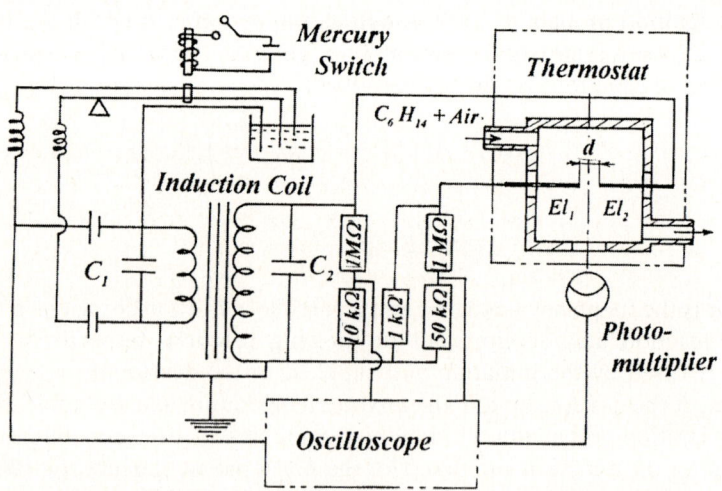

Fig. 8.4. Scheme of experimental equipment for spark ignition

8.2 Application of the Stochastic Ignition Theory to Spark Ignition

a pair steel electrodes El_1 and El_2 having 1.0 mm diameter. The vessel is placed in a thermostat to keep the mixture temperature constant at an arbitrary value.

The experiments are carried out under atmospheric pressure at different temperatures from 20°C to 70°C, different primary current and potential, and with a different gap distance d between both the electrodes from 0.8 to 2.0 mm, observing the luminescence of spark and ignition by a photomultiplier through a plexiglas window.

An example of the oscillograms recording the electric discharge potential V_2, current I_2, and luminescence L of the spark and ignition is shown in Fig. 8.5. In a direct photograph of the luminescence by spark and ignition taken on a rotating film shown in Fig. 8.6, we recognize that the luminescence of the spark produced

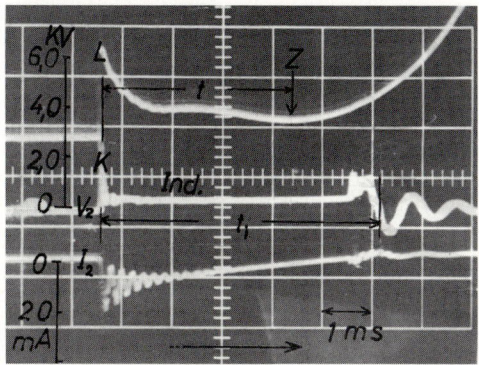

Fig. 8.5. Oscillogram at a spark ignition (Z) in a stoichiometric n-hexane–air mixture. L: luminescence from spark and ignition, V_2: discharge potential, and I_2: discharge current

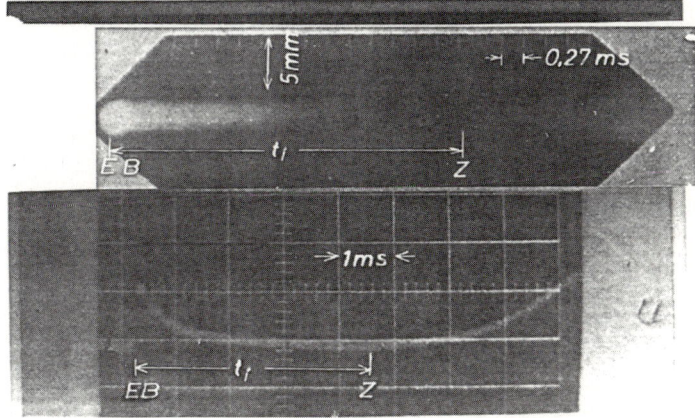

Fig. 8.6. Direct photograph of spark and ignition luminescence taken on a rotating film (above) and an oscillogram of both the luminescence observed by a photomultiplier (below)

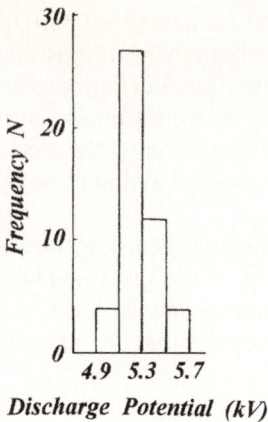

Fig. 8.7. Histogram of discharge potential

by the dielectric breakdown decreases with time because of flame propagation, therefore, the minimum luminescence point Z should be taken as the instant of ignition, as shown below in Fig. 8.6. The period from the spark beginning to the ignition should be the ignition induction period t. An experiment is repeated more than 60 times under the same conditions in order to obtain the histogram of the ignition induction period.

The electric discharge itself is a stochastic phenomenon, but in these experiments carried out here the fluctuation of discharge potential is so small as shown in Fig. 8.7 that its influence on the fluctuation of the ignition induction period does not come into question.

Besides, changing the electric potential of the primary circuit and setting a condenser of 500 pF into the secondary circuit, the duration period t_i of the inductance component, potential V_k and current I_k of the capacity component, average values V_{im} and I_{im} of those of inductance component can be varied as illustrated in Fig. 8.8.

8.2.3 Action of Capacity and Inductance Components

Probability of Ignition by Discharge Spark

An example of the histograms of the ignition induction period observed in the experiments of spark ignition is illustrated in Fig. 8.9 (left). Normalizing it, the probability density $q(t)$ is obtained. Then, according to (3.11), a diagram of ln $P(t)$ with respect to the ignition induction period t is obtained, as shown also in Fig. 8.9 (right) from which the ignition nucleus growing period τ can also be estimated.

The fundamental equation (3.13) allows us to deduce the overall ignition probability $m\mu(t)$ of m mole mixture per unit time. Here m is the mole number of the mixture excited by the spark, but kept almost constant in this case having

8.2 Application of the Stochastic Ignition Theory to Spark Ignition

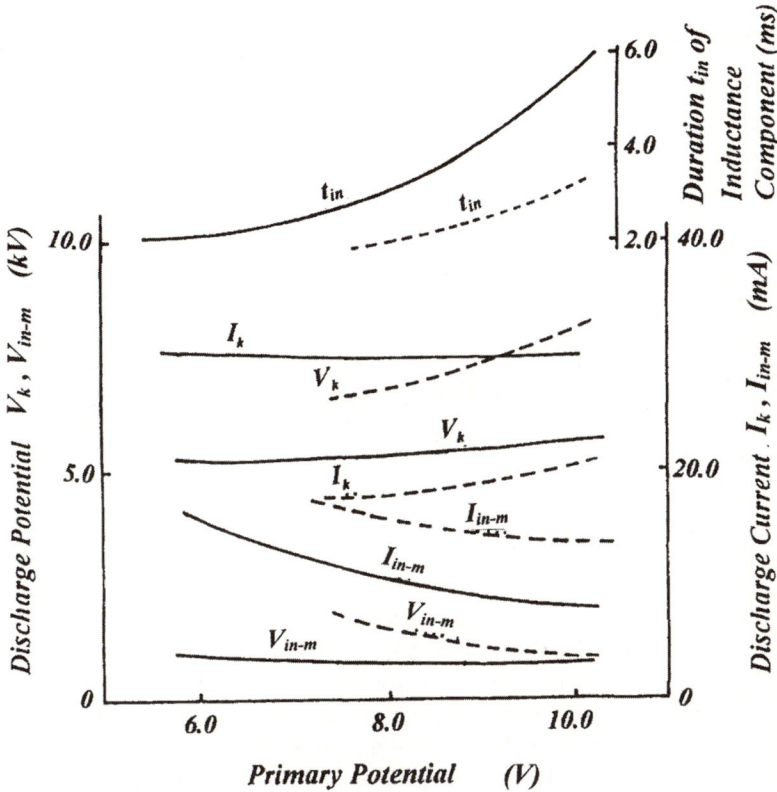

Fig. 8.8. Discharge potential V, current I, and duration t_i of inductance component in relation to the primary potential. Subscript k means capacity component, i: inductance component, and m average value. *Solid lines*, without condenser; *broken line*, with a condenser of 500 pF in the secondary circuit

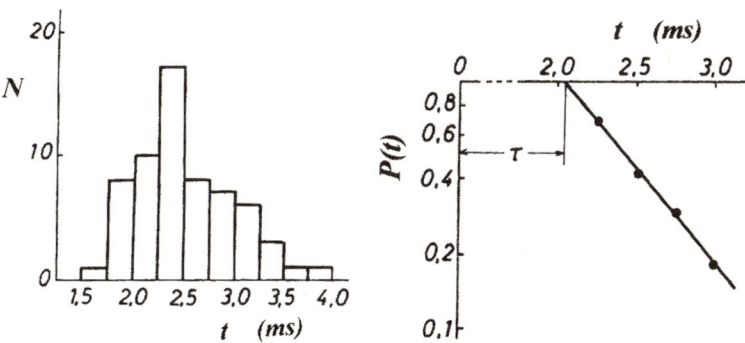

Fig. 8.9. Histogram of inductance period t of spark ignition (left) and $P(t)$ in relation to t (right)

8 Ignition by Electric Sparks

Table 8.1. Summarized experimental results

primary potential (V)	primary current (A)	mixture temperature (°C)	overall ignition probability $m\mu$ (ms^{-1})	ignition nucleus growing period τ(ms)	average ignition induction period t_m(ms)
10.0	3.5	70.0	1.90	1.17	1.70
		20.0	0.88	3.40	4.50
6.0	2.0	70.0	1.80	2.15	2.70
		20.0	1.80	2.10	2.65
10.0*	3.5	70.0	2.42	1.40	1.81
		20.0	0.96	2.45	3.49
8.0*	2.5	70.0	1.82	1.85	2.40
		20.0	0.52	1.90	3.83

*In the case of the secondary circuit having a condenser of 500 pF

a constant gap distance between the electrodes. In the ln $P(t) - t$ diagram we always obtain a straight line. This means that the ignition probability is constant with time, which is therefore expressed simply by μ.

The experimental results of the spark ignition using the electrodes having a constant gap distance of 0.8 mm under different primary potentials are summarized in Table 8.1 together with the average ignition induction period $t_m = \tau + 1/(m\mu)$, while the diagram of logarithms of the overall ignition probability $m\mu$ and ignition nucleus growing period τ in relation to the reciprocal mixture temperature $1/T$ are illustrated in Fig. 8.10.

Energies Supplied from the Spark to Ignition

From the relations of ln $m\mu$ and ln τ to $1/T$ in Fig. 8.10, we can calculate the values of $(E_1 + E_2 - W - e_1 - e_2)$ and $(E_2 - W - e_2)$ according to (8.5) introduced from (8.3):

$$m\mu = mA\exp\left[-\frac{E_1+E_2-W-e_1-e_2}{RT}\right] \quad (8.5)$$

and (8.4). As we already know the values of effective activation energies $(E_1 + E_2 - W)$ and $(E_2 - W)$ of the spontaneous ignition in the stoichiometric n-hexane–air mixture obtained from the shock tube experiments in Sect. 6.3, we can further calculate the values of e_1 and e_2, too. The results are listed in Table 8.2.

e_1 is the energy supplied to the initiation reaction, while e_2 that to the development reaction. Considering the discharge process shown in Fig. 8.2, e_1 should be supplied from the capacity component, while e_2 from the inductance component. At the ignition by the external energy, more energy than lost by diffusion must be given to the ignition reaction.

The energy supplied from the external source per unit time, i.e., the power of each component, therefore, plays an important role for the ignition much more

8.2 Application of the Stochastic Ignition Theory to Spark Ignition

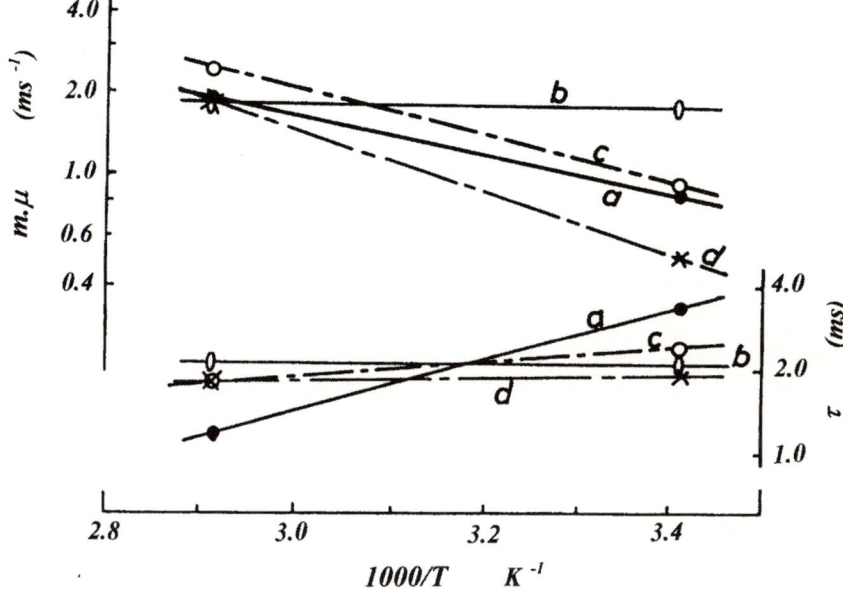

Fig. 8.10. Logarithms of overall spark ignition probability $m\mu$ and ignition nucleus growing period τ with respect to the reciprocal of initial mixture temperature T. (**a**) at primary voltage $V_{pr} = 10$ V, (**b**) at $V_{pr} = 6$ V, (**c**) at $V_{pr} = 10$ V with 500 pF in the secondary circuit, and (**d**) at $V_{pr} = 8$ V with 500 pF

Table 8.2. Activation energies and those supplied to ignition from the spark

primary voltage (V)	$E_1 + E_2 - W - e_1 - e_2$ (kJ mol^{-1})	$E_2 - W - e_2$ (kJ mol^{-1})	e_1 (kJ mol^{-1})	e_2 (kJ mol^{-1})
10.0	13.0	18.0	75.0	11.3
6.0	0.0	0.0	71.1	29.3
10.0*	15.9	4.6	60.0	25.1
8.0*	22.6	0.0	48.5	29.3

*With the secondary circuit having a condenser of 500 pF

than the whole energy, as the energy loss by the diffusion increases with time. On the other hand, we can approximately estimate the power L_k of the capacity component and that L_i of the inductance component from the discharge potential and current recorded in the oscillograms like that shown in Fig. 8.2.

The relations of L_k and L_i to e_1 and e_2 are shown in Table 8.3 and illustrated in the diagram of Fig. 8.11. From this diagram we can conclude that the energy supplied from the spark to the ignition is proportional to the spark power.

8 Ignition by Electric Sparks

Table 8.3. Spark powers and energies supplied from the spark

primary voltage (V)	mean power of capacity component L_k (W)	spark energy supplied to the initiation reaction e_1 (kJ mol^{-1})	mean power of inductance component L_i (W)	spark energy supplied to the development reaction e_2 (kJ mol^{-1})
10.0	42	75.0	6.4	11.3
6.0	42	71.1	15	29.3
10.0*	40	60.0	14	25.1
8.0*	26	48.5	26	29.3

*With a condenser of 500 μμF in the second circuit

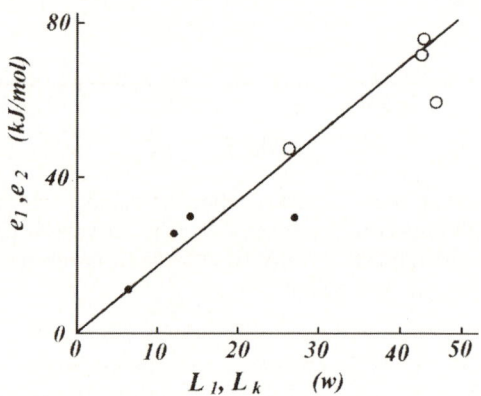

Fig. 8.11. Energies e_1 and e_2 supplied from the spark to the ignition with respect to the powers L_k and L_i of capacity and inductance components, respectively

Summary of the Experimental Results

Summarizing the experimental results, we can conclude as follows:

1. At the first stage of spark ignition, an energy proportional to the power of the spark is supplied to the initiation reaction. In the case using induction coils, the energy is supplied from the capacity component to the initiation reaction, but the value never exceeds that of the effective activation energy for the spontaneous ignition. If the energy given to the mixture from the capacity component is high enough, an ignition can take place, but the energy from the capacity component is never given to the development reaction stage proceeding later.
2. As the inductance component appears after the capacity component, the energy from the inductance component is supplied to the development reaction of the ignition. If no capacity component is produced, a part of the

energy from the inductance component can be supplied to the initiation reaction, too.
3. If the energy supplied from both the components to the ignition is higher than the effective activation energies for the spontaneous ignition ($E_1 + E_2 - W$) and ($E_2 - W$), the ignition is independent of the mixture temperature, but the supplied energy is less than the effective energies for the spontaneous ignition, the mixture temperature has an influence on the ignition corresponding to the energy difference. The higher the temperature, the more easily it ignites.

The results have not much difference to those obtained by the classical theory, but the energy supplied from the spark to the ignition is quantitatively estimated, while only the whole energy supplied to the circuit can be estimated by the classical theory.

8.2.4 Gap Distance and Ignition[58]

Ignition by Electrodes having Different Gap Distances

The experiments of spark ignition in a stoichiometric *n*-hexane mixture under atmospheric pressure at different temperatures are carried out just like those described in Chap. 8.2 using the same apparatus shown in Fig. 8.1. A pair of steel electrodes of 1.0 mm in diameter are set in the combustion chamber, so that the distance d between both the electrodes can be varied to an arbitrary length from 0.8 to 2.0 mm.

The experimental results are summarized in Fig. 8.12, in which the logarithms of the overall ignition probability $m\mu$ and ignition nucleus growing period τ in relation to the reciprocal mixture temperature $1/T$ are illustrated.

From the results the frequency factor mA, effective activation energies ($E_1 + E_2 - W - e_1 - e_2$) and ($E_2 - W - e_2$), consequently supplied energies e_1 and e_2 can be estimated according to (8.4) and (8.5). The results are listed in Table 8.4 together with the power L_k and L_i of capacity and inductance components, respectively.

As e_1 and e_2 supplied from the spark to the ignition are almost independent of the gap distance in this case, the difference of the overall ignition probability $m\mu$ at different gap distances d depends only on mA.[60, 61]

From the results described in this table, we obtain a straight line mA in relation to the gap distance d, as shown in Fig. 8.13, that is, mA is proportional to d. As the frequency factor A is, however, almost constant because of the constant mixture density, the diagram in Fig. 8.13 suggest that m is proportional to d and is expressed as follows:

$$m = C_a + C_g d. \tag{8.6}$$

As illustrated in Fig. 8.14, the mixture excited by the spark is not only that between both the electrodes, but also of the hemispherical spaces S_1 and S_2 around the electrodes.

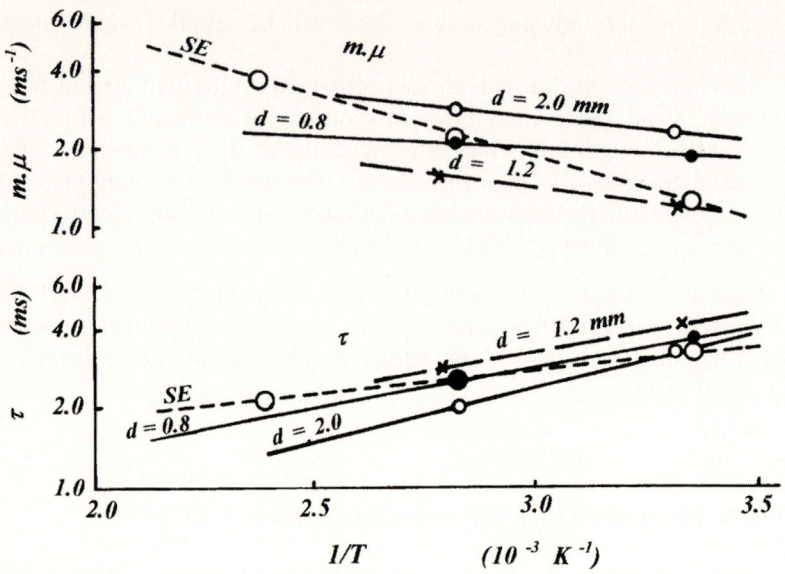

Fig. 8.12. Logarithms of overall spark ignition probability $m\mu$ and ignition growing period τ at different d with respect to the reciprocal of initial mixture temperature $1/T$. Lines denoted by *SE* represent results of special electrodes explained in "Quantity Effect on Spark Ignition"

Table 8.4. Spark powers and energies supplied to ignition from spark in relation to gap distance

gap distance d (mm)	mean power of capacity component L_k (W)	e_1 (kJ mol⁻¹)	mean power of inductance component L_i (W)	e_2 (kJ mol⁻¹)	mA (ms⁻¹)
0.8	42	75.0	10.4	23.0	3.7
1.2	50	74.0	8.0	22.6	5.3
2.0	40	75.6	8.8	21.3	8.6

The mixture volume V_e heated by the spark is, thus, expressed by the following equation:

$$V_e = \pi r^2 d + S_1 + S_2, \tag{8.7}$$

where r is the radius of the spark column. Using the molar volume V_m of the mixture,

$$m = \frac{V_e}{V_m} = \frac{\pi r^2 d + S_1 + S_2}{V_m}. \tag{8.8}$$

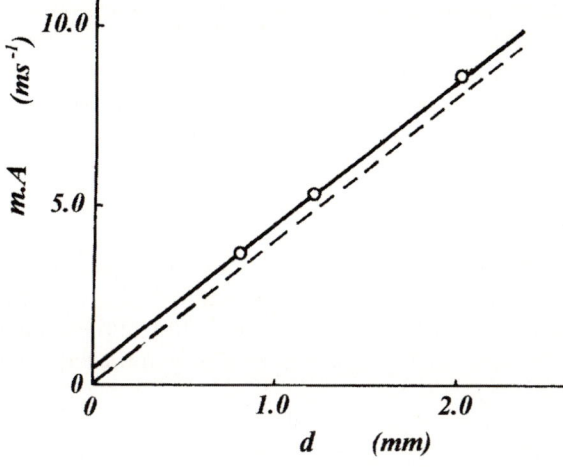

Fig. 8.13. Frequency factor mA in relation to gap distance d

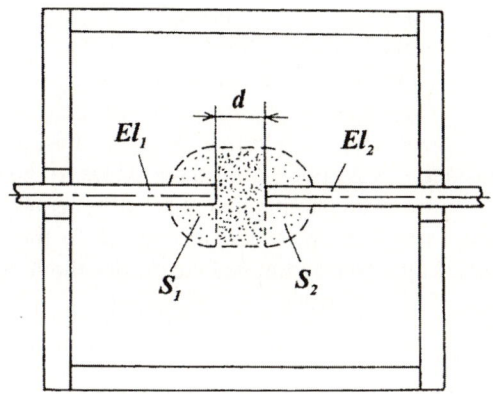

Fig. 8.14. Space (*dark area*) excited by a spark in a test chamber. El_1, El_2: electrodes of steel

Comparing this equation with (8.6), we obtain the following relations:

$$C_a = (S_1 + S_2)/V_m, \tag{8.9}$$

$$C_g = (\pi r^2)/V_m. \tag{8.10}$$

The overall probability $m\mu$ of ignition by the electrodes having a gap distance d is expressed by the following equation:

$$m\mu = (C_a + C_g d) A \exp\left(-\frac{E_1 + E_2 - W - e_1 - e_2}{RT}\right). \tag{8.11}$$

8 Ignition by Electric Sparks

The overall ignition probability $m\mu$ is obviously proportional to mixture quantity m excited by the spark, which is also proportional to the gap distance d. The most effect of gap distance on the spark ignition is caused not by the cooling effect of the electrodes, but by the quantity effect of the mixture between both the electrodes.

Minimum Ignition Energy

The minimum ignition energy at the spark ignition means the electric discharge energy at the ignition limit. In the stochastic ignition theory, there is no ignition limit. The so-called ignition limit, therefore, should be the state where the mixture heated by the spark has a certain constant ignition probability, just like at the spontaneous ignition.

Besides, the spark ignition is usually investigated at room temperature, the ignition limit is always observed in the mixture at room temperature. According to the stochastic ignition theory, the ignition limit is the point having a certain constant ignition probability $m\mu$ in relation to the reciprocal mixture temperature $1/T$, shown in Fig. 8.15.

In this, $\ln m\mu - 1/T$ diagram we can draw many straight lines through the point Z having different slopes corresponding to the mA, consequently to $(C_a + C_g d)$. From the slope of each line the effective activation energy $(E_1 + E_2 - W - e_1 - e_2)$, consequently $(e_1 + e_2)$ supplied from the spark to the ignition can be calculated according to (8.11). There is no straight line of a slope having negative angle, as it means that some energy is supplied from the mixture to the spark. The calculated energy $(e_1 + e_2)$ means the supplied energy at the ignition limit, i.e., minimum ignition energy, which is illustrated in relation to the gap distance d in Fig. 8.16. At the spark ignition by flanged electrodes the hemispherical spaces

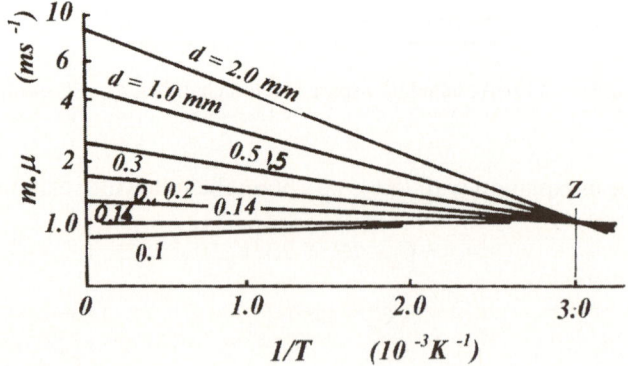

Fig. 8.15. Logarithm of overall probability $m\mu$ of spark ignition between the electrodes having different gap distances d with respect to the reciprocal mixture temperature $1/T$. All lines through the point Z at room temperature having a certain ignition probability (=1.0 ms^{-1})

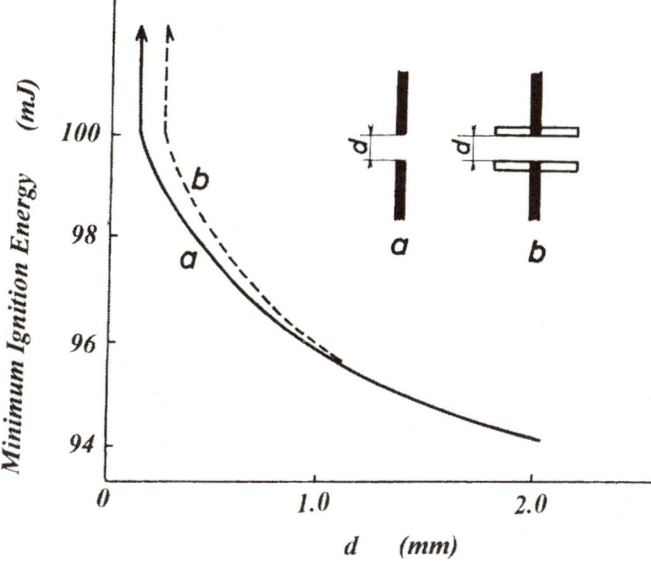

Fig. 8.16. Minimum ignition energy supplied from spark to the ignition estimated from the diagram in Fig. 8.15 in relation to the gap distance d. (**a**) by normal electrodes, (**b**) by glass disk flanged electrodes

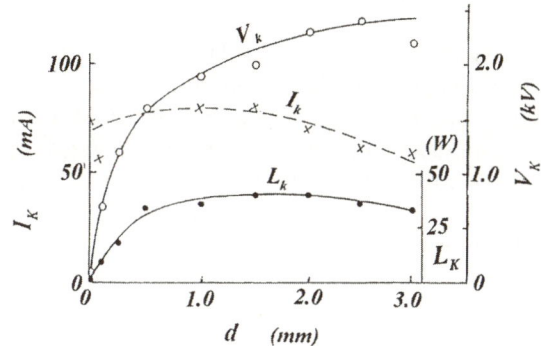

Fig. 8.17. Discharge current I_k and power L_k of capacity component with respect to the gap distance d

S_1 and S_2 do not exist and m is smaller than that by normal electrodes. The minimum ignition energy by flanged electrodes, thus, is expressed by the curve b in Fig. 8.16.

On the other hand, the relation between the gap distance and the spark characteristics comes into question. In Fig. 8.17 the discharge potential V_k, mean current I_k, and power L_k of the capacity component of an electric spark in relation to the gap distance d are illustrated. This diagram suggests that the spark power has a maximum value at $d = 1.8$ mm. The current through a longer gap distance, consequently the power decreases, as the electric resistance between the electrodes increase because of ion diffusion increase.

In a region of electrode gap having a distance less than a finite value (about 1.8 mm), the overall ignition probability $m\mu$ also decreases with the decrease of the gap distance d, since the mixture quantity excited by the spark as well as the spark power decreases. In the other region of the electrode gap having longer distance, the overall ignition probability also decreases, as the spark power decreases and supplies less energy to the ignition.

In Fig. 8.3 of Sect. 8.1.2. the minimum ignition energy at the spark ignition in relation to the gap distance is presented by Lewis. In the diagram, however, the whole energy $CV^2/2$ at the ignition supplied from the circuit to the discharge but not to the ignition is taken as the minimum energy. The energy supplied to the ignition is $(e_1 + e_2)$, which depends not on the whole circuit energy, but is proportional to the spark power.

Considering the relation between the spark power and circuit energy shown in Fig. 8.17, the relation between the minimum energy at the spark ignition and the gap distance illustrated in Fig. 8.16 should be close to that presented in Fig. 8.3 by Lewis.

For the minimum energy of spark ignition the quantity effect of ignition and the characteristics of the spark play the most important role.

Quantity Effect on Spark Ignition

As already described, the overall ignition probability is expressed by $m\mu$ and proportional to the mole number m of the mixture excited by the spark, the more the overall ignition probability, the more easily the mixture can be ignited.

In order to confirm the quantity effect on the ignition, a spark ignition using a couple of special electrodes as shown in Fig. 8.18a is examined. One of the electrodes has a flat disc form on which a thin mica plate is placed, so that the spark should take a long detour path of about 8 mm.

The spark ignition in a stoichiometric n-hexane–air mixture at different mixture temperatures is observed and the results are shown in Fig. 8.18b, where the overall ignition probability is expressed by $\mu(n)$–s and that by a normal one by $\mu(n)$–N, or that and τ with SE in Fig. 8.12 in "Ignition by Electrodes having Different Gap Distances." Comparing the results, mA at the ignition by the long-path spark much larger than the normal one, while the energy $(e_1 + e_2)$ supplied from the sparks is less, namely $e_1 = 67$ kJ mol^{-1} at the power of the capacity component $L_k = 42$ W and $e_2 = 25$ kJ mol^{-1} at the power of the inductance component $L_i = 24$ W are observed, as calculated from the result in Fig. 8.12.

This suggests that the mixture quantity m excited by the spark increases by the detour-spark path, while the energy $(e_1 + e_2)$ supplied from the spark decreases by the ion diffusion along the longer spark path. The slope of the straight line of ln $m\mu$ against $1/T$ is steeper, that is, the mixture can be ignited

8.2 Application of the Stochastic Ignition Theory to Spark Ignition

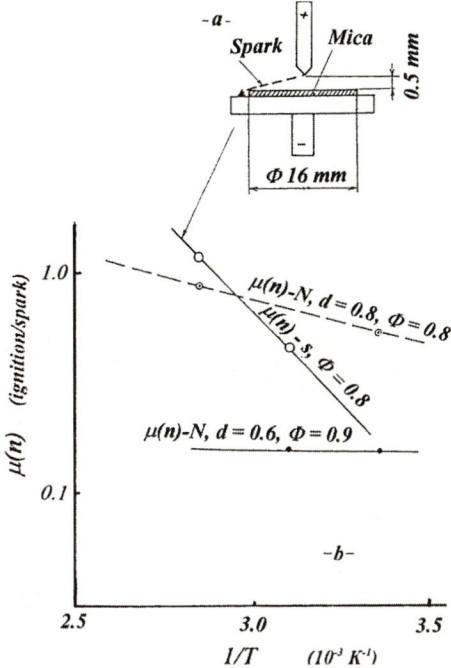

Fig. 8.18. Special long-spark electrodes (**a**) and overall ignition probability $\mu(n)$–s is (ignition frequency/spark) by the special electrodes and that by a normal one (**b**) having different gap distance d, equivalence Φ: ratio of the mixture

more easily at higher temperature by such a long-path spark, but has a reverse tendency at lower temperature.

Remark 4: $\mu(n)$ is different from $\mu(t)$. In a combustible mixture ignited not easily by an electric spark, the mixture can be ignited by a spark of electrical discharge repeated many times. Such an experiment can be repeated many times under the same condition. From such a repeated experiment, a frequency N of ignition by the spark at nth discharge, that is, a histogram of ignition by the spark of the nth discharge. Normalizing the histogram we can obtain its probability density $q(n)$. From this $q(n)$ $P(n) = \int_n^\infty q(n)dn$ can be obtained, from which we can calculate the probability $\mu(n)$ just like $m\mu(t)$ according to (3.13), replacing t by n, $\mu(n)$ means, therefore, the probability how many times the mixture can be ignited by one spark.

Spark Ignition in a Flowing Mixture

The quantity effect is also observed at the spark in a flowing mixture. Figure 8.19 represents logarithms of the overall ignition probability $m\mu$ and the ignition nucleus growing period τ at the spark ignition in a stoichiometric n-hexane–air

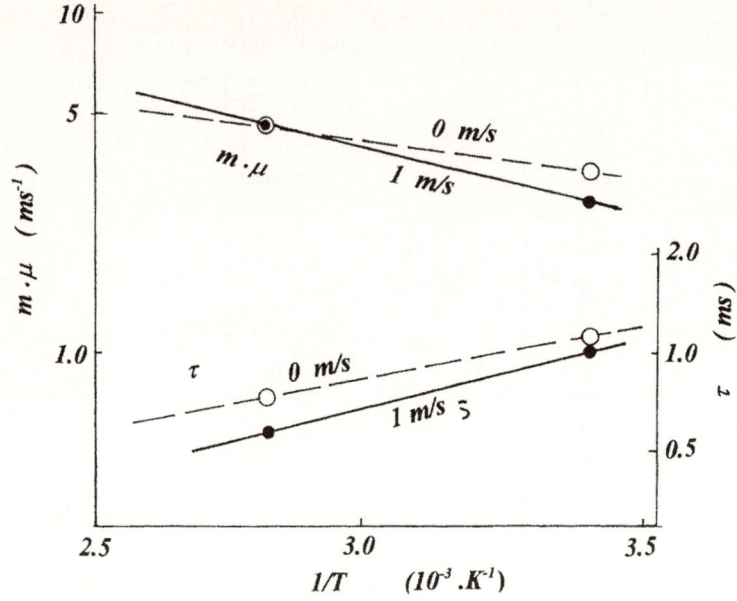

Fig. 8.19. Logarithms of overall ignition probability $m\mu$ and ignition nucleus growing period τ in a flowing (*solid line*) and standing (*broken line*) stoichiometric n-hexane–air mixture with respect to the reciprocal mixture temperature $1/T$

mixture flowing with a velocity of 1.0 m s⁻¹ as well as those in the standing mixture as a function of the reciprocal mixture temperature $1/T$. At the spark ignition of a flowing mixture, the mixture moving between the electrodes is excited by the spark during the electric discharge, consequently, the mixture quantity m excited by the spark becomes larger, but the specific energy $(e_1 + e_2)$ supplied from the spark to the unit mixture quantity becomes less than those at the spark ignition in the standing mixture. The slope of ln $m\mu$ vs. $1/T$ at the ignition in the flowing mixture is steeper than at the ignition in the standing mixture. The flowing mixture, thus, can be more easily ignited by the spark than the standing one at higher temperature, while a reverse tendency is observed at lower temperature.

If the mixture flow velocity, however, is too high, the spark cannot be formed or continues not long enough. The mixture, therefore, can be ignited very hard.

Detour-spark Ignition Plug[62, 63]

According to the quantity effect introduced from the stochastic ignition theory, the overall ignition probability of a combustible mixture, that is, the ignition ability of the mixture is proportional to the quantity of the mixture having an energy high enough for ignition. Applying this quantity effect, a detour-spark plug for spark-ignition engine is developed.

8.2 Application of the Stochastic Ignition Theory to Spark Ignition

The detour-spark plug has a long central electrode insulated by a thin ceramic tube and two or more earthed electrodes set near the root of the central insulator tube keeping a small gap (0.1–0.2 mm) from the central insulator and electrode, as shown in Fig. 8.20. The electric discharge spark takes place not along the shortest path between the central electrode and earthed one, but between the top edge of the central electrode and that of the earthed outer one beyond the central insulator, as the photograph in Fig. 8.20 shows.

By this method, a very long spark, as long as several times that of a conventional spark plug, can be obtained without any increase of the discharge potential excessively, as illustrated in Fig. 8.21.

In order to compare the ignition probability by the detour-spark plug with that by a conventional one, several experiments of the ignition in n-hexane–air mixtures having different equivalence ratios Φ (fuel concentration ratio to the stoichiometric) by two kinds of plug are carried out according to the same method as already explained. The results are represented in Fig. 8.22, in which the logarithms of the overall ignition probability $m\mu$ are shown in relation to the reciprocal mixture temperature $1/T$ for different equivalence ratios Φ of the mixture. The overall ignition probability by the detour-spark plug is much higher than that by the conventional one, especially in the lean mixture region.

We applied the detour-spark plugs of the same type as that mentioned earlier to a spark-ignition engine for a passenger automobile. The test engine has 4 cylinders with overhead valves, a swept volume of 1,300 cc, and a compression ratio of 8.0. Figure 8.23 represents the performance curves of the engine driven at 3,000 rpm engine speed and a boost pressure (inlet mixture gas pressure) of 150 mm Hg below the atmospheric pressure by the conventional spark plugs (broken lines) and those by the detour-spark plugs (solid lines) as functions of the equivalence ratio Φ of the inlet gasoline–air mixture.

The results suggest that the engine can be driven by the detour-spark plugs stably, keeping the fuel consumption relatively low, even if the mixture is leaner

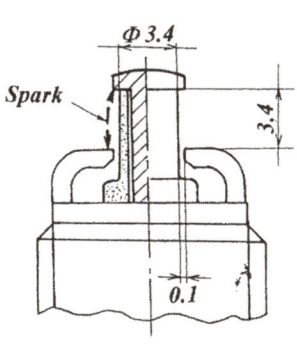

Fig. 8.20. Detour-spark plug

8 Ignition by Electric Sparks

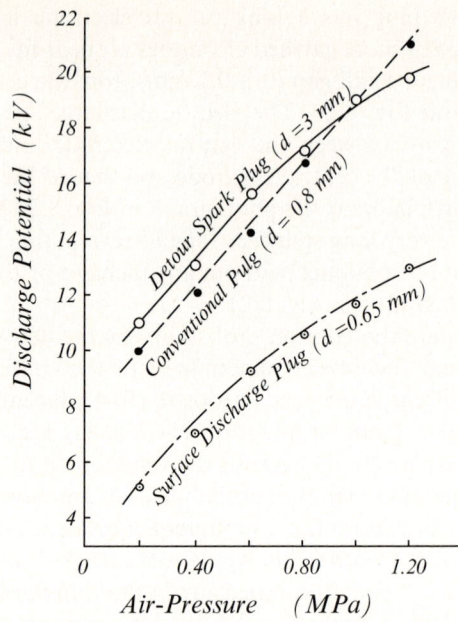

Fig. 8.21. Discharge potential of different spark plugs in relation to the gas pressure

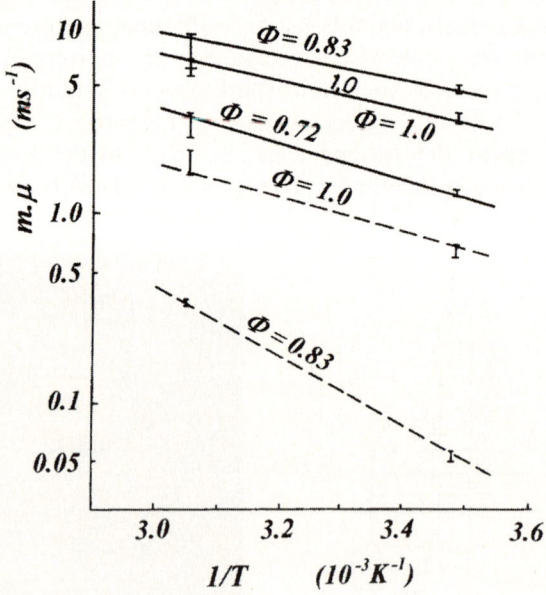

Fig. 8.22. Logarithms of overall ignition probability $m\mu$ of detour (*solid line*) and conventional (*broken line*) spark plugs in *n*-hexane–air mixture having different equivalence ratio Φ

8.2 Application of the Stochastic Ignition Theory to Spark Ignition

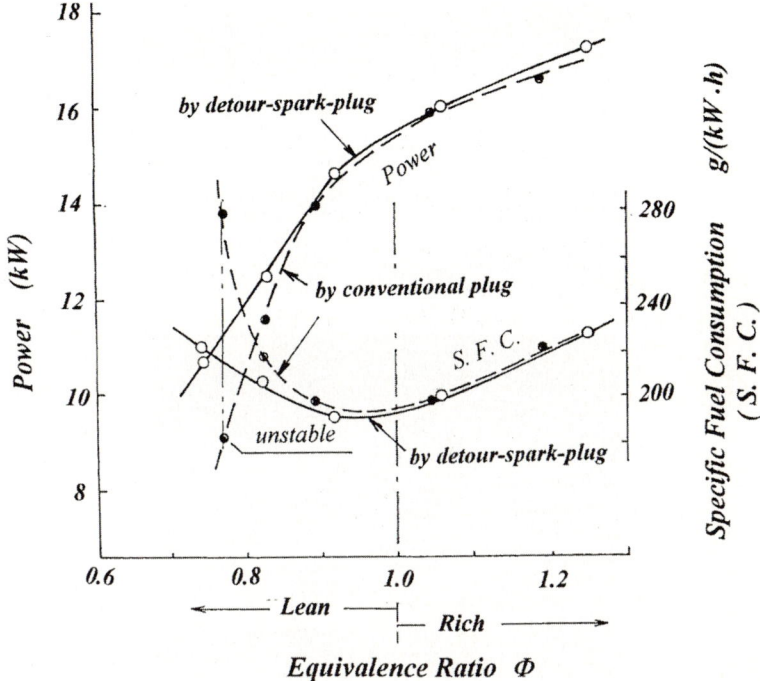

Fig. 8.23. Performance (power) and fuel consumption (*S. F. C.*) curves of a 4 cycle spark-ignition engine driven by conventional (*broken line*) and by detour-spark (*solid line*) plugs in relation to the fuel equivalence ratio Φ

than the equivalence ratio $\Phi = 0.75$, while it is driven unstably by the conventional plugs, if the mixture is leaner than $\Phi = 0.85$.

In Fig. 8.24 the pressure diagram measured by a piezoelectric pressure transducer in a cylinder of the engine driven by the conventional spark plug (above) and that by the detour-spark plugs (below) is shown. From such pressure diagram we can calculate the heat release rate (combustion rate) $dQ/d\alpha$ during the combustion in the cylinder in relation to the crank angle α of the engine, as also illustrated in Fig. 8.24.[64] Both the diagram of pressure variation and heat release rate suggest that the combustion velocity at ignition by the detour-spark plug is much higher than that by the conventional plug. The reason for it should be attributed to a larger fire ball at the ignition by the detour-spark plug because of the larger mixture quantity excited by the spark.

8 Ignition by Electric Sparks

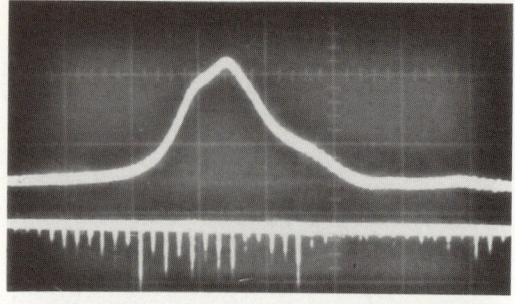

Driven by a conventional spark plug *(3.0 bhp)*

Driven by a detour-spark plug *(4.0 bhp)*

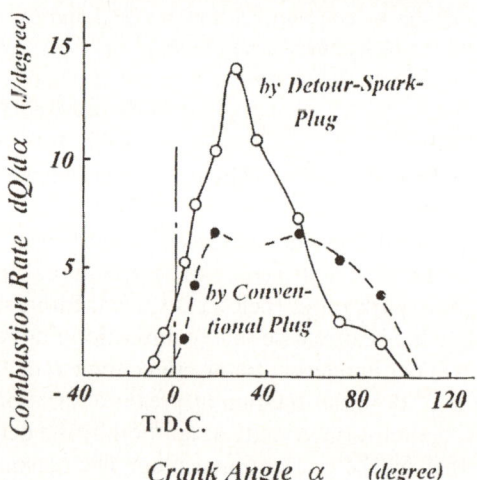

Fig. 8.24. Above: Indicator diagram (pressure–time) in a combustion chamber of a spark-ignition engine driven by a conventional plug (above) and by detour-spark plug (below). 1 MPa (div.)$^{-1}$ and 5 ms (div.)$^{-1}$. Below: Combustion rate (heat release rate) dQ/dα at detour plug drive *(solid line)* and that at conventional plug drive *(broken line)*, α: crank angle

9

Nonequilibrium State

In an irreversible process a heterogeneous and nonequilibrium state appears even for a short period. The heterogeneous state has been explained already in the previous chapters. Now we have to make clear, under which condition such a nonequlibium state really appears. In order to confirm the nonequilibrium state, the nonequilibrium temperature during the process should be detected. In this chapter the methods to measure the temperature, the measured results, and their meanings are explained.

At the nonequilibrium state the system is composed of many small parts having different temperatures and densities always changing their values, as explained already in Sect. 3.2. As the temperature can be measured only under the assumption of an equilibrium state, it is very difficult to measure such a temperature which changes always spatially and temporally. In the small parts having very high temperature, however, the molecules are ionized, and many ions and free electrons having high temperature are produced. The free electrons keep their temperature for a relatively long time, while the molecules and ions are cooled by collisions with cold molecules moving near the high temperature particles. Applying the methods developed in plasma physics, the temperature of the free electrons keeping their temperature for a relatively long time or that of free electrons and ions continuously produced during the measuring period can be detected.

Changing the equivalence ratio of a mixture of propane–air, its combustion velocity can be varied, while the ion density in the flame can be measured by a double probe method explained later. In Fig. 9.1 the combustion velocity (flame propagation velocity minus flow velocity of the mixture) in relation to the ion density in the flame is illustrated.[65] As this diagram shows, the combustion velocity is proportional to the ion density. This means that the ion density is proportional to the density of high temperature particles, as the combustion velocity must be proportional to the density of the particles having an energy higher than the activation energy. Therefore, observing the behavior and density of ions and free electrons, the state of nonequilibrium can be investigated.

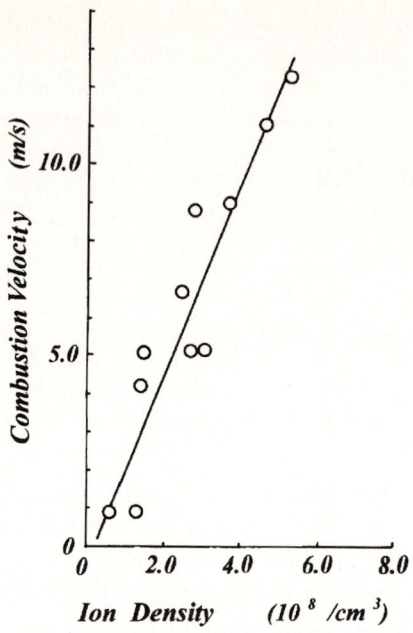

Fig. 9.1. Relation between combustion velocity and ion density in a propagating flame

As mentioned already, some methods developed in plasma physics are convenient to observe the ionization in the combustion systems. One of them, the probe method is often applied to measure the ionization in flames because of its simplicity. A thin metal needle having a high electroconductivity is inserted into the plasma. From the relation between the potential given to the needle and the current through the needle the temperatures and densities of ions and free electrons are calculated. In this case, however, the correct values cannot be measured because of cooling effect and disturbance of the plasma by needle, but with the measured results we can compare the different states with each other.

On the contrary, there is a nonintruding method, laser light scattering method. By this method we can measure the ionization in a small part of the plasma without any disturbance. Though the scattered light, however, is so weak that we have much difficulty to detect it, this method is very effective for the diagnostic of plasma having very high temperature and ion density.

In order to apply the methods, the density of free electrons as well as ions must be higher than a certain value. These methods, therefore, can be applied not to the cold mixture before combustion, but only to the flame or the heated gases behind combustion waves.

In this chapter, first, a calculation method of the adiabatic combustion temperature behind a combustion wave under assumption of an equilibrium state, then applications of a probe method and a laser light scattering method to propagating combustion and detonation waves, and the measured results are explained.

9.1 Adiabatic Combustion Temperature at Equilibrium

In order to compare the temperature at the nonequilibrium with the combustion temperature under the assumption of an equilibrium state, the combustion temperature at an equilibrium state should be estimated. In this chapter a calculating method of the so-called adiabatic combustion temperature under the assumption of an equilibrium state without any heat loss by radiation and conduction is introduced in brief.[66]

9.1.1 Reaction Process and Dissociation

As an example, we discuss the combustion reaction of a stoichiometric mixture of methane and oxygen during a perfect combustion. According to the following reaction:

$$CH_4 + 2O_2 \Leftrightarrow CO_2 + 2H_2O \tag{9.1}$$

CO_2 and H_2O are produced. But at the high temperature observed after combustion, a part of CO_2 and H_2O is dissociated as follows:

$$CO_2 \Leftrightarrow CO + O_2/2 - 289.8 \text{ kJ} \tag{9.2}$$

$$H_2O \Leftrightarrow H_2 + O_2/2 - 285.9 \text{ kJ} \tag{9.3}$$

Assuming that the dissociation degree of CO_2 is α that of H_2O β, the combustion reaction should be expressed as follows:

$$CH_4 + 2O_2 = (1-\alpha)CO_2 + \alpha CO + (1-\beta)H_2O + \beta H_2 + \frac{1}{2}(\alpha+\beta)O_2 \tag{9.4}$$

As the reaction heat depend on α and β it is necessary to know the correct values of dissociation degrees α and β to calculate the combustion temperature.

Now we want to explain here a general calculation method of the adiabatic combustion temperature in a mixture of hydrocarbon fuel. Assuming first a combustion reaction in a mixture composed of 1 mol hydrocarbon C_nH_m, k mole oxygen, and l mole A element which is independent of the reaction, the following reaction proceeds:

$$C_nH_m + k O_2 + l A \rightarrow x_1CO_2 + x_2CO + x_3H_2O + x_4O_2 + x_5H_2 + lA \tag{9.5}$$

In order to calculate the 5 unknown x_1, x_2, x_3, x_4, and x_5, we need 5 equations involving these 5 unknowns. From the equation of mass conservation, the following relations of mole number between before and after combustion are obtained:

$$C : n = x_1 + x_2, \tag{9.6}$$

9 Nonequilibrium State

$$H: m = 2x_3 + 2x_4, \tag{9.7}$$

$$O_2: k = x_1 + \frac{1}{2}x_2 + x_4 + \frac{1}{2}x_3. \tag{9.8}$$

The dissociation degrees of CO_2 and H_2O are listed in Table 9.1. On the other hand, there is an equilibrium constant K based on the law of mass action. As long as the combustion temperature is not extremely high and only the dissociation process shown by (9.2) and (9.3) come into question, the equilibrium constants are shown as follows:

$$K_1 = \frac{P_{CO}\sqrt{P_{O_2}}}{P_{CO_2}}, \tag{9.9}$$

$$K_2 = \frac{P_{H_2}\sqrt{P_{O_2}}}{P_{H_2O}}, \tag{9.10}$$

where P_{CO}, P_{O_2}, etc. mean the partial pressure of CO, O_2, etc. respectively. The equilibrium constants K_1, K_2 depend on the gas temperature, as shown in Table 9.2.

Let the total mole number of the combustion gas be X, then

$$X = x_1 + x_2 + x_3 + x_4 + x_5 + l, \tag{9.11}$$

Table 9.1. Dissociation degree (%)[67]

CO_2: $CO_2 = CO + (1/2)O_2 - 289.8$ kJ							
P (atm)		1.0	5.0	10.0	20.0	50	100
T (K)	log K						
3,500	+0.201	85.2	59.7	52.4	45.3	36.4	30.4
3,250	−0.105	61.8	45.0	38.3	32.1	25.0	20.5
3,000	−0.463	44.6	29.9	24.7	20.3	15.4	12.5
2,750	−0.899	26.8	16.8	13.6	11.0	8.24	6.85
2,500	−1.421	13.2	7.98	6.41	5.09	3.87	3.06
2,250	−2.062	5.32	3.11	2.47	1.96	1.15	1.15
2,000	−2.862	1.56	0.91	0.72	0.57	0.42	0.34
H_2O: $H_2O = H_2 + (1/2)O_2 - 285.9$ kJ							
P (atm)		1.0	10	20	50	100	
T (K)	log K						
3,500	−0.717	33.5	17.5	14.2	10.7	8.6	
3,250	−1.008	23.2	11.6	9.29	6.98	5.58	
3,000	−1.349	14.6	7.09	5.68	4.22	3.40	
2,750	−1.749	8.20	3.90	3.12	2.31	1.85	
2,500	−2.229	4.03	1.90	1.52	1.12	0.89	
2,250	−2.813	1.68	0.78	0.62	0.46	0.36	
2,000	−3.543	0.55	0.25	0.20	0.15	0.12	

9.1 Adiabatic Combustion Temperature at Equilibrium

Table 9.2.[68] Equilibrium constants

Temperature (K)	$O_2/2 =$ $O-495.3$ kJ	$H_2/2 =$ $H-433.2$ kJ	$O_2/2 + H_2/2 =$ $OH-286.2$ kJ	$O_2/2 + H_2 =$ $H_2O + 285.9$ kJ	$CO + O_2/2 =$ $CO_2 + 289.8$ kJ
	K_1	K_2	K_3	K_4	K_5
4,000	1.584	1.694	1.571	1.732	
3,800	1.055	1.175	1.480	2.586	
3,600	6.730×10^{-1}	7.856×10^{-1}	1.386	4.003	
3,400	4.091×10^{-1}	5.014×10^{-1}	1.294	6.654	
3,200	2.327×10^{-1}	3.022×10^{-1}	1.191	1.165×10^{1}	
3,000	1.228×10^{-1}	1.703×10^{-1}	1.078	2.192×10^{1}	
2,800	5.948×10^{-2}	8.879×10^{-2}	9.712×10^{-1}	4.485×10^{1}	
2,600	2.559×10^{-2}	4.182×10^{-2}	8.569×10^{-1}	1.044×10^{2}	
2,400	9.575×10^{-3}	1.742×10^{-2}	7.369×10^{-1}	2.760×10^{2}	
2,200	3.024×10^{-3}	6.227×10^{-3}	6.192×10^{-1}	8.682×10^{2}	
2,000	7.535×10^{-4}	1.812×10^{-3}	5.002×10^{-1}	3.436×10^{3}	
1,800	1.386×10^{-4}	4.027×10^{-4}	3.856×10^{-1}	1.838×10^{4}	
1,600	1.681×10^{-5}	6.203×10^{-5}	2.777×10^{-1}	1.491×10^{5}	
1,400	1.118×10^{-6}	5.391×10^{-6}	1.812×10^{-1}	2.190×10^{6}	
1,200	3.048×10^{-8}	2.346×10^{-8}	1.025×10^{-1}	7.792×10^{7}	
1,000	1.986×10^{-10}	2.788×10^{-9}	4.566×10^{-2}	1.132×10^{10}	6.710×10^{25}
800	1.068×10^{-12}	3.740×10^{-12}	1.345×10^{-2}	1.190×10^{13}	2.521×10^{34}
600	3.591×10^{-19}	6.473×10^{-17}	1.765×10^{-3}	4.223×10^{18}	3.439×10^{51}
400	6.027×10^{-30}	2.204×10^{-26}	2.769×10^{-5}	1.707×10^{29}	1.234×10^{69}
298.16	5.452×10^{-41}	4.867×10^{-38}	4.002×10^{-7}	1.097×10^{40}	

If the total pressure of the combustion gas is expressed by P,

$$P_{CO_2} = Px_1/X,$$
$$P_{CO} = Px_2/X,$$
$$P_{H_2O} = Px_3/X,$$
$$P_{O_2} = Px_4/X,$$
$$P_{H_2} = Px_5/X,$$
$$P_A = Pl/X, \tag{9.12}$$

therefore

$$K_1 = \frac{x_2}{x_1}\sqrt{x_4}\sqrt{\frac{P}{X}}, \tag{9.13}$$

$$K_2 = \frac{x_5}{x_3}\sqrt{x_4}\sqrt{\frac{P}{X}}. \tag{9.14}$$

Thus, from (9.6–9.8), (9.13), and (9.14), x_1, x_2, x_3, x_4, and x_5 can be calculated. In the practical calculation, however, it is convenient to presume an approximate value of X and calculate each value of x, putting the presumed X into the equations. The calculation is repeated to obtain a reasonable value of X.

The dissociation degrees are expressed as follows: $\alpha = x_2/(x_1 + x_2)$ and $\beta = x_5/(x_3 + x_5)$.

9.1.2 Reaction Heat and Adiabatic Combustion Temperature

From the reaction process, the combustion heat can be calculated by the following two different methods:

1. From the dissociation degrees and the heat absorbed at each reaction, the total absorbed heat Q_{dis} by the dissociation during the combustion can be obtained. The difference between the total reaction heat Q_b in the perfect combustion and the absorbed heat Q_{dis} by dissociation, that is, $\Delta Q = Q_b - Q_{dis}$ is the real combustion heat.
2. From the difference between both the enthalpies of formation of each gas component before and after the combustion the reaction heat can be calculated, using an enthalpy table of gases.

From the real reaction heat, the combustion temperature can be obtained, considering the combustion process, namely applying the isobaric heat capacity C_p in an isobaric combustion, or the isochoric heat capacity C_v in an isochoric combustion. This calculation is carried out under the assumption of equilibrium after the combustion. Considering the heat loss during the combustion the real temperature should be lower than the adiabatic combustion temperature.

9.1.3 Adiabatic Combustion Temperature of Propane–Oxygen Mixture

As an example, we try to calculate the adiabatic combustion temperature of the propane–oxygen mixture often applied to produce detonation waves.

Assuming that

1. The homogeneous mixture remains in a closed vessel
2. The vessel is kept in an adiabatic state, i.e., no heat is transmitted outside through the vessel wall
3. The combustion gas can be treated as an ideal gas
4. The combustion is dissociated corresponding to the gas temperature raised by the combustion

The following five chemical reactions proceed under an equilibrium state:

1. $O_2/2 \underset{}{\overset{k_1}{\rightleftharpoons}} O$
2. $H_2/2 \underset{}{\overset{k_2}{\rightleftharpoons}} H$
3. $O_2/2 + H_2 \underset{}{\overset{k_3}{\rightleftharpoons}} OH$

9.1 Adiabatic Combustion Temperature at Equilibrium

4. $O_2/2 + H_2 \overset{k_4}{\Leftrightarrow} H_2O$
5. $CO + O_2/2 \overset{k_5}{\Leftrightarrow} CO_2$

If the equilibrium constant of each reaction is K_1, K_2, K_3, K_4, and K_5, as shown in the reactions, and the partial pressure of each component P_{O_2}, P_O, P_{H_2} ... are expressed as follows:

$$K_1 = \frac{P_O}{\sqrt{P_{O_2}}} \cdots, \tag{9.15a}$$

$$K_2 = \frac{P_H}{\sqrt{P_{H_2}}}, \tag{9.15b}$$

$$K_3 = \frac{P_{OH}}{\sqrt{P_{O_2} P_{H_2}}}, \tag{9.15c}$$

$$K_4 = \frac{P_{H_2O}}{P_{H_2}\sqrt{P_{O_2}}}, \tag{9.15d}$$

$$K_5 = \frac{P_{CO_2}}{P_{CO}\sqrt{P_{O_2}}}. \tag{9.15e}$$

The equilibrium constants K_1, K_2, K_3, K_4, and K_5 are listed in Table 9.2
The combustion reaction is thus expressed as follows:

$$C_3H_8 + 5O_2 \rightarrow x_1O_2 + x_2O + x_3CO + x_4CO_2 + x_5H + x_6OH + x_7H_2 + x_8H_2O \tag{9.16}$$

Therefore, mole number of molecules containing

$$C : x_3 + x_4 = 3 \tag{9.16a}$$

that containing

$$O : 2x_1 + x_2 + x_3 + 2x_4 + x_6 + x_8 = 10 \tag{9.16b}$$

that containing

$$H : x_5 + x_6 + 2x_7 + 2x_8 = 8. \tag{9.16c}$$

Whole mole number of the molecules participating in the combustion reaction $X = \Sigma x_i$ and partial pressure of element i, $P_i = Px_i/X$, where P is the total pressure of the combustion gas. Putting $P/X = A^2$ and applying the relations described above to (9.15a–e),

$$x_2 = \frac{K_1\sqrt{x_1}}{A}, \tag{9.17a}$$

9 Nonequilibrium State

$$x_3 = \frac{x_4}{AK_3\sqrt{x_1}}, \qquad (9.17b)$$

$$x_5 = \frac{K_2\sqrt{x_7}}{A}, \qquad (9.17c)$$

$$x_6 = K_3\sqrt{x_1}\sqrt{x_7}, \qquad (9.17d)$$

$$x_8 = K_4 A \sqrt{x_1}\, x_7 \qquad (9.17e)$$

are obtained. Putting further $x_1 = B^2$, $x_7 = C^2$ and applying (9.17c–e) to (9.16c), we obtain the following equation:

$$\frac{CK_2}{A} + K_3 BC + 2C^2 + 2K_4 ABC^2 = 8 \qquad (9.18a)$$

or

$$2(1 + K_4 AB)C^2 + \left(\frac{K_2}{A} + K_3 B\right)C - 8 = 0. \qquad (9.18b)$$

From this equation, we obtain

$$C = \frac{-\left(\frac{K_2}{A} + K_3 B\right) + \sqrt{\left(\frac{K_2}{A} + K_3 B\right)^2 + 64(1 + K_4 AB)}}{4(1 + K_4 AB)}. \qquad (9.19)$$

1. Assuming an approximate value of T, A is estimated
2. Assuming an approximate value of B, $x_1 \sim x_8$ are calculated from (9.17a)–(9.19)
3. Applying thus calculated $x_1 \sim x_8$ to (9.16c), we can check, if the values are correct or not. If not, the calculation is repeated, till the correct values are obtained, substituting another value of T and B
4. According to the results of the calculation, the chemical composition of the combustion gas can be defined
5. The energy balance at the temperature T should also be checked. The heat Q produced by the reaction (9.16) should be equal to the sum of the heat difference between the whole formation enthalpy $\Sigma x_i H_i$ of each component of combustion gas in the right term of (9.16) at the initial temperature T_0 and that of the mixture in the left term before combustion and the heat Q is to raise the temperature of the combustion gas from the initial temperature T_0 to T, that is

$$Q = \Sigma x_i H_i \text{ after combustion} - \Sigma x_j H_j \text{ before combustion} = C_x(T - T_0),$$

where x_i and x_j are mole number of each component of the combustion gas and that of the mixture before combustion, respectively, H_i and H_j formation enthalpy of each component of the gas after and before combustion, respectively, C whole capacity of the combustion gas. Several examples of formation enthalpy are listed in Table 9.3.

Table 9.3. Formation enthalpy (kJ/mol)

Temperature (K)	H	H_2	O	O_2	HO	H_2O	CO	CO_2	CH	CH_2	CH_4	C_2H_6	C_3H_8	C_6H_{14}	C_8H_{18}
298.15	217.999	0	249.173	0	38.987	−241.826	−110.527	−393.522	594.128	386.392	−74.85	−84.68	−103.85	−198.8	−250.0
1,000	222.248	0	252.682	0	38.230	−247.857	−111.983	−394.623	593.285	382.022					
2,000	226.898	0	255.299	0	36.685	−251.575	−118.896	−396.784	589.723	376.844					
3,000	229.790	0	256.741	0	35.194	−253.024	−127.457	−400.111	586.510	371.127					
4,000	231.509	0	257.496	0	33.136	−254.501	−137.537	−405.251	581.937	363.650					

9 Nonequilibrium State

Table 9.4. Adiabatic combustion temperature T and mole number x

	$T(K)$	x_1	x_2	x_3	x_4	x_5	x_6	x_7	x_8
at isochoric combustion	3,349	0.873	0.490	1.193	1.087	0.491	1.028	0.590	2.650
at isobaric combustion	28.06	0.889	0.473	1.815	1.185	0.546	0.844	0.585	2.719

The adiabatic combustion temperature T and x for the stoichiometric propane–oxygen mixture are listed in Table 9.4.

9.2 Investigation of Flame by Probe Method

The combustion temperature of gas in a diffusion flame under atmosphere is usually observed to be 1,500–2,000 K. Though under such a relatively low temperature, the gas in the flame is remarkably ionized. The reason for it is not yet clear, though some chemical ionization processes are proposed. The gas is anyhow in a plasma state where ions, free electrons, and neutral particles coexist. The state of the gas, therefore, can be investigated applying different plasma diagnostics. One of them is the probe method using one or more metallic needles inserted into the plasma. Representative examples of the probe method are Langmuir probe method and double probe method.[69]

9.2.1 Electron Temperature and Ion Temperature

The ionized gas having a temperature lower than several thousand Kelvin can usually never be perfectly ionized plasma and has not only free electron and ions, but also neutral particles. In an equilibrium state the temperature of these three particles have the same temperature, but in a nonequilibrium state, these three temperatures have different values. The definition of temperature is, therefore, defined according to the concept of gas kinetics, namely, the particle temperature T is expressed as follows:

$$\frac{1}{2}kT = \frac{1}{2}mw_x^2, \tag{9.20}$$

where k is Boltzmann's constant, m particle mass, and w_x the average translation velocity of the particle in x-direction. The electron temperature T_e as well as ion temperature T_i is then defined as follows:

$$T_e = \frac{m_e w_e^2}{k}, \tag{9.21}$$

$$T_i = \frac{m_i w_i^2}{k}, \tag{9.22}$$

9.2 Investigation of Flame by Probe Method

where m_e and w_e are mass and average translation velocity in one direction of the free electrons and m_i and w_i are those of ions, respectively.

Inserting a terminal of a metallic needle having a high electroconductivity into a plasma, giving a low potential near the earth to the other terminal, the former has a negative potential to the plasma. The free electrons near the needle terminal, therefore, are repelled and remove far from the terminal surface, then the ion density near the terminal surface increases. The region near the needle terminal having almost no free electrons is called ion sheath, in which the potential drops by closing to the terminal surface and has a potential V lower than the plasma potential, as schematically illustrated in Fig. 9.2.

Free electrons overcoming this potential barrier V and reaching the terminal surface must have an energy higher than eV, namely only the electrons having a velocity w_e shown in the following equation,[70]

$$\frac{1}{2} m_e w_e^2 \geq eV, \tag{9.23}$$

can reach the probe surface, where e is the elementary charge. As long as each velocity of free electrons, ions, and neutral particles have a Maxwell's distribution, the energy has a Boltzmann's distribution. The number n_e of the free electrons satisfying the condition shown in (9.23), therefore, is proportional to $\exp(-eV/kT_e)$. If F is the effective area of probe surface for collecting the free electrons, the probe current i_e transported by the free electrons is expressed as follows:

$$i_e = eFn \sqrt{\frac{kT_e}{2\pi m_e}} \exp\left(-\frac{eV}{kT}\right), \tag{9.24}$$

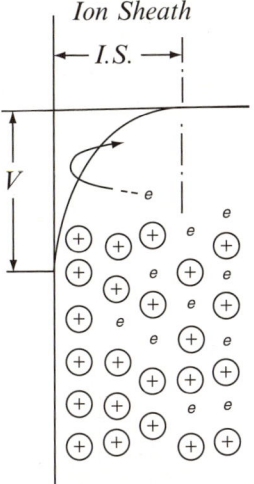

Fig. 9.2. Ion sheath on probe surface

where e is the elementary charge and n the density of the whole particles.

As the ions have a much larger mass, such a small potential in the ion sheath cannot play any role in the ions. The ion current i_i through the probe is, therefore,

$$i_i = eFn_i \sqrt{\frac{kT_i}{2\pi m_i}}. \tag{9.25}$$

9.2.2 Langmuir Probe Method[71]

As in Fig. 9.3 illustrated, two electrodes are set in both the ends of a homogeneous plasma, keeping the potential between both the electrodes relatively high (50–150 V). A third electrode, i.e., a probe having a potential of V lower than that between the other electrodes is inserted into the plasma, measuring the current i through the probe. The probe current i should be the difference between the current i_e transported by free electrons and that i_i transported by ions, namely

$$i = i_e - i_i. \tag{9.26}$$

According to (9.24) and (9.25) the relation between the probe potential V and current i is obtained, as illustrated in Fig. 9.4. From (9.24) we obtain the following equation:

$$\frac{d \ln i}{dV} = -\frac{e}{kT_e}. \tag{9.27}$$

According to (9.27), the electron temperature T_e can be calculated from the diagram of Fig. 9.4. From this diagram the saturated electron current i_{es} and ion current i_{is}, then the electron density n_e and ion density n_i can be calculated according to the following equations:

$$n_e = \frac{i_{es}}{Fe\sqrt{\frac{kT_e}{2\pi m_e}}}, \tag{9.28}$$

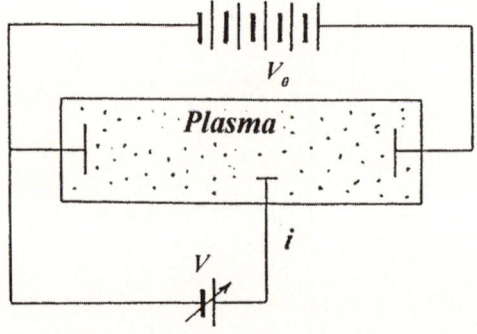

Fig. 9.3. Langmuir probe

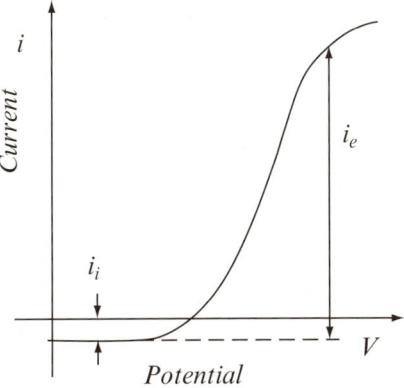

Fig. 9.4. Characteristic of Langmuir probe

$$n_i = \frac{i_{is}}{Fe\sqrt{\frac{kT_i}{2\pi m_i}}}, \qquad (9.29)$$

The probe potential V_f at $i = 0$ is called "floating potential" and expressed by the following equation:

$$V_f = \frac{kT_i}{2e} \ln \frac{m_i T_e}{m_e T_i}. \qquad (9.30)$$

Many measured results of ionization in flames using a Langmuir probe have been reported.[71] Because of high probe potential a thin probe having a diameter less than 1.0 mm can be applied, but we cannot expect to obtain any correct values, as the influence of the high potential to flame is not unavoidable and the flame is not homogeneous against the assumption of the measurement.

9.2.3 Double Probe Method[72–74]

As in Fig. 9.5 illustrated, two electrodes are inserted into a plasma to face with each other keeping the distance about 2–5 mm.

Varying the potential V given between both the electrodes within several volts, a current i through the probe circuit is measured. Theoretically the probe circuit is not earthed and it is called *floating double probe*.

The relation between the probe potential V and probe current i, so-called probe characteristic is illustrated in Fig. 9.6, where a straight line at $V = 0$ and $i = 0$ is observed, while the probe current i is saturated at a large potential in both the plus and minus sides and shows a straight line. This saturated current i_1 or i_2 is transported only by ions.

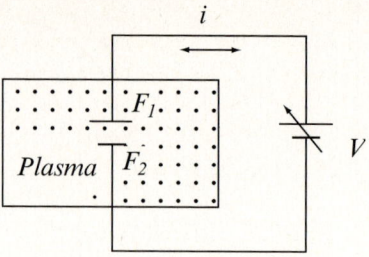

Fig. 9.5. Double probe

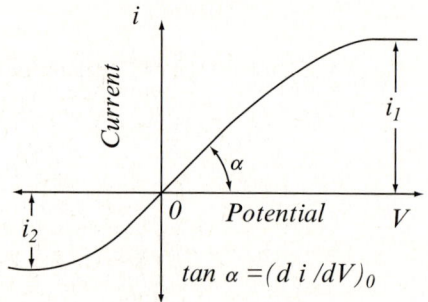

Fig. 9.6. Characteristic of double probe. Probe current vs. probe potential

From the double probe characteristic the following equation is obtained:

$$T_e = \frac{e}{k} \frac{i_1 i_2}{(i_1+i_2)\left(\frac{di}{dV}\right)_0} \tag{9.31}$$

in which T_e is the electron temperature and (di/dV) at the turning point of the characteristic curve where $V = 0$.

As the current is proportional to the effective area of the electrode, the characteristic curve has a symmetric form and $i_1 = i_2$, if the form and size of the both electrodes are the same, respectively. A probe potential V_f where the probe current i begins to saturate for increasing or decreasing probe potential V is equal to the plasma potential and therefore should be the floating potential.

The double probe theory described above can theoretically be available for a floating double probe isolated from the earth. In order to carry out a high speed measurement, it is often necessary to use an oscilloscope which has to be earthed In this case the current passes through the earth, but nevertheless (9.31) can be applied to the probe characteristic obtained by an earthed system. Only the turning point does not fall on the point of $V = 0$. In order to find the turning point easily, it is convenient to have a perfectly symmetric probe characteristic curve and for it a double probe of electrodes having the same form and size must be applied.

In Fig. 9.7 an example of probe characteristic curves obtained in a flame by an oscilloscope.

$$T_e = \frac{e}{k} \frac{i_s}{2(di/dV)_0},$$

where $(di/dV)_0 = \tan \alpha$

$$V_f = \frac{kT_e}{2e} \ln \frac{m_i T_e}{m_e T_i},$$

$$N_i = \frac{i_s}{Fe\sqrt{kT_i/(2\pi m_i)}}.$$

The symmetry center should be taken as the turning point of the probe characteristic obtained by a floating double probe at $V = 0$. With increase of probe potential in both the plus and minus sides, the thickness of the ion sheath, consequently the effective area of electrodes edges increase, too. The saturated current is not parallel to the i-axis, but shows a straight line inclined to i-axis, as shown in Fig. 9.7. In order to obtain the saturated current i_s, both the straight lines in the plus and minus sides are elongated to cross with the line of V = constant passing through the turning point, then the distance between both the cross points in the diagram is to be $2i_s$.

In this case, as both the electrodes have the same form and size, $i_1 = i_2 = i_s$. Equation (9.31), therefore, is rewritten to

$$T_e = \frac{e}{k} \frac{i_s}{2\left(\dfrac{di}{dV}\right)_0}. \tag{9.32}$$

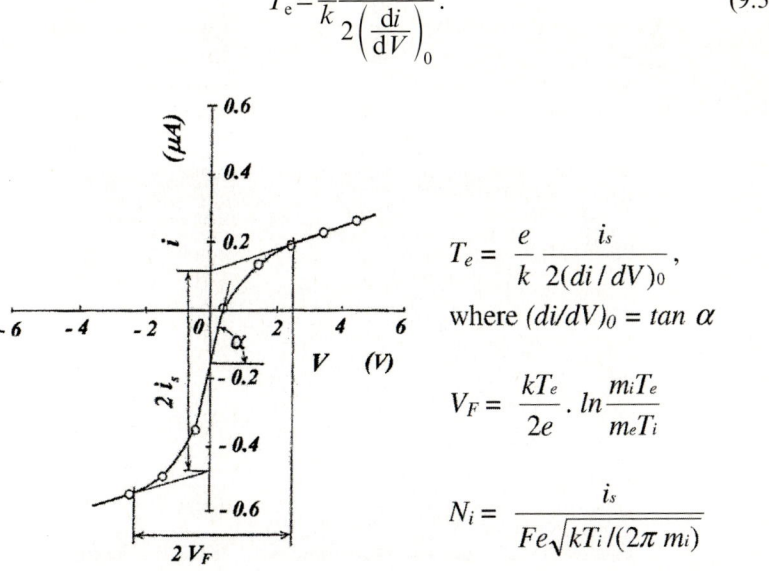

Fig. 9.7. Characteristic of an earthed double probe having the same electrodes. Probe current i vs. probe potential V

9.2.4 Investigation of Ionization in a Standing Flame by a Double Probe Method

As schematically illustrated in Fig. 9.8, a double probe composed of two brass bars having a diameter of 3 mm, insulated by a silica (silicon oxide) tube is inserted into a standing flame of an *n*-hexane–air mixture above a Bunsen burner.[75]

Giving a sawtooth potential as shown in the oscillogram of Fig. 9.9 to one electrode of the probe, the probe current corresponding to the potential through the other electrode is observed by an oscilloscope as also shown in Fig. 9.9. From the results observed in Fig. 9.9, the relation between the probe potential and current, i.e., probe characteristic can be obtained, from which saturated current i_{s1}, i_{s2} and slope $(di/dV)_0$ at the turning point, and consequently the electron temperature T_e can be obtained according to (9.32). The results are listed in the following Table 9.5.

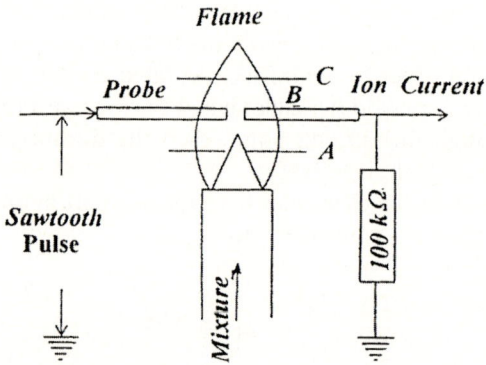

Fig. 9.8. Measurement of ionization in a standing flame by a double probe method

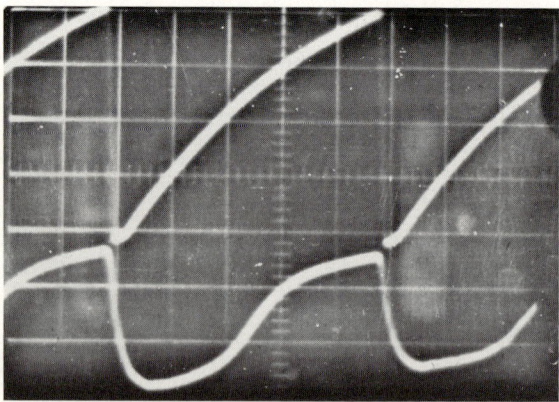

Fig. 9.9. Sawtooth probe potential (above, 1.0 V per div.) and probe current (below, 0.05 µA per div.), 2 ms per div.

9.2 Investigation of Flame by Probe Method

Table 9.5. Results measured by a double probe method in a standing flame of n-hexane–air mixture above a bunsen burner

measured position, height above the burner h (mm)	distance between both the electrodes d (mm)	floating potential V_f (V)	di/dV (µA V^{-1})	saturated current $i_{s1} = i_{s2}$ (µA)	electron temperature T_e (K)	gas temperature measured by a thermocouple T_g (K)
24	2.0	0.5	0.030	0.009	1,820	1,843
24	2.0	0.6	0.060	0.021	1,990	1,843
10	2.0	0.8	0.130	0.038	1,700	1,803
10	2.0	0.7	0.073	0.027	2,130	1,803
10	3.5	0.75	0.060	0.024	2,270	1,803
10	3.5	0.55	0.066	0.019	1,670	1,803

The results suggest that the electron temperature T_e measured by the double probe method is approximately equal to the gas temperature T_g in the flame measured by a thermocouple. Because of the cooling effect by the probe, however, the temperature measured by the double probe method may be lower than the real value.

9.2.5 Investigation of Ionization in a Propagating Flame by a Double Probe Method

In order to avoid the cooling effect of the probe, the double probe method is applied to measure the ionization in a flame propagating in a tube. As the flame passes by the probe within a short time, neither the flame is cooled too much by the probe, nor the probe is heated too much by the flame.

As shown in Fig. 9.10,[76] a pair of electrodes of a double probe are set face to face into a channel of plexiglas having a square cross-section of 2 cm × 2 cm and a length of 30 cm. Each of the electrodes of copper has a diameter of 2 mm. Each side of them is insulated by a silica tube, while only their end faces are exposed to the flame to collect the ions and free electrons. The gap distance between both the electrodes is kept to be 3 mm.

A stoichiometric methane–air mixture filled in the channel under atmospheric pressure and room temperature is ignited at an end of the channel, from which a flame propagates along the channel with a velocity of 20–30 m s^{-1}. Giving a sawtooth potential of 1.0–25 kHz having a range of ±15 V to one of the electrodes, a prove current corresponding to the probe potential is observed by an oscilloscope as shown in Fig. 9.11,[77] in which the sawtooth potential as well as the luminescence from the flame is observed by a photomultiplier set at the position of the probe. The ionization of the gas in the flame having a length of 10–20 cm is measured by the double probe during the passing period of the flame.[78]

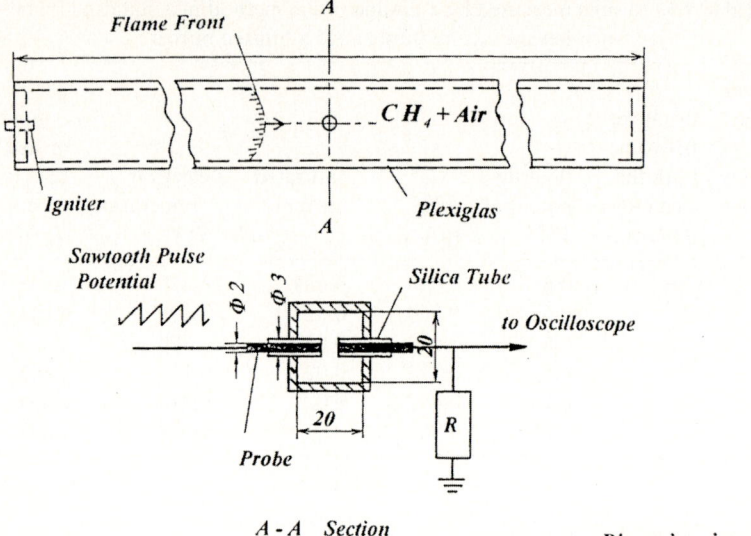

Fig. 9.10. Sketch of measurement system of ionization in a flame propagating in a stoichiometric methane–air mixture in a tube using a double probe method

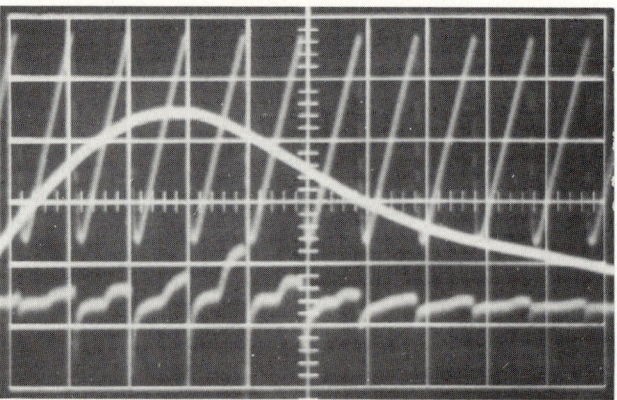

Fig. 9.11. Probe potential of sawtooth pulse (above, 5 V per div., 25 kHz), luminescence from the flame (middle) and probe current (below, 0.4 µA per div.) in a flame propagating in a stoichiometric methane–air mixture

From the probe potential and current recorded in the oscillogram of Fig. 9.11, we obtain probe characteristics at different positions of the propagating flame, from which further the electron temperature T_e can be calculated according to (9.32) and is represented in Fig. 9.12 with respect to the time t after passage of the flame front at the probe position or to the distance l behind the flame front, the atomic or molecular weight of ions calculated according to (9.30)

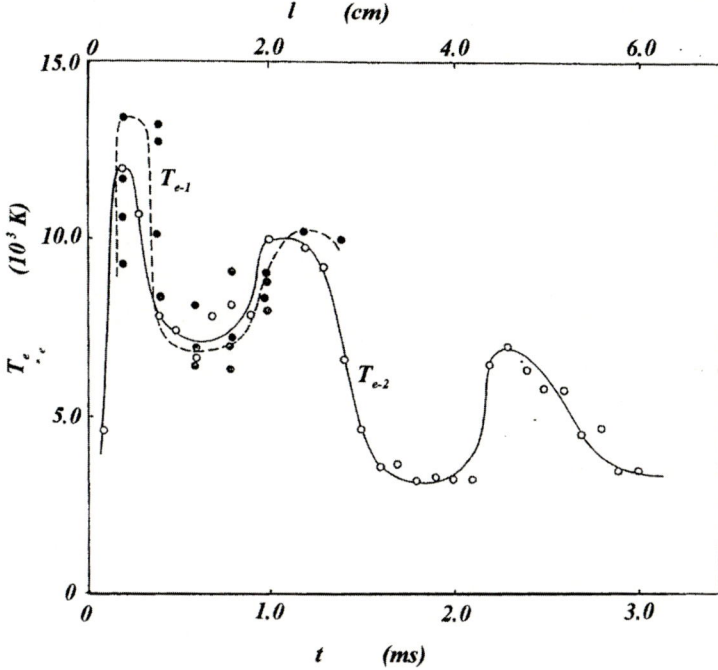

Fig. 9.12. Electron temperature T_e in a flame propagating in a stoichiometric methane–air mixture measured by the double probe method using a sawtooth pulse potential vs. the distance l behind the flame front and time t after the passage of the flame front at the measuring point

is represented in Fig. 9.13 and ion density n_i calculated according to (9.29) in Fig. 9.14. In this case $T_e = T_i$ is assumed.

Though the temperature behind the flame front in an equilibrium state is estimated to be about 2,000 K, the electron temperature just behind the flame front measured by the double probe method is so high to be 10,000–15,000 K, namely 5–7 times the equilibrium temperature. This suggests that there is a nonequilibrium state behind the flame front, where the particles having a temperature higher than 10,000 K and those having a temperature lower than 2,000 K coexist, then both the particles of high temperature and those of low temperature are mixed with each other in the course of time, and the nonequilibrium state closes to an equilibrium state. About 100 mm behind the flame front, thus, the electron temperature drops to about 2,000 K and has almost the same value as that in the equilibrium state.

As long as the double probe method using a high frequency sawtooth potential is applied to a propagating flame changing the state behind its front steadily and little during the period of one sawtooth potential, probe characteristic can be used to calculate the electron temperature, ion density, and so on. This method,

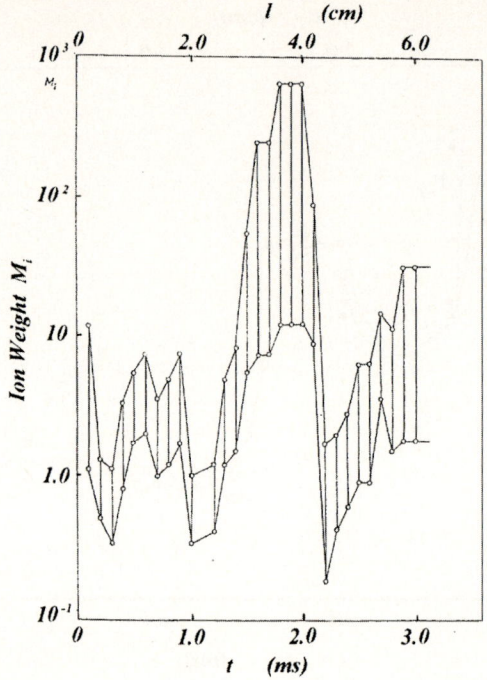

Fig. 9.13. Atomic or molecular weight of ion M_i in a flame propagating in a stoichiometric methane–air mixture with respect to the distance l or time t

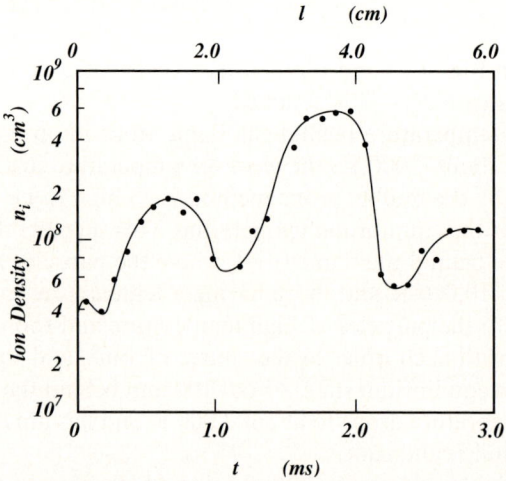

Fig. 9.14. Ion density n_i in a flame propagating in a stoichiometric methane–air mixture with respect to the distance l behind the flame front and time t after passage of the flame front at the measuring point

however, cannot be applied to turbulent flames and detonation waves in which the states change very rapidly. For such unstable flames or plasmas a method applying a high frequency multistep potential is developed.[78]

Giving a multistep potential composed of 6–10 steps having 5–100 kHz to one of the electrodes of a double probe, a stepwise probe current corresponding to each potential step can be recorded in an oscillogram, in which the current corresponding to the same step potential as a line composed of many dots, that is, the same number of curve as that of the given potential step are recorded.

In Fig. 9.15, an experimental apparatus for investigating the ionization in a propagating flame by a double probe method using a high frequency multistep probe potential is schematically illustrated. Into the middle of a channel of plexiglas having a square cross-section of 15 mm × 15 mm and a length of 500 mm in which a stoichiometric propane–air mixture is filled under atmospheric pressure and room temperature a double probe is set. During a flame produced by a spark ignition at a channel end propagates through the channel with a velocity of 3 m s^{-1}, a probe potential having six steps and 5 kHz is given to the one electrode of the probe, as shown in an oscillogram in Fig. 9.16. Then we obtain six curves of probe current corresponding to the six step potential, as shown in Fig. 9.17.

From such a group of current we can obtain a probe characteristic at an arbitrary position behind the flame front or time after the passage of the flame front at the measuring point. An example of such probe characteristics is shown

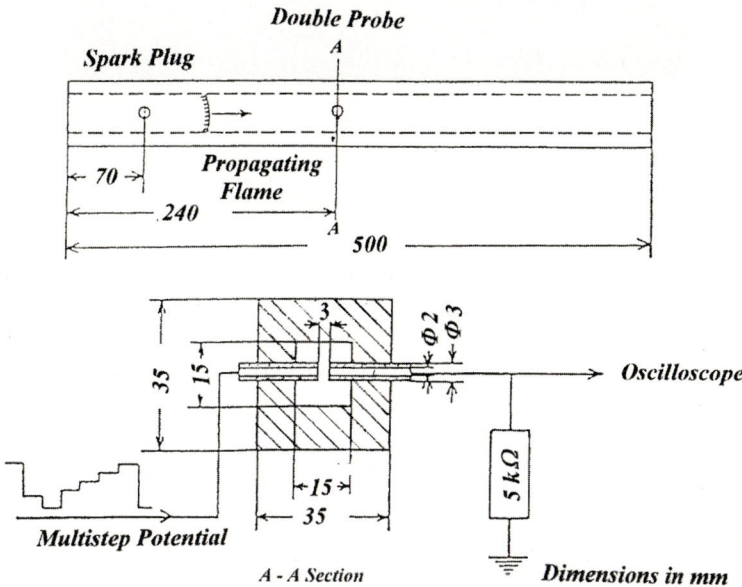

Fig. 9.15. Measurement of ionization in a propagating flame by a double probe method applying a high frequency multistep potential

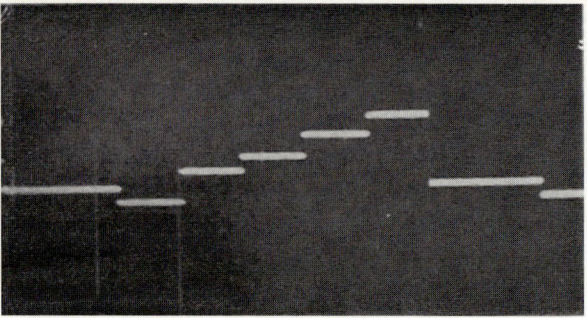

Fig. 9.16. High frequency multistep probe potential 5 V per div., 40 μs per div.

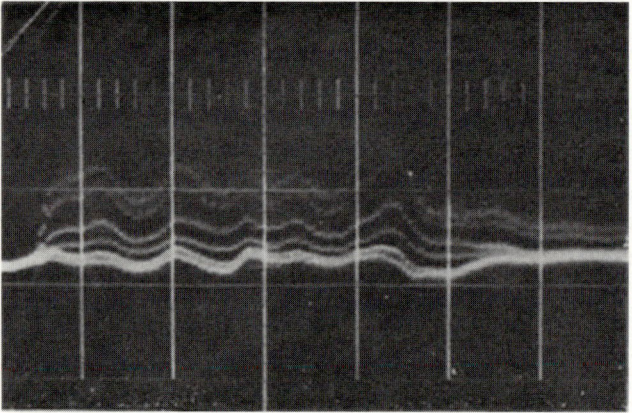

Fig. 9.17. Probe current corresponding to multistep potential of double probe measured in a flame propagating in a mixture of $C_3H_8 + 10O_2$, 0.2 μA per div., 4 ms per div.

in Fig. 9.18. From such probe characteristic we can calculate the electron temperature in the flame according to (9.32). In Fig. 9.19 the electron temperature T_e in a flame propagating in a stoichiometric propane–air mixture measured by the double probe method using a high frequency multistep potential is illustrated with respect to the time after the passage of the flame front at the measuring point.

The results also suggest that just behind the flame front the electron temperature T_e is measured to be 6,000 to 8,000 K, so high as 3–4 times the adiabatic combustion temperature and some parts having high temperature and those having low temperature coexist in a heterogeneous state behind the flame front. Comparing the results with those measured in a standing flame, the cooling effect of the probe at the measurement in a propagating flame is much less than those in a standing flame.

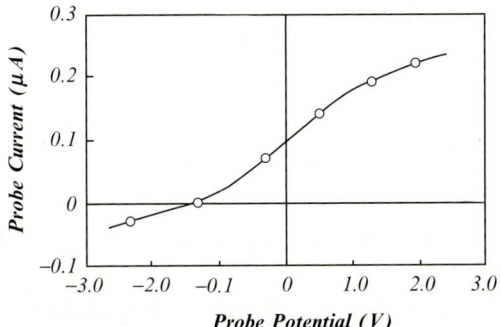

Fig. 9.18. Probe characteristic obtained by a double probe method using a multistep probe potential

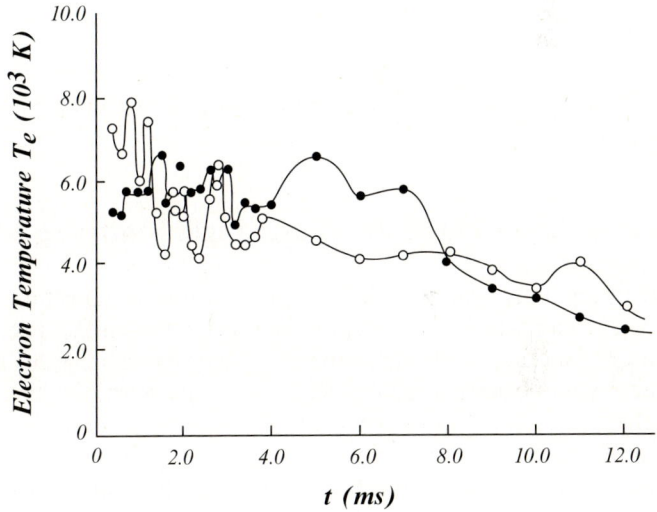

Fig. 9.19. Electron temperature T_e in a flame propagating in a mixture of $C_3H_8 + 10O_2$ measured by a double probe method using a high frequency multistep probe potential with respect to time t after passage of the flame front at the measurement point. Obtained from two different experiments

Nevertheless we still observe a remarkable cooling effect of the probe at the measurement using a probe method. The electron temperature T_e is measured by the double probe method in a flame propagating in a stoichiometric propane–air mixture under different pressures. In Fig. 9.20 the difference between the measured electron temperature T_e and the initial mixture temperature T_g is represented with respect to the mixture pressure P.

With increase of the mixture pressure, the electron temperature decreases and closes to the combustion temperature. This means, the cooling effect of the probe still plays an important role on the ionization measurement in gases.

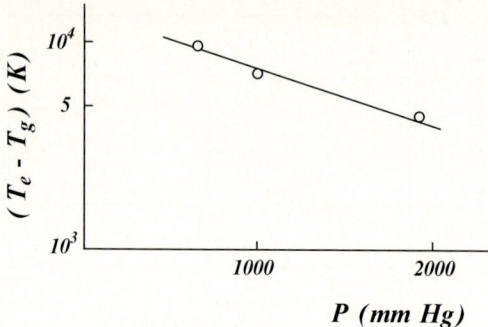

Fig. 9.20. Difference between both the measured electron and initial mixture temperatures $(T_e - T_g)$ in relation to the initial mixture pressure P

In order to avoid such a cooling effect of the probe, we applied a nonintruding method, i.e., a laser light scattering method developed in an area of plasma physics.

9.3 Investigation of Flame by a Laser Light Scattering Method

Let light be incident on a gas particle, the light is scattered, reflecting or refracting on it. As such a particle in a gas is moving with a velocity corresponding to its temperature, especially with such a mean velocity shown in (9.20) at an equilibrium state. A monochromatic light scattered by gas particles, therefore, has Doppler effect[79] and shows a distribution of spectrum having a center at the wave length of the incident light corresponding to the velocity distribution of the particles, i.e., to the gas temperature. This is called "Rayleigh scattering," from which the gas temperature can be calculated.

At such a light scattering, the particle size effectively playing for the scattering is called scattering cross-section. Gas molecules usually have a small scattering cross-section, for example, oxygen or nitrogen molecules have a scattering cross-section $\sigma_R = 4.08 \times 10^{-26}$ cm², while a free electron has a scattering cross-section $\sigma_{Th} = 0.67 \times 10^{-24}$ cm², that is, larger than 50 times that of molecules. The light scattered by free electrons is much stronger than by molecules.

The scattered light is in general very weak to be detected, but applying a strong laser beam to the light source and a suitable detecting method, the scattered light can be measured. The plasma diagnostics using laser light scattering have been developed, theoretically by Salpeter and experimentally by Kunze, Kegel, and others. The method is available for the diagnostics of strong plasma having an electron density higher than 10^{13} per cm³. In the next chapter the theory is briefly explained, while the details are described in the referred books and papers.

9.3.1 Laser Light Scattering Method[80-82]

Let the wave number vector of incident light k_0, that of scattered light k_s, difference of both the vector $k = k_s - k_g$, velocity vector of electron v, the Doppler shift ω, deviation by Doppler effect, is expressed by $\omega = k \cdot v$. In plasma there are two sorts of electrons, one of them is of perfectly free electrons, and the other those correlated to ions. The light scattered by the free electrons is called "electron component," while the other by those correlated to ions "ion component." The electron component is formed by freely moving electrons and has a wider distribution, as the electrons have much less mass and higher velocity than ions, while the ion component has a more narrow distribution, as the electron correlated to ions moving together with the ions and have therefore much less velocity.

Let the total intensity integrated over the intensity of each wavelength of scattered light over the whole wavelength region be S, that of electron component S_e and that of ion component S_i, then formally the relation is expressed by the following equation:

$$S = S_e + S_i. \tag{9.33}$$

These total intensities depend on a characteristic parameter α expressed by the following equation:

$$\alpha = \frac{1}{k\lambda_D} \approx \frac{\lambda_0}{4\pi D \sin\frac{\theta}{2}}, \tag{9.34}$$

where α for the electron component is expressed by α_e, λ_0 is the wavelength of the incident light, and D_e or D_i is Debye length expressed as follows, if it is formed by free electrons:

$$D_e = \sqrt{\frac{kT_e}{(4\pi n_e e^2)}} \tag{9.35}$$

and α for the ion component is expressed by α_i and Debye length formed by ions is

$$D_i = \sqrt{\frac{kT_i}{(4\pi n_e Z e^2)}}, \tag{9.36}$$

where θ is the scattering angle, k Boltzmann's constant, T_e the electron temperature, T_i the ion temperature, n_e and n_i the density of free electrons and ions, respectively, e the elementary charge, and Z the charge of ion. That is

$$S_e = 1/(1+\alpha_e^2),$$
$$S_i = (Z\alpha_e^4) \Big/ \left[(1+\alpha_e^2)\left\{1+\alpha_e^2+\alpha_e^2 Z \frac{T_e}{T_i}\right\}\right]. \tag{9.37}$$

The relation of S_e and S_i to α is shown in Fig. 9.21. The total intensity S_e in the electron component depends only on α_e, whereas S_i in the ion component depends not only on α_e, but also on the temperature ratio T_e/T_i. At $\alpha = 0$, $S_e = 1.0$, and

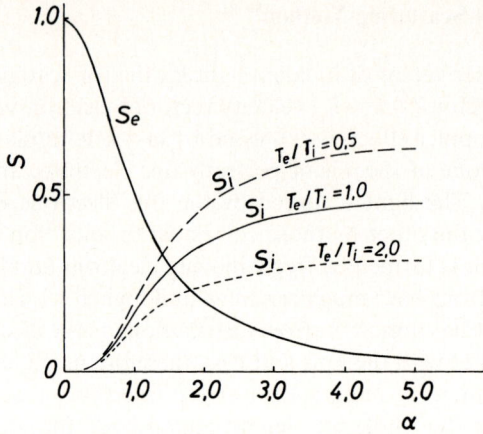

Fig. 9.21. S, S_e, and S_i in relation to α

$S_i = 0$, the scattering is caused only by free electrons and is Thomson scattering having a Gaussian profile.

1. At $\alpha \approx 0$:
The halfwidth $\Delta\lambda_\eta$, having the half intensity of the maximum one, is expressed as follows:

$$\Delta\lambda_\eta = 4\lambda_0 \sin\frac{\theta}{2}\sqrt{\frac{kT_e}{m_e c^2}\ln 2}, \qquad (9.38)$$

where θ is the scattering angle, m_e the electron mass, and c the light velocity. According to this equation the electron temperature T_e can be calculated from $\Delta\lambda_\eta$, but the electron density n_e cannot be obtained without any absolute value of the scattering intensity.

2. $\alpha \gg 0$:
The method proposed by Kegel[83] is introduced. Under an assumption that the translation velocity of free electrons is much faster than that of ions, the intensity I_s of he scattered light is approximately expressed by the following equation:

$$I_s \approx \frac{n_e}{k_s}\sqrt{\frac{m_e}{2kT_e}}F_\alpha(X) + Z\left(\frac{\alpha_e^2}{1+\alpha_e^2}\right)^2 \frac{n_e}{k_s}\sqrt{\frac{m_i}{2\pi kT_t}}F_\beta(Y), \qquad (9.39)$$

where k_s is the wave number of scattered light,

$$F(X) = \frac{\exp(-X^2)}{\left\{1+\alpha_e^2 - \alpha_e^2 g(X)\right\}^2 + \pi\alpha_e^4 X^2 \exp(-2X^2)}, \quad X = (\Delta\omega/k_s)\sqrt{\frac{m_e}{2kT_e}},$$

$\Delta\omega$ deviation from the angular frequency, ω_0 of the incident light beam and

$$g(X) = 2X\exp(-X^2)\int_0^X t^2\,dt$$

9.3 Investigation of Flame by a Laser Light Scattering Method

$F_\beta(Y)$ is obtained by replacing α_e with $\beta = \sqrt{\alpha_i^2/(1+\alpha_e^2)}$ and X with $Y = (\Delta\omega/k_s)\sqrt{m_i/2KT}$ in the equation of $F_\alpha(X)$ described above. Here m_i is ion mass.

The first term of (9.39) is of the electron component, while the second term of the ion component. The state of the free electrons is obtained from the electron component, and that of ions from the ion component.

It is, however, not easy to estimate the temperatures and densities of free electrons and ions according to (9.39). As the spectral profile of scattered light is defined by the characteristic parameter α, Kegel proposed a method to calculate the values described above from the experimentally obtained distribution curves of scattered light as follows:

1. The spectral profiles of ruby laser beam having a wavelength $\lambda_0 = 6{,}948$ Å scattered in a hydrogen plasma having a temperature $T_0 = 100{,}000$ K under the scattering angle $\theta = 90°$ for different α together with the electron density n_0 are theoretically obtained according to (9.39) and a set of the profiles is proposed by Kegel. Some examples of the spectral profiles in the Kegel's graphs are shown in Fig. 9.22.
2. Comparing the spectral profiles of scattered light experimentally obtained using the same laser under a scattering angle θ with those in the Kegel's graphs and finding out one profile having the same form, the characteristic parameter α_e of the electron component of the profile experimentally estimated.
3. Characteristic length $\Delta\lambda$ of the scattered light experimentally obtained, for example the halfwidth, is compared with that of the profile $\Delta\lambda_0$ of the spectral profile in the graphs having the same α_e.

According to the following relations:

$$T_e = T_0 \left(\frac{\Delta\lambda}{\Delta\lambda_0}\right)^2 \left(\frac{\sin\dfrac{\theta_0}{2}}{\sin\dfrac{\theta}{2}}\right)^2, \qquad (9.40)$$

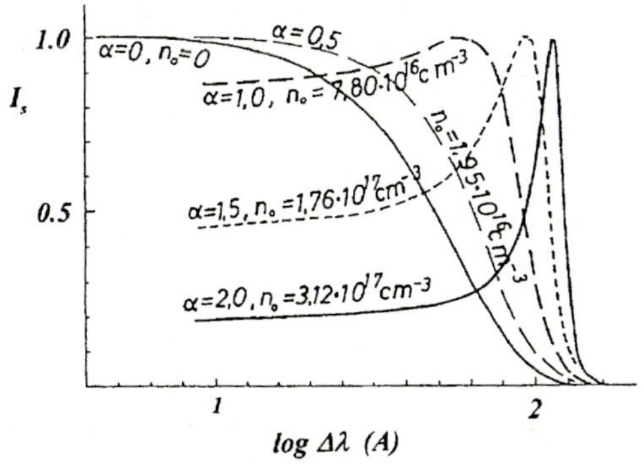

Fig. 9.22. Intensity I_s of ruby laser beam scattered in a hydrogen plasma of 10^5 K having different α with respect to logarithm of wavelength $\Delta\lambda$ deviated from the wavelength of the laser beam

$$n_e = n_0\left(\frac{\Delta\lambda}{\Delta\lambda_0}\right)^2, \tag{9.41}$$

the electron temperature T_e and electron density n_e of the plasma experimentally observed can be calculated.

In the same way, ion temperature T_i and density n_i can be obtained, taking β as the characteristic parameter instead of α_e. Looking for a profile having the same form as that of ion component in the Kegel's graphs, α_e is replaced by β. Let the characteristic length of both the profiles be $\Delta\lambda_0$ and $\Delta\lambda$, then according to the following equation:

$$T_i = T_0\left(\frac{m_i}{m_e}\right)\left(\frac{\Delta\lambda}{\Delta\lambda_0}\right)^2 \left(\frac{\sin\frac{\theta_0}{2}}{\sin\frac{\theta}{2}}\right)^2, \tag{9.42}$$

the ion temperature T_i can be calculated. In an isothermal plasma, $\alpha_i^2 = Z\alpha_e^2$. Therefore,

$$n_i = n_0\left(\frac{\Delta\lambda}{\Delta\lambda_0}\right)^2 \frac{m_i}{m_e}(1+\alpha_e^2), \tag{9.43}$$

from which also n_i is obtained.

9.3.2 Notices to be Considered at the Measurement

At the measurement by the laser light scattering method, it has to be considered that the scattered light is so weak and the width of its profile is so narrow.

In order to have a detectable scattered light, a very strong laser producing a intensive incident light beam must be prepared, but at the same time, the influence of the laser beam on the plasma must be avoided as much as possible, as the plasma absorbs the energy of the laser beam and a breakdown may take place in the plasma. The power of the laser, therefore, must be controlled not to generate a breakdown, while the radiation duration of the laser must be as short as several nanoseconds and finish, before the plasma is heated by the laser beam. For this a Q-switched giant pulse ruby laser having a power of 1–30 MW may be the most suitable. Also using a rotating prism is recommended to avoid power fluctuation.

Some filters and polarizer are applied to minimize the emission and light from the flame as well as from outside the measurement system. Some consideration for preventing the stray light and useless reflection are also necessary. A scattering angle near 90° is recommended because of easy analysis of the scattered light, if it is possible.

A spectroscope having a high resolving power and sensitivity should be prepared to take the spectra of scattered light. The most used spectroscope at present is a monochromator having a diffraction grating. The spectra passed through the spectroscope are caught by a photographic film having a very high sensitivity. With the former method the same experiment must be repeated many times, as the intensity at only one wavelength can be observed by one measurement, while the latter the whole spectra can be taken at once, but it is very difficult to obtain such a sensitive film. Using an image intensifier, however, it is today not so difficult to take the spectra directly on a film. The spectra taken on the film can be analyzed by a microphotometer, by which the light intensity at each wavelength, consequently the spectra of the scattered light can be obtained. The measurement error is estimated to be ±30%, but the accuracy is high enough to assess the nonequilibrium state.

9.3.3 Investigation of the Ionization in a Propagating Flame Using the Laser Light Scattering Method[84]

As the laser light scattering method can be applied only to strongly ionized plasma having an ionization degree higher than an extent value. A flame propagating in a stoichiometric propane–oxygen mixture ($C_3H_8 + 5O_2$) is thus investigated by the laser light scattering method.

In Fig. 9.23 the experimental apparatus and measurement system are schematically illustrated. In order to suppress the transition to detonation waves, a flat cylindrical combustion chamber of steel having an inner-diameter of 300 mm and inner-width of 25 mm is prepared. A stoichiometric propane–oxygen mixture

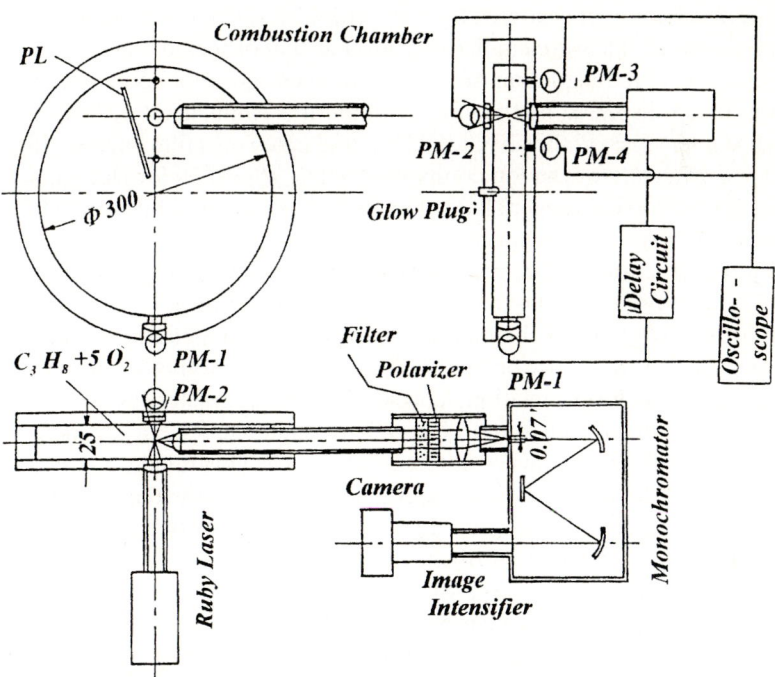

Fig. 9.23. Experimental apparatus for measurement of ionization in a flame propagating in a stoichiometric propane–oxygen mixture using s laser light scattering method

filled in the combustion chamber under a pressure of 200 torr (27 kPa) and room temperature is ignited by an electrical spark at the center, from which a divergent flame propagates radially with a velocity of about 75 m s^{-1}. From a Q-switched giant pulse ruby laser a beam having a power of 25 ± 5 MW and a half intensity duration of 20 ns is emitted to focus onto a center of the mixture in a position at a radial distance of 100 mm from the chamber center. The emission of the laser beam is triggered by the light emission of the propagating flame before its arrival at the measuring point through a photomultiplier PM-1. Its triggering time is regulated by a delay circuit in such a way that the laser beam can be emitted at an arbitrary instant before or after the flame front passes the measuring point. The laser beam incident instant is observed by a photomultiplier PM-2, while the flame propagation by those PM-3 and PM-4, which are all at the same time recorded on an oscilloscope, so that the measured position can be verified.

At the same time, the light scattered by the charged particles in the flame at an angle of 90° to the incident laser beam is introduced into a monochromator (Shimazu having a grating line number of 1,200 per mm and two concave mirrors of 500 mm focus length) through a condenser lens at the top of the tube having an inner-diameter of 10 mm inserted into the chamber, a filter and polarizer. In the chamber a steel plate PL is set between both the chamber walls on the radial direction to reduce the stray light from the flame to the monochromator as much as possible.

The distribution of scattered light dispersed by the monochromator has to be recorded over a wide wavelength range around that of incident beam all at once. Because of a large fluctuation of the measurement position and laser beam, the scattered light dispersed by the monochromator is stimulated to 22,000 times by an image intensifier (Philips XX 1389 HG) and taken on a film having a high sensitivity of *ISO-400* for the ion component, while *ISO-1000* for electron component, since the ion component has a narrow dispersion range but strong intensity, while the electron component has a wide range but weak intensity. In Fig. 9.24 an example of scattered light recorded directly on a film through the image intensifier after the dispersion by the monochromator is shown.

In order to obtain a diagram of intensity vs. wavelength, each spectrum recorded on the film must be analyzed by a microphotometer and then transformed into the linear relation, considering the characteristic of the image intensifier and that of the film. In Fig. 9.25 the characteristic of the image intensifier for the light beam having different intensity is illustrated.

In Fig. 9.26 an example of the laser light scattered in the flame and analyzed by a microphotometer, that is, a diagram of diffuse density with respect to the wavelength deviated from that of the incident laser beam.

In this diagram, we observe the electron component having a wide range below the ion component which has a narrow range and strong intensity. The symmetrical center of the electron component is shifted to the shorter wavelength by about 10 Å from that of the laser beam. Considering the sensitivity of the film, characteristic of the image intensifier and the luminescence of the flame, we can transform the diagram to the linear scale of light intensity with respect to the wavelength deviated from that of incident laser beam.

9.3 Investigation of Flame by a Laser Light Scattering Method

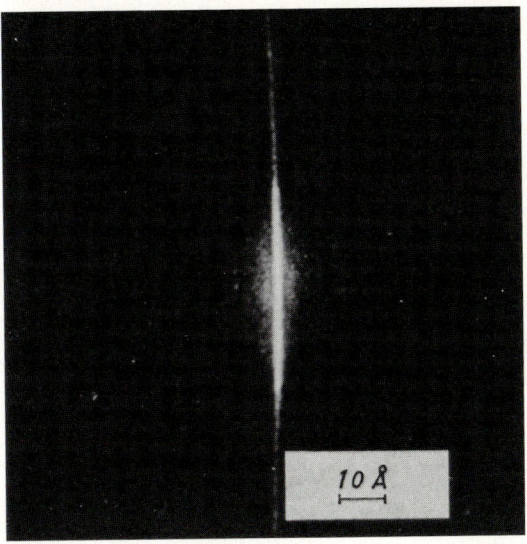

Fig. 9.24. Laser beam scattered in a flame taken on a photographic film

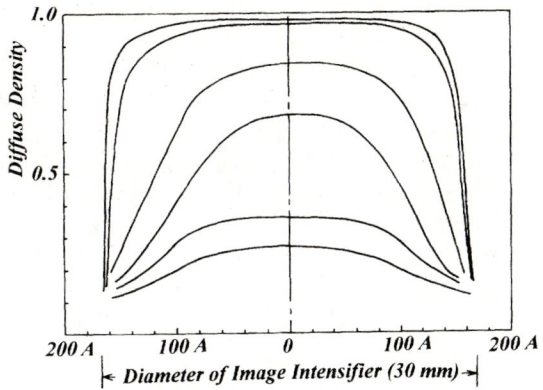

Fig. 9.25. Characteristic of the image intensifier. The wavelength on the horizontal coordinate means the corresponding deviation of the wavelength dispersed by the monochromator

Electron Component

From the transformed spectral distribution of the electron component we can estimate the characteristic number α_e (0.5–0.9) then, further the electron temperature T_e and density n_e, comparing these parameters with those proposed by Kegel and according to (9.40) and (9.41). In Figs. 9.27 and 9.28, the thus obtained electron temperature and density are illustrated with respect to the distance from the visible flame front. The results suggest

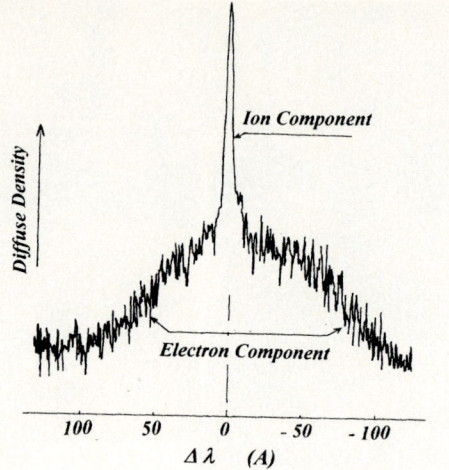

Fig. 9.26. Spectral distribution of the scattered light analyzed by a microphotometer

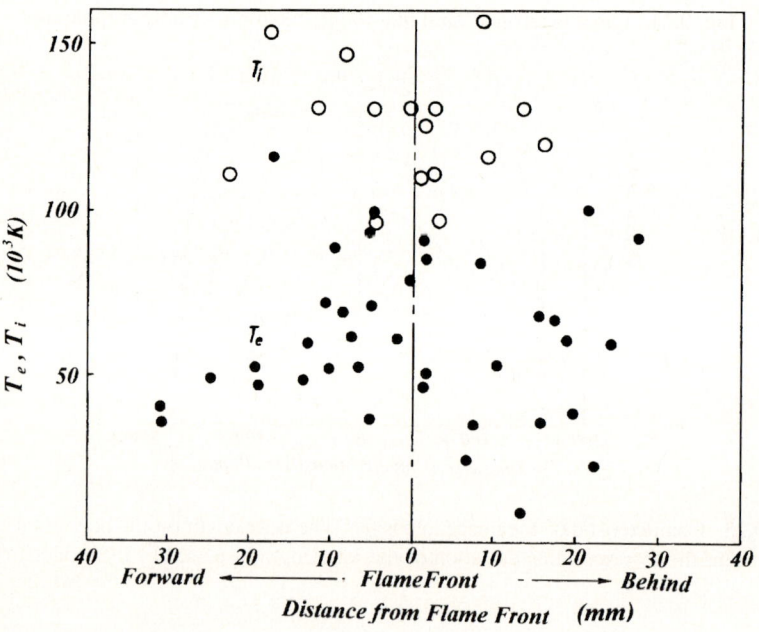

Fig. 9.27. Electron temperature T_e (*black circles*) and ion temperature T_i (*white circles*) with respect to the distance from the flame front

1. The ionization proceeds not only behind the flame, but also in front of it
2. The electron temperature is very high and most of the measured results have a value from 30,000 K to 10,000 K
3. The electron density has a value from 10^{15} to 10^{17} per cm^3

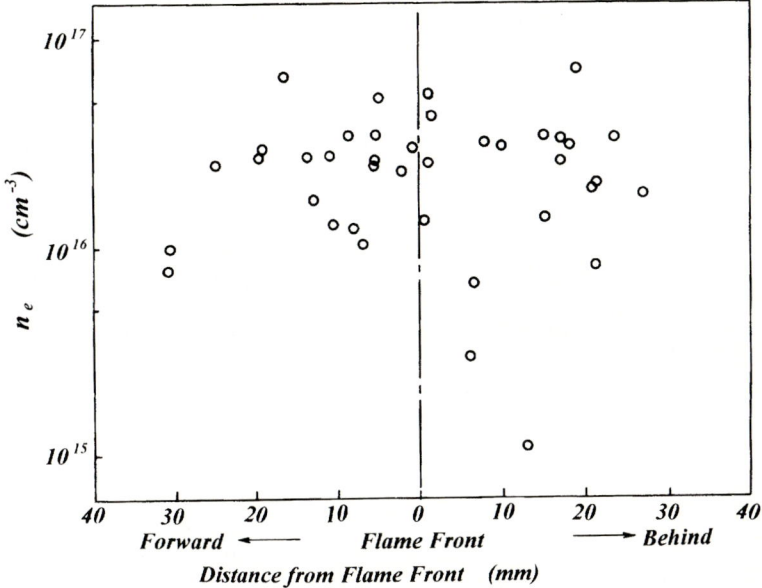

Fig. 9.28 Electron density n_e with respect to the distance from the flame front

Considering that the theoretical adiabatic combustion temperature is estimated to be 2,700–3,000 K at isobaric and to be 3,200–3,500 K at isochoric combustion, the electron temperature measured by the laser light scattering method is 6–30 times the theoretical combustion temperature and seems to be much too high. Considering, however, that the electron temperature measured by the double probe method shows also a temperature higher than 10,000 K, the results observed by the laser light scattering method are also reasonable ones.

Ion Component

A laser beam focused on a gas composed of different kinds of ions is scattered differently by each kind of ion. Each spectral distribution profile, therefore, is different according to the mass of each kind of ion. The smaller the mass, the wider the profile. Consequently, the laser light scattered by different kinds of ions is composed of different spectral distributions like a pagoda of several storeys.

For example, if we analyze a photograph of the spectrum of an ion component composed of H-ions and C-ions having a temperature of 100,000 K taken on a film with a microphotometer, we obtain a spectral distribution profile illustrated in Fig. 9.29, which shows a pagoda form composed of two profiles made by C-ions and H-ions. The top Gaussian curve in this spectral distribution means the spectrum of the laser beam itself having an arbitrary intensity reflected on the wall of the combustion chamber and observed through the measurement system together with the scattered light. The second storey is the

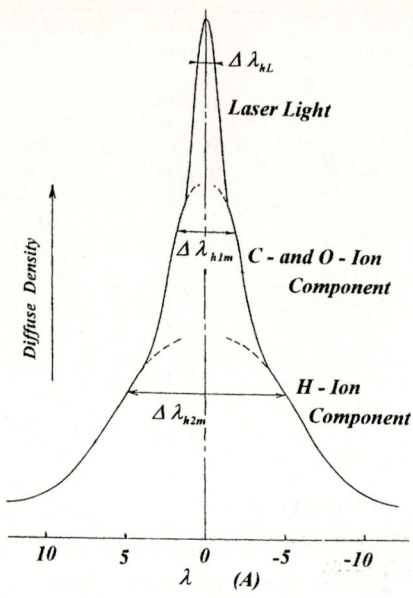

Fig. 9.29. Spectrum composed of H-ion and C-ion components having a temperature of 100,000 K. $\Delta\lambda_\eta$: halfwidth

C-ion component, while the third one the H-ion component. Because of the film characteristic, the intensity is expressed not in a linear scale, but approximately in a logarithmic one. In order to obtain the real spectral distribution of the scattered light, the width of the spectrum of the laser beam itself observed through the same measurement system must be considered. In the diagram of Fig. 9.29, for example, the real halfwidth of the spectral distribution of the scattered light is the difference between the observed value $\Delta\lambda_{h1m}$ or $\Delta\lambda_{h2m}$ and the halfwidth $\Delta\lambda_{hL}$ of the laser beam observed on the same film, that is, the real halfwidth of the scattered light $\Delta\lambda_{h1} = \Delta\lambda_{h1m} - \Delta\lambda_{hL}$ or $\Delta\lambda_{h2} = \Delta\lambda_{h2m} - \Delta\lambda_{hL}$.

In Fig. 9.30 an example of the spectral distribution of the ion component experimentally observed in the flame and analyzed by the microphotometer is illustrated. We indeed observe in this diagram a pagoda form of three bell-like profiles just like those in Fig. 9.29. Corresponding to the film characteristic and its exposure, the form of the spectral distribution is somewhat different, but such a pagoda form is always observed. After transforming the diagram into the linear scale, considering the film sensitivity and comparing it with that of the laser beam shown in Fig. 9.31, the top bell-shaped profile $S_i - L$ is the spectrum of the laser beam itself reflected on the window, lens or wall of the chamber without any scattering.

The other bell-shaped profiles $S_i - 1$ and $S_i - 2$ are the spectral distribution of the light scattered by two kinds of charged particles having different masses. As these profiles have Gaussian distributions and the characteristic number β

9.3 Investigation of Flame by a Laser Light Scattering Method 171

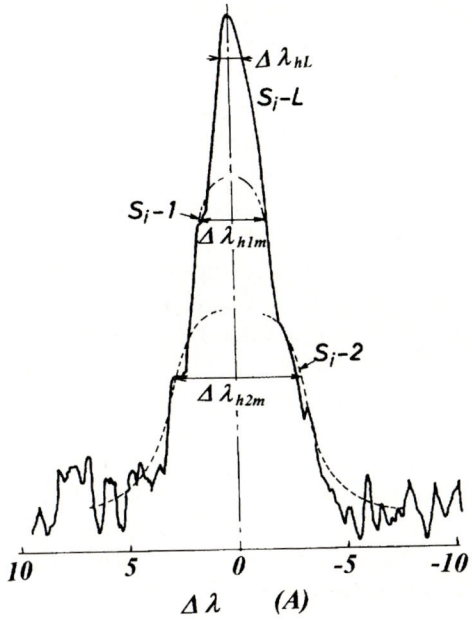

Fig. 9.30. Spectrum composed of two ion components of laser beam scattered in a flame propagating in a propane–oxygen mixture

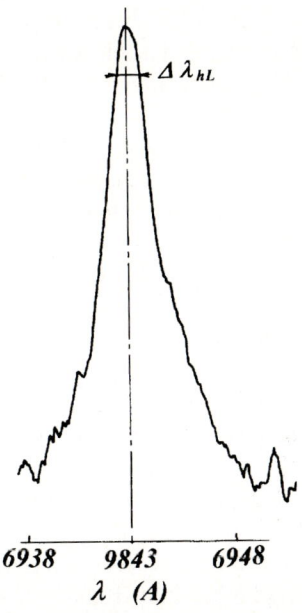

Fig. 9.31. Experimentally obtained spectrum of laser beam analyzed by microphotometer

is almost equal to zero, they can be treated as a spectrum of Thomson scattering. Applying the real halfwidth of the scattered light $\Delta\lambda_{h1} = \Delta\lambda_{h1m} - \Delta\lambda_{hL}$ or $\Delta\lambda_{h2} = \Delta\lambda_{h2m} - \Delta\lambda_{hL}$ into (9.38) where T_e is replaced by the ion temperature T_i and m_e by the ion mass m_i, we can calculate the ion temperature, if the mass of each ion is known. Comparing the width of both the profiles, the mass of the profile S_i–2 is smaller than that of S_i–1.

If the profile S_i–2 were caused by electrons, the temperature of the charged particles is estimated to be about 200 K, which is much too low and impossible to occur in the flame. Assuming the hydrogen ion as the next smallest particles, we obtain the ion temperature in the flame also shown in Fig. 9.27 with white circles. The ion temperature has a value from 100,000 to 150,000 K. Besides these results, no spectral distribution is obtained in 30% of the total experiment, just like in the case of electron temperature. The measurement error estimated to be ±20%.

Assuming further that the ion temperatures of both the profiles S_i–1 and S_i–2 are the same, we can calculate the mass of the charged particles having the profile S_i–1 according to (9.38). The calculated results suggest that the atomic weight of the ions forming the profile S_i–1 is 6 to 15. This means that the charged particles must be carbon or oxygen ions.

Discussion on the Measurement Results

As already mentioned, the center of the spectral distribution profile of the electron component is shifted to the shorter wavelength by about 10 Å. This means that the free electrons are driven in the direction to the monochromator with a velocity having an order of 10^5 m s^{-1}. The electrical field made by the laser beam can drive the free electrons in such a way, but not the ions, as they have a much larger mass.

As the neutral particles in this flame have a density of about 10^{17}–10^{18} per cm^3, the Rayleigh scattering by the neutral particles must be considered. The temperature of the neutral particles is, however, lower than the adiabatic combustion temperature of 3,000 K and the halfwidth of its spectrum is estimated to be less than 0.4 Å. The spectral distribution of the Rayleigh scattering, therefore, cannot be distinguished from that of laser beam itself and plays no role in the calculation of the ion temperature in this case.

The laser beam focused in the flame can heat the gas, but has energy less than 10^{-2} eV and its influence on the measured results is less than 0.1%. The breakdown in the gas caused by the focused laser beam can be observed by the photomultiplier PM-2 in Fig. 9.23, if it takes place, but it never occurs in the experiments. Even if it took place, the scattered light is never affected by it because of its induction period.

Considering that the electron temperature in a flame measured by the double probe method shows a value higher than 10,000 K, although the free electrons are cooled by the probes, the results measured by the laser light scattering method must be reasonable ones.

Nonequilibrium State

The experimental results suggest as follows:

1. Considering that either the electron temperature or the ion temperature near the flame front has a value of 10,000–150,000 K which is 3–50 times higher than the adiabatic combustion temperature, the gas near the flame front must be in a distinct nonequilibrium.
2. The ion temperature is higher than the electron temperature. As the ions as well as the free electrons lose their energy by the endothermic reaction at the ionization their temperatures decrease. As the ionization occurs in some high temperature regions, the ions get some energy from the high temperature particles surrounding them by collisions and soon heated up to the temperature of the particles surrounding the ions, while the free electrons colliding with the particles surrounding them can be heated not so quickly as the ions because of less coefficient of heat transmission, as the electrons have much smaller mass than the ions. An electron temperature less than ion temperature is, thus, observed in a flame.
3. An electron density of 10^{16} per cm^3 to 10^{17} per cm^3 is observed in the whole space where the flame propagates, but about in 70% of the space. The average electron density therefore must be much lower than it. In about 30% of the total experiments the laser beam having a focus diameter d_e of 0.2–0.3 mm cannot be incident to the part of the flame having a temperature higher than 10,000 K. The flame is composed of many parts having different temperatures like a mosaic, keeping a nonequilibrium state. Considering the probability that the laser beam can never be incident to the high temperature part, the average electron density is, thus, estimated to be about 10^{15} per cm^3.
4. Also in the front of the flames a lot of particles having extremely high temperature are observed, where sometimes a weak or ultraviolet luminescence is observed. This suggests that some combustion reactions are initiated by such high temperature particles supplied from the flame front and the flame is driven by the combustion reaction occurring in the front of the flame. Because of their mass, ions or atoms having high temperatures can play some role for combustion only in their diffusion area having several millimeter from the flame front. On the contrary the free electrons can fly so far as 30 mm from the flame front without any large energy loss, transporting the energy to the fuel and oxygen molecules, initiating the combustion reaction in the fresh mixture area in the front of the flame. In the explosion in a hydrogen–oxygen mixture described in Sect. 6.2.2, or in the ignition in an octane–air mixture reported in Sect. 7.1.1, the energy loss coefficient at collision is estimated to be an order of 10^{-3}–10^{-4}. This also suggests that the energy of combustion is transported by electrons. The flame propagation is, thus, mainly driven by free electrons having a temperature higher than 10,000 K.
5. It is difficult to explain only by the nonequilibrium state in the irreversible phenomena why the ions and free electrons have so high temperature as

100,000 K in the flame. In order to explain the phenomena, following two-step nonequilibrium process may be available,
(a) High energy particles, mainly free electrons having an energy of 2–3 eV are emitted from the flame front into the fresh mixture in front of the flame and excite the molecules and heat them to an order of 10,000 K
(b) In the mixture having a temperature higher than 10,000 K, a nonequilibrium state appears, in which some areas have a temperature of an order of 100,000 K, while the other areas have much lower temperature.

9.4 Nonequilibrium and Heterogeneous State behind Shock Waves

The shock wave is an irreversible process, in which some heterogeneous and nonequilibrium states must be observed. It is already well known that some ions, free electrons, and neutral particles having different temperatures exist in a short time behind a shock wave. The time from the initiation of a shock wave to the instant at which all particles behind the shock wave reach an equilibrium state is called "relaxation time."[85]

The ionization of gases, especially argon gas, behind shock waves has been investigated theoretically and experimentally mainly by means of the probe or microwave methods and many papers on this subject have been published.

In this chapter the nonequilibrium state of argon behind reflected shock waves experimentally observed in a shock tube, measuring the temperatures of free electrons and neutral particles by the laser light scattering method or a spectroscopic temperature measurement is explained.

9.4.1 Spectroscopic Temperature Measurement Method[86, 87]

Under an assumption of an equilibrium state a spectroscopic method is applied to measure the temperature of argon behind reflected shock waves in a shock tube. First, the method is explained.

Each atom in a gas emits or absorbs several spectral lines having different proper wavelengths corresponding to the energy level of each electron contained in the atom. The intensity I of each spectral line depends on the temperature of the gas and is expressed by the following equation under an assumption of equilibrium:

$$I = \frac{hc}{4\pi\lambda} g_n A_{nm} \frac{N(T)}{U'(T)} \exp\left(-\frac{E_n}{kT}\right), \tag{9.44}$$

where h is Planck's constant, c the light velocity, λ wavelength of the spectral line, g_n statistical weight, A_{nm} transition probability, $N(T)$ gas density, $U(T)$ partition function, E_n energy of the spectral line at n-level, k Boltzmann's constant, and T gas temperature at the equilibrium.

9.4 Nonequilibrium and Heterogeneous State behind Shock Waves

As N and U are constant in the gas at a constant state,

$$I = \text{const.} \times \frac{g_n A_{nm}}{\lambda} \exp\left(-\frac{E_n}{kT}\right). \tag{9.45}$$

Therefore,

$$\ln\left(\frac{\lambda I}{g_n A_{nm}}\right) = -\frac{E_n}{kT} + \text{const.} \tag{9.46}$$

the gas temperature T can be calculated from the slope Ψ of $\ln\{\lambda/(g_n A_{nm})\}$ against E_n according to the relation $\tan \Psi = 1/kT$, if g_n and A_{nm} are known and I of several spectral lines having different wavelength λ are measured.

9.4.2 Temperature in Argon Gas Behind Shock Waves[85]

Using a shock tube having an inner-diameter of 49 mm, shock waves propagating in argon gas are produced, as shown in Fig. 9.32. The tube is divided with a polyester film into two segments of about 1 and 3 m length, respectively. Hydrogen gas is charged as a driver gas in the 1 m long tube at a pressure between 15 and 35 atm., while the remaining 3 m long tube is filled by argon of 99.996% under an arbitrary pressure between 5 and 20 torr and room temperature of 20°C. In order to keep the residual impurity in the test gas less than 10^{-7}%, the low pressure tube is cleaned up with the fresh gas twice before each experiment.

The shock waves are produced by breaking the polyester film, propagate in the low pressure argon gas, and are reflected from the end wall of the tube. The propagation velocity of the incident shock waves is measured by three piezo-electric pressure transducers P-1, P-2, and P-3 set on the low pressure tube, as shown in Fig. 9.32. Temperatures in argon gas behind reflected shock waves are measured under the following two conditions

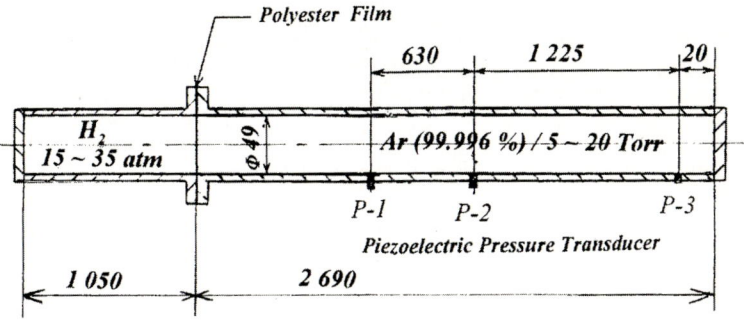

Fig. 9.32. Shock tube for temperature measurement in argon behind shock waves, dimensions in mm

	Mach number of	theoretical gas dynamic temperature	argon density behind (kg m^{-3})
1.	4.8	5,300 K	0.18
2.	6.7	10,000 K	0.10

Measurement of Electron Temperatures

The electron and ion temperatures in the shock heated argon are measured by the laser light scattering method described in Chap. 8. In Fig. 9.33 the arrangement of the measuring apparatus is illustrated. Just like in Chap. 8, a Q-switched ruby giant pulse laser beam having 20 ± 2 MW, 20 ns duration of half value intensity and a window on the axis of the shock tube behind the reflected shock waves. The diameter of the focus is estimated to be about 0.2 mm.

The emission of the laser beam is triggered by the pressure of the incident shock front through a piezoelectric pressure transducer. The triggering time is regulated by a delay-circuit in such a way that the laser beam can be emitted at an arbitrary instant after the passage of the reflected shock front at the measuring position. Thus, the laser beam can be emitted to an arbitrary position behind the reflected shock waves. Both the laser light and the luminescence of the reflected shock waves are recorded on the same oscilloscope through a photomultiplier PM-2 in Fig. 9.33, so that the distance of the observed position from the reflected shock front can be measured.

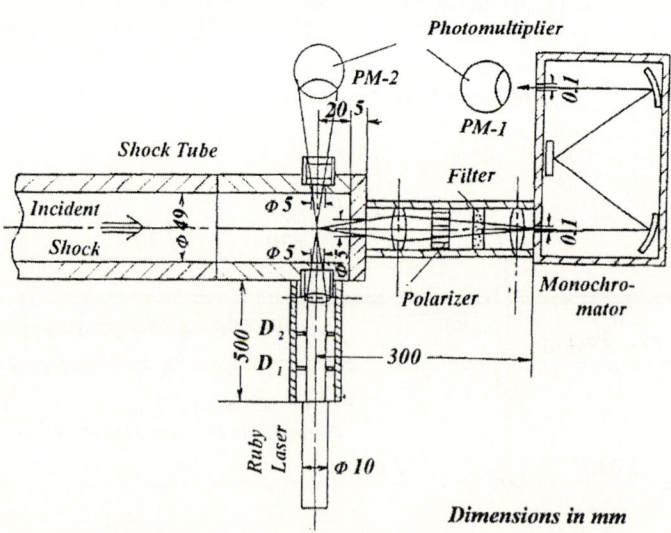

Fig. 9.33. Arrangement of the experimental apparatus applying the laser light scattering method

9.4 Nonequilibrium and Heterogeneous State behind Shock Waves

At the same time, the light scattered by the charged particles in argon behind the reflected shock waves in a direction at an angle of 90° to the incident laser beam is observed through a sapphire window at the end plate of the shock tube and a monochromator. The intensity of spectral lines of the scattered light having different wavelengths near that of the incident laser beam is measured by a photomultiplier P-1 and oscilloscope for the electron component. Light trap diaphragm D_1, D_2, a filter, and a polarizer are set in the light path in order to eliminate the stray light and the luminescence of the shock waves, as shown in Fig. 9.33.

Since each experiment with the photomultiplier permits to measure only one wavelength, the measurement must be repeated several times at the same position by changing the measured wavelength to obtain the spectrum of the scattered light. The same measurement for each spectral line is repeated more than five times and its average value is used to estimate the temperature.

Electron Temperature

In Fig. 9.34 an example of the spectra of the electron component scattered in the shock heated argon gas, measured with the photomultiplier PM-1 is illustrated. Comparing the relative profiles of the spectra obtained from the experiments with those in Kegel's graphs, the characteristic parameter α expressed by (9.34) can be estimated. In Fig. 9.34, the theoretically possible profiles of the scattered light corresponding to the maximum and minimum values of α are shown. From the values of α and the absolute characteristic wavelength of the spectra (halfwidth $\Delta\lambda_h$ for example) we can obtain the electron temperature T_e and density n_e according to (9.40) and (9.41). Figure 9.35 illustrates the measured electron temperature T_e and Fig. 9.36 the electron density n_e in argon

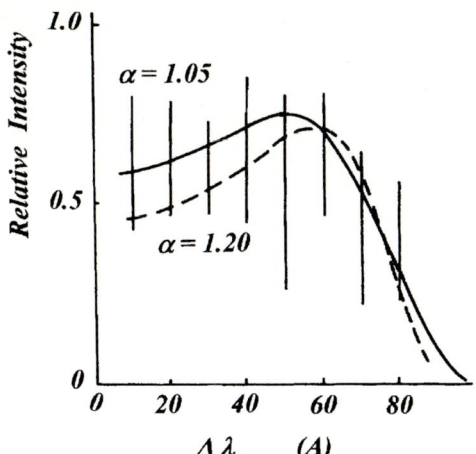

Fig. 9.34. Electron component of the laser beam scattered in argon at 50 μs after arrival of the reflected shock front. Mach number of the incident shock: 6.7

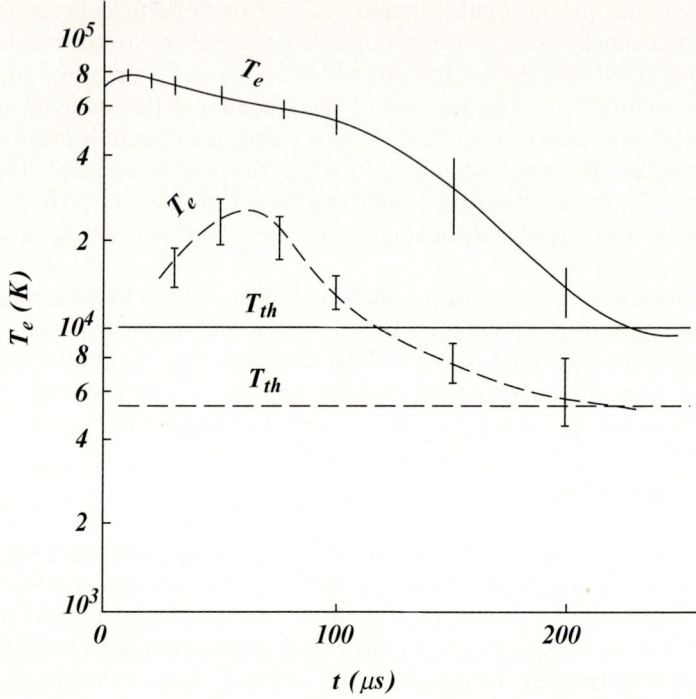

Fig. 9.35. Measured electron temperature T_e and theoretical gas dynamic temperature T_{th} with respect to the time t after arrival of the reflected shock front at the measuring point. Mach number of the incident shock is 6.7 for the *solid line* and 4.8 for the *broken line*.

behind the reflected shock waves with respect to the time t after passage of the reflected shock front at the measuring point.

The experimental results suggest that the electron temperature is just behind the reflected shock front much higher than the theoretical gas dynamic temperature T_{th}, but it approaches the theoretical one during about 200 μs after the arrival of the reflected shock front at the measuring point.

Ionization Degree

The ionization degree $\Lambda_i = n_e/(n_e + n_g)$ of plasma in an equilibrium state can be calculated, applying the following Saha's equation[70]:

$$r_i^2 = \frac{3 \times 10^{21}}{n_0} T^{3/2} \exp\left(-\frac{V_i}{T}\right), \tag{9.47}$$

where $r_i^2 = (n_e n_i)/n_0^2$, n_e the electron density, n_i the ion density, n_0 the density of neutral particles, V_i the ionization potential, and T the temperature of the plasma. The theoretical ionization degree Λ_{th} of argon at $T = 10{,}000$ K is esti-

9.4 Nonequilibrium and Heterogeneous State behind Shock Waves

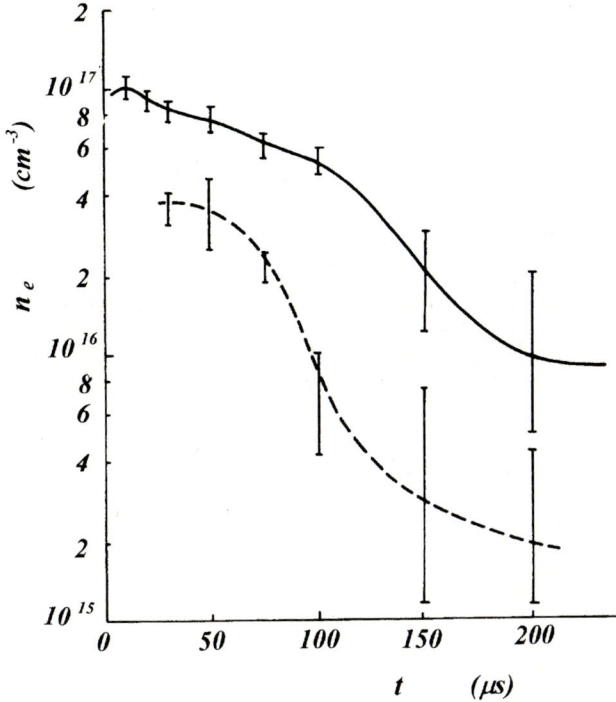

Fig. 9.36. Electron density n_e in argon with respect to the time t after arrival of the reflected shock front. Mach number of the incident shock is 6.7 for the *solid line* and 4.8 for the *broken line*

mated to be approximately 3×10^{-2}, but that measured behind reflected shock waves experimentally is different. Under the assumption of $n_e = n_i$ the ionization degree Λ_t of argon behind the reflected shock waves is calculated from the results obtained by laser light scattering method. The ionization degree Λ_t experimentally obtained behind the reflected shock waves having a gas dynamic temperature of 10,000 K is illustrated in Fig. 9.37 together with the theoretical one Λ_{th} in relation to the time t after the arrival of the reflected shock front at the measuring point. Just behind the shock front the measured value of the ionization degree is much higher than the theoretical one, but then approaches it during about 200 μs. This means that the ionized argon is in a nonequilibrium state just behind the reflected shock waves and then approaches an equilibrium state.

Spectroscopic Temperature[86, 87]

Using the monochromator and photomultiplier set at the position of the laser used for the scattering method, as shown in Fig. 9.38 the spectroscopic temperature of argon behind reflected shock waves is measured by observing the intensity of

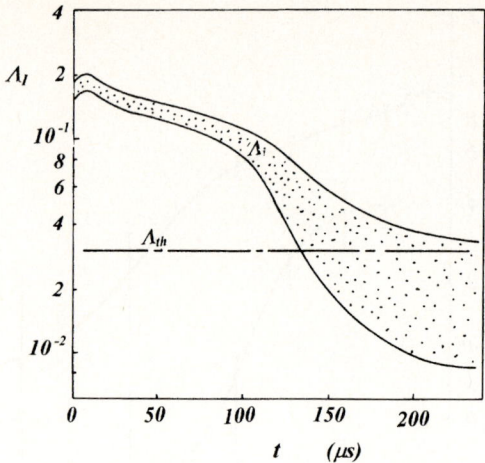

Fig. 9.37. Ionization degree Λ_i measured in argon gas the spectroscopic temperature behind reflected shock waves with respect to the time t after reflected shock waves the arrival of the reflected shock front. Λ_{th}: theoretical one in an equilibrium state

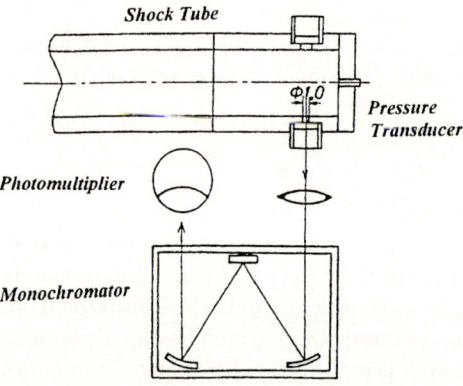

Fig. 9.38. Experimental apparatus for measuring the spectroscopic temperature behind reflected shock waves

several ArI lines under the assumption that the shock heated gas is in an equilibrium state. The wavelengths of measured ArI lines are 5912.1 Å, 6043.2 Å, 5373.6 Å, and 6052.7 Å and their transition probabilities A_{nm} statistical weights g_n and energy levels E_n proposed by Desai and Corcoran are taken, which are listed in Table 9.6.[88]

9.4 Nonequilibrium and Heterogeneous State behind Shock Waves

Table 9.6. Transition probability A_{nm}, statistical weight g_n and energy level E_n

λ (A)	5,912.1	6,043.2	5,373.6	6,052.7
A_{nm} (10^{-5} s^{-1})	15.24	30.6	5.75	2.19
g_n	3	7	5	5
E_n	121,011.98	123,832.50	124,692.02	120,619.8

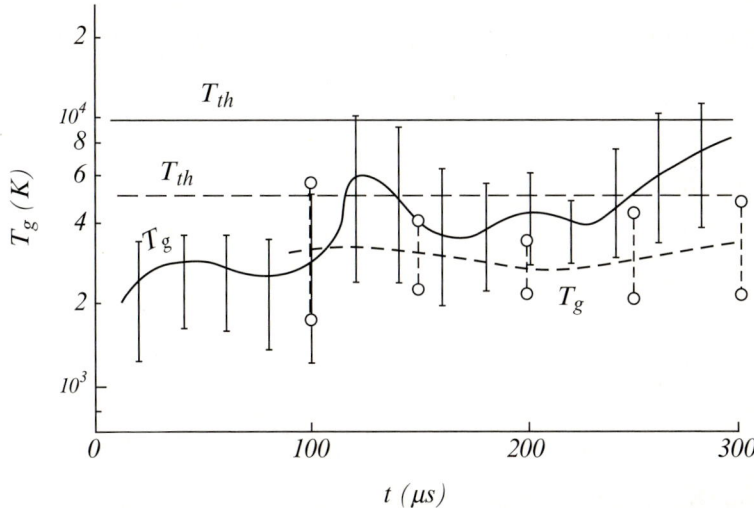

Fig. 9.39. Measured spectroscopic temperature T_g and theoretical gas dynamic temperature T_{th} in argon behind reflected shock waves with respect to the time t after arrival of the reflected shock front at the measuring point. Mach number of the incident shock: 6.7 for the *solid line* and 4.8 for the *broken* one

As the fluctuation of the measured result is relatively large, the same measurement is repeated several times and the average value of the measured results is taken into account. From the results the spectroscopic temperature is calculated according to (9.46). The spectroscopic temperature T_g thus experimentally obtained with respect to the time t after the arrival of the reflected shock front at the measuring point is illustrated in Fig. 9.39.

In contrast to the electron temperature the spectroscopic temperature T_g is lower than the theoretical gas dynamic temperature T_{th}. As the spectroscopic temperature is obtained under the assumption of equilibrium state, it may not be equal to the gas temperature, but qualitatively shows the tendency of the neutral gas temperature, that is, the gas temperature is lower than the theoretical gas dynamic temperature during a period from 200 to 300 μs after the arrival of the reflected shock front at the measuring point.

9.4.3 A Stochastic Phenomenon Behind Shock Waves

The induction period of light emission from argon behind the reflected shock wave fluctuates in a fairly large range. A example of the histogram of emission induction period t obtained from many experiments under the same condition is shown in Fig. 9.40.

According to (3.11) the relation $\ln P(t) - t$ can be obtained from such histogram of the emission induction period as shown in Fig. 9.41, from which further

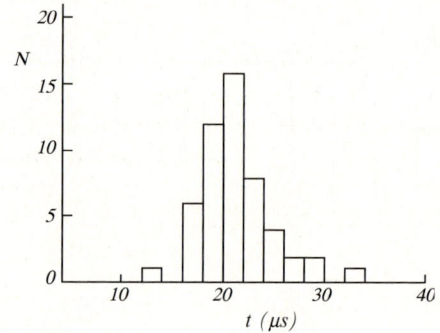

Fig. 9.40. Histogram of induction period of light emission in argon behind reflected shock waves at the gas dynamic temperature of 4,200 K and density of 0.33 kg m^{-3}, Mach number of the incident shock: 4.8

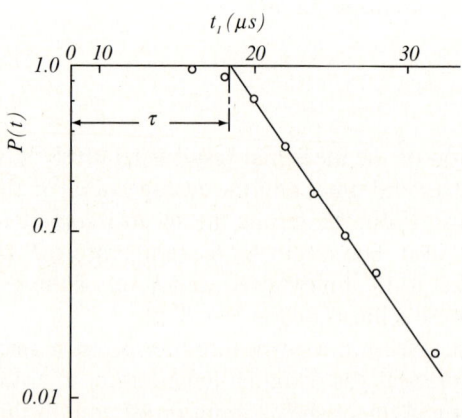

Fig. 9.41. $\ln P(t)$ against the time t after arrival of reflected shock front

9.4 Nonequilibrium and Heterogeneous State behind Shock Waves

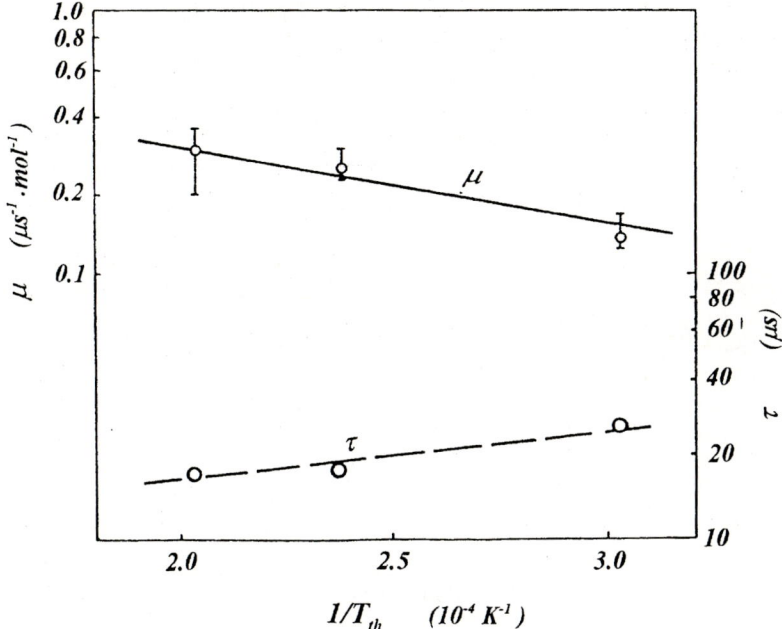

Fig. 9.42. Logarithms of the light emission probability μ and that of growing period τ with respect to the reciprocal gas dynamic temperature $1/T_{th}$ in argon having a density of 0.33 kg m^{-3} behind reflected shock wave

the probability μ of light emission and the emission nucleus growing period τ are obtained according to (3.13).

Logarithms of μ and τ with respect to the reciprocal gas dynamic temperature $1/T_{th}$ are illustrated in Fig. 9.42. The light emission from argon gas behind shock waves is a stochastic phenomenon, as observed always in irreversible phenomena.

9.4.4 Nonequilibrium and Heterogeneous State Behind Shock Waves

Summarizing the measured results we can illustrate a diagram as shown in Fig. 9.43, where the temperature T_e and density n_e of the free electrons, and the spectroscopic temperature T_g as well as the theoretical gas dynamic temperature T_{th} behind the reflected shock waves whose incident shock has a Mach number of 6.7 are presented in group with respect to the time t after the arrival of the reflected shock front at the measuring point. In order to know the ionization process and state behind reflected shock waves, we have to discuss on the spectroscopic temperature, electron temperature, and ionization degree.

(1) *Spectroscopic temperature.* The measured spectroscopic temperature is much lower than the theoretical gas dynamic temperature. As explained

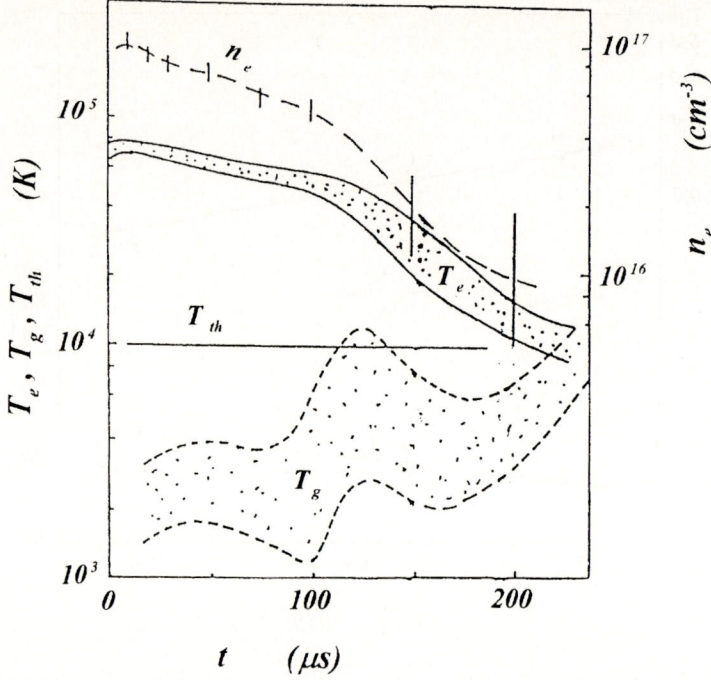

Fig. 9.43. Measured electron temperature T_e and spectroscopic temperature T_g in argon behind reflected shock waves with respect to the t after arrival of the reflected shock front. T_{th} is theoretical gas dynamic temperature. Mach number of the incident shock : 6.7

above, the light emission in a shock heated gas is a stochastic phenomenon in which a fairly large fluctuation occurs with a certain probability. This light emission consists of two processes: the initiation and the development. During the period of a few hundred microseconds from the beginning of emission to the time in which the whole gas emits light, the gas is not in a homogeneous state. The gas contains two parts: one of which has a high temperature and emits light, while the other has a low temperature and absorb the light. The measured spectroscopic temperature does not mean the average gas temperature, but also suggests that there are some parts having a temperature lower than the average one.

(2) *Electron temperature and ionization*. The electron temperature measured by the laser light scattering method is much higher than the theoretical gas dynamic temperature as shown in Fig. 9.43. On the other hand the ionization degree of argon experimentally obtained behind reflected shock waves is at first also much higher than the theoretical one, then approaches to the theoretical value during about 200 μs just like the spectroscopic and electron temperature, as shown in Fig. 9.37. This means the ionized argon gas behind reflected shock waves is in a nonequilibrium state during a certain

9.4 Nonequilibrium and Heterogeneous State behind Shock Waves

period at the beginning of the ionization. If the gas behind reflected shock waves is heterogeneous during the first period and composed of many parts having different temperatures, as described above, the laser light is scattered by both the electrons in the high temperature parts and those of the low temperature parts. The ionization, however, depends strongly on the temperature and most of free electrons are produced in the high temperature parts. The electron temperature as well as the ionization degree estimated from the scattered light, thus, expresses mainly those in the high temperature parts, in which all the temperatures of gas, ions, and electrons have almost the same value. The measured results showing the nonequilibrium state, therefore, also suggest that there must be some very high temperature parts in a shock heated gas in which the gas is ionized.

Because of the heterogeneous state the relaxation time in the ionization of gas behind shock waves means the period in which the heterogeneous state becomes homogeneous, mainly through diffusion between the parts having different temperatures. The relaxation time, therefore, is very long in comparison with that caused by simple collisions between the neutral particles, ions and electrons, as the experimental results show. The recombination, dissociation, and energy exchange by the direct collisions proceed so rapidly that they can contribute little to such a long relaxation time in the ionization of gas behind shock waves.

Such nonequilibrium and heterogeneous states in gas behind shock waves play a very important role for detonation waves or interaction between shock and combustion waves, as explained in next following chapters.

10
Interaction Between Combustion and Pressure or Shock Waves

A flame propagating in a combustible gaseous mixture is called a combustion wave. Such combustion waves are also observed in flames of gas or oil burners in which the fuels are mixed with air by diffusion.

As the gas behind the combustion wave is heated by the combustion and expanded, producing a flow in the mixture gas, the combustion wave is driven by the flow of the mixture gas. Applying the three conservation laws to the flowing gas in front of as well as behind the combustion wave under the assumption of an equilibrium state, the mechanism of the combustion wave has been investigated, discussed, and tried to explain. As the propagation velocity of the combustion wave in a laminar flow is much lower than that in a turbulent flow, the theory of each case has been developed.

As ignitions as well as combustion are irreversible phenomena, the mechanism of the combustion wave must be investigated and discussed, considering that the gas near the flame is in a nonequilibrium and heterogeneous state. Besides, at the combustion, not only the flow, but also pressure and shock waves are produced and play important roles for the combustion and propagation of the flame.

In this chapter, after the classical propagation theory of combustion waves is briefly explained, the application of the irreversible theory and the interaction between the combustion and pressure or shock waves are explained, based on some experimental results.

10.1 Propagation of Combustion Waves[89, 90]

The classical propagation theory of combustion waves is constructed mainly based on heat and mass transfer. First we want to analyze a plane combustion wave propagating in a combustible mixture through a tube having a unit area.

As illustrated in Fig. 10.1 we assume a plane flame propagating through a tube filled by a combustible mixture having temperature of T_1, density ρ_1, and

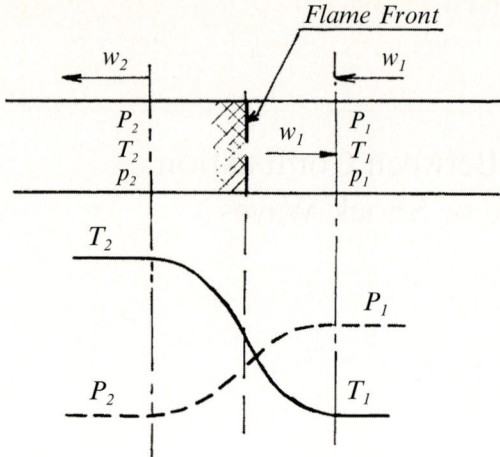

Fig. 10.1. A model of flame propagation in a combustible mixture through a tube

pressure P_1 with a velocity of w_1, behind which the combustion gas has the temperature of T_2, density ρ_2, and pressure P_2. The combustion gas behind the flame flows into the opposite direction of the flame propagation. Considering a coordinate system moving with the combustion wave, the fresh mixture flows into the combustion wave with a velocity of w_1, while the combustion gas flows out with a velocity w_2. Assuming a steady state, that is, the combustion wave propagates with a constant velocity, three fundamental equations of flow representing the combustion waves as follows:

$$\rho_1 w_1 = \rho_2 w_2 \tag{10.1}$$

$$\rho_1 w_1^2 + P_1 = \rho_2 w_2^2 + P_2 \tag{10.2}$$

$$\frac{1}{2} w_1^2 + E_1 + \frac{P_1}{\rho_1} + \Delta Q = \frac{1}{2} w_2^2 + E_2 + \frac{P_2}{\rho_2}, \tag{10.3}$$

where E is the energy and ΔQ the combustion heat.
From these equations we obtain the following relations:

$$P_2 - P_1 = \rho_1 w_1 (w_1 - w_2), \text{ then}$$

$$\frac{P_2 - P_1}{\frac{1}{\rho_1} - \frac{1}{\rho_2}} = \rho_1^2 w_1^2 \tag{10.4}$$

and

$$E_2 - E_1 - \Delta Q = \frac{1}{2}(P_1 + P_2)\left(\frac{1}{\rho_1} - \frac{1}{\rho_2}\right). \tag{10.5}$$

These relations are also applied to detonation waves, as later explained, and (10.5) express the so-called Hugoniot-curve for detonation waves, as shown in diagram of pressure vs. density in Fig. 10.2. The state of the gas behind the combustion wave falls on the point A' between F and G on the Hugoniot-curve in Fig. 10.2. According to (10.4), the propagation velocity w_1 of the combustion wave can be estimated from the slope of the straight line drawn from the initial mixture state A to the point A' on the Hugoniot-curve, as also illustrated in Fig. 10.2. As the combustion gas is heated by the combustion heat, expands and flows, its pressure is lower than that of the initial mixture.

The propagation velocity of combustion waves is often measured in a flame standing on a Bunsen burner. Assuming a standing flame having a conical form, a height h of the flame and radius r of the burner exit, as shown in Fig. 10.3, the area F of the combustion wave is expressed as follows:

$$F = \pi r^2 / \sin \alpha = \pi r \sqrt{r^2 + h^2}. \tag{10.6}$$

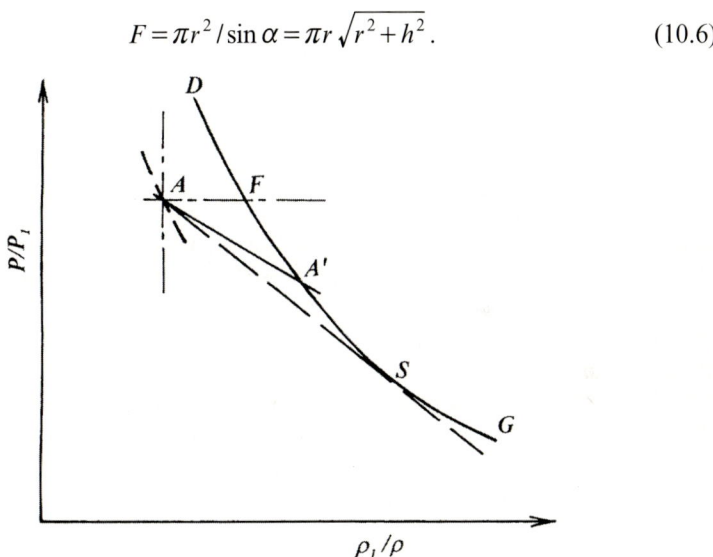

Fig. 10.2. Hugoniot curve in a combustion wave

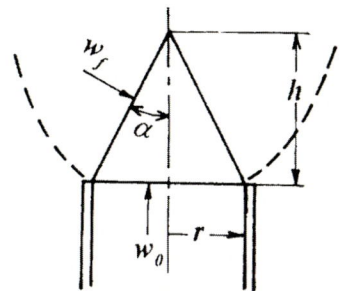

Fig. 10.3. Flame propagation velocity w_f and mixture flow velocity w_0 in a flame having a conical form in a Bunsen burner, 2α: angle of the flame cone, r: radius of the burner exit, h: flame height

The propagation velocity w_f of the combustion wave:

$$w_f = \frac{V_0}{\pi r \sqrt{r^2 + h^2}}, \tag{10.7}$$

where V_0 is the mixture volume passing through the burner per unit time (cm³ s⁻¹). This equation is, however, only applicable for flames having a conical surface. Flame whose form can be clearly recognized is called a laminar flame. According to the following equation,

$$w_f = w_0 \sin \alpha, \tag{10.8}$$

the propagation velocity w_f of the combustion wave can be calculated. In this equation w_0 is the average flow velocity of the mixture at the exit of the Bunsen burner.

In a mixture having a higher flow velocity, or flowing out from a burner having a larger diameter, the flame cannot keep its conical form, but obtains unstable one. Such flames are called turbulent flames. The propagation velocity of a turbulent flame is expressed by the mixture volume passing the burner per unit time and cross-section area.[91]

The propagation velocity of a laminar flame depends on the density of activated particles, diffusion among the particles, and heat transfer. It is, therefore, different according to the sorts of fuel, mixture ratio with air, but usually has a value between several decimeters and several meters. A pressure drop is also rarely observed in flames propagating with a velocity less than 10 m s⁻¹.

The turbulent flame propagates much faster than the laminar one and its propagation velocity depends on the Reynolds number Re of the flow in which the combustion wave propagates, as illustrated in Fig. 10.4. The reason for it has been attributed to the increase of the contact surface between the flame surface

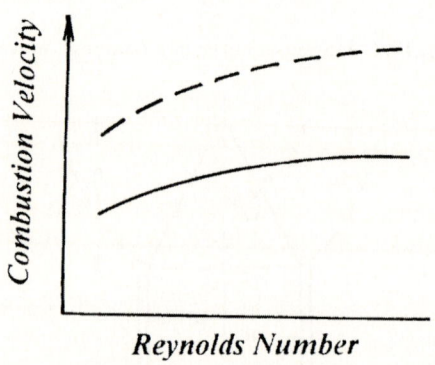

Fig. 10.4. Propagation velocity of a turbulent flame

and the unburned gas, consequently, to the increase of diffusion and heat transfer among the particles, namely the unburned mixture is heated much faster.[92] One says, the propagation velocity of a turbulent flame is nearly proportional to $\sqrt{Re}$.

The relations explained earlier can be applied to flames propagating in a combustible mixture through a tube, as long as both tube ends are open. The thermal and hydraulic influences should also be considered. The propagation velocity of a turbulent combustion wave seems to be much higher than that of a laminar one, but for it the difference between both definitions must be considered. The propagation velocity of a laminar flame is calculated as a mixture flow through a unit area of the flame surface, while that of a turbulent flame as that the unit cross-section of the tube, that is, the area through which the turbulent combustion wave passes is smaller than that of the laminar one.

10.2 Flame Propagation as an Irreversible Phenomenon

10.2.1 Theoretical Treatment

The flame propagation velocity w_f (m s^{-1}) in a combustible mixture through unit area (m^2) is expressed by the combustion velocity w_m (kg s^{-1}) as follows:

$$w_m = \rho w_f, \qquad (10.9)$$

where ρ is the mixture density. On the other hand, the flame propagation should be a phenomenon in which some energy released in the reaction zone behind flame is supplied to the fresh mixture in the front of the flame and ignites it one point after another. The combustion velocity, therefore, should be proportional to the ignition probability in the fresh mixture. In this case, however, the energy supplied to the fresh mixture is transported from the reaction zone behind the flame not only by high-temperature molecules, atoms, and radicals, but also by free electrons. The molecules, atoms, and radicals have such a large mass and size that they collide with the particles in the fresh mixture easily, lose their energy at once and can scarcely fly further than the usual diffusion zone, while the ions having anomalously high-temperature can play a more important role for combustion in the diffusion zone, but can fly not further. The free electrons, however, having much less energy loss in a collision, can penetrate into the fresh mixture ten times further than the diffusion zone and excites the particles in the fresh mixture much more than the molecules, atoms, and radicals supplied from the reaction zone, as the electrons have much smaller mass, but much higher velocity and temperature, 5 to 30 times that of the molecules, atoms, and radicals in the flame. Therefore, the free electrons play an important role for the ignition in the fresh gas much more than the other particles.

Besides, as the high-temperature zone exist in the flame front never homogeneously, because of irreversible phenomena, the high-temperature particles from the flame front heterogeneously permeate into the fresh mixture. The fresh mixture facing the flame front can be heated never homogeneously and has never a homogeneous temperature distribution. The classical combustion theory explained in the previous section, therefore, cannot be applied to the flame propagation. It is more reasonable to assume that the fresh mixture having an initial low-temperature obtains some energy Q_f (J mol^{-1}) from the reaction zone of the flame.

Using the effective activation energy of spontaneous ignition $E_i = E_1 + E_2 - W$ in (3.15), the ignition probability μ_f in the fresh gas facing the flame front can be expressed as follows,

$$\mu_f = A \exp\left(-\frac{E_i - Q_f}{RT}\right). \tag{10.10}$$

The combustion velocity can be expressed by an equation proportional to this probability like (6.10). As the collision frequency Z is proportional to the concentrations of oxygen and fuel molecules, the combustion velocity w_m is expressed by the following equation:

$$w_m = \rho^2 C_o C_f \cdot \left(\frac{C_o}{C_f}\right)^n \exp\left(-\frac{E_i - Q_f}{RT}\right). \tag{10.11}$$

where ρ is the density of the mixture, C_o and C_f are the concentration of oxygen and fuel, respectively, and n the reaction order. Simplifying this equation, we obtain the next equation:

$$w_m = A_F \rho^2 \exp\left(-\frac{E_i - Q_f}{RT}\right). \tag{10.12}$$

Connecting this with (10.9), $w_f = A_F \rho \exp\{-(E_i - Q_f)/RT\}$ is obtained. Assuming the mixture as an ideal gas, $w_f \cdot T$ is proportional to $A_F \exp\{-(E_i - Q_f)/RT\}$, $w_b = w_f T/T_0$ is called the normalized flame propagation velocity, where $T_0 = 273$ K, which is expressed by the equation:

$$w_b = w_f \frac{T}{T_0} = A_F \exp\left(-\frac{E_i - Q_f}{RT}\right), \tag{10.13}$$

where A_F depends on the mixture density at T_0, the concentration of fuel and oxygen, and a quantity effect.

10.2.2 Experiments of Flame Propagation[93]

As the mixture flowing out from the tube of a Bunsen burner burns, diffusing into the air, the mixture ratio cannot be kept constant. In order to investigate a combustion wave propagating in a mixture keeping the mixture ratio constant, we investigate a combustion wave standing in a homogeneous combustible mixture flowing through a tube, balancing the flame propagation velocity with the flow velocity. By this method we can obtain the relation among the flame

10.2 Flame Propagation as an Irreversible Phenomenon

propagation velocity, fuel concentration, temperature, and Reynolds number of the mixture flow.

As illustrated in Fig. 10.5, a channel of steel is prepared having a square section of 10×10 mm in which a wedge form barrier is set to change the cross-section area of the channel. One side of the channel where the wedge barrier is set is covered by pyrex-glass, so that the flame state can be observed from outside. *n*-hexane–air mixtures having different mixture ratios are introduced from the right-side into the channel decreasing first and then increasing the cross-section area along the path having the wedge barrier with a velocity from 0.5 to 3.5 m s^{-1}. The mixture is ignited by a spark plug set at the down stream end of the channel, from which a flame propagates against the mixture flow, but is decelerated and stands still somewhere in the channel narrows by the wedge barrier where the flame propagation velocity balances with the flow velocity.

The apparatus is set in a thermostat to regulate the mixture temperature from 300 to 600 K. Seven thermocouples having different distance with each other are set in the channel where the propagating flame should stand still, so that the temperature as well as the position of the flame can be observed. In Fig. 10.6 a photograph of one of the standing flames in the channel is shown.

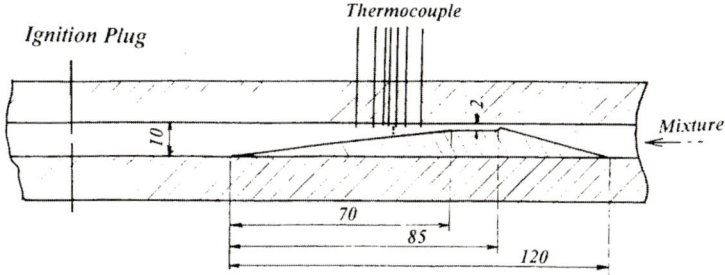

Fig. 10.5. Experimental apparatus for the observing standing flames in a mixture flow

Fig. 10.6. A photograph of a standing flame in a flow of a stoichiometric *n*-hexane–air mixture

In such a standing flame, the flame propagation velocity must be equal to the flow velocity. Assuming that the cross-section area of the channel where the flame is standing is S (m²) and mixture flow volume V (m³ s⁻¹), then the flame propagation velocity should be:

$$w_f = \frac{V}{S} \text{ (m s}^{-1}\text{)}. \tag{10.14}$$

The Reynolds number Re can be calculated from the velocity, size of the section area of the channel, and kinematic viscosity. As the channel has a rectangular cross-section area at the standing flame, the Reynolds number of transition from laminar flow to turbulent one is estimated to be 600. The experiments are carried out in a range of Reynolds number from 200 to 900, that is, from the area of laminar flow to that of turbulent flow under the initial mixture temperature of 378, 423, and 533 K.

In Fig. 10.7 the flame propagation velocity w_f in a stoichiometric n-hexane–air mixture measured in this experiment is illustrated with respect to the Reynolds number Re with initial mixture temperature T_1 as parameter. Applying (10.13) to the relation in Fig. 10.7, we obtain the normalized flame propagation velocity $w_b = w_f T/T_0$ in relation to the reciprocal mixture temperature $1/T$ with parameter Re. The result is illustrated in Fig. 10.8. According to (10.13) we can calculate the values of $(E_i - Q_f)$ and A_F for mixtures having different equivalence ratios ϕ from such diagrams of $\ln w_b$ with respect to $1/T$. As E_i of n-hexane–air

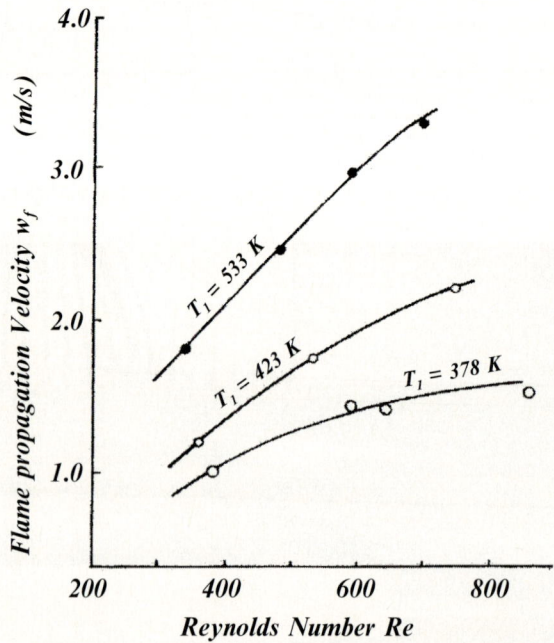

Fig. 10.7. Flame propagation velocity w_f with respect to Reynolds number Re of the stoichiometric n-hexane–air mixture flow, T_1, initial mixture temperature

10.2 Flame Propagation as an Irreversible Phenomenon

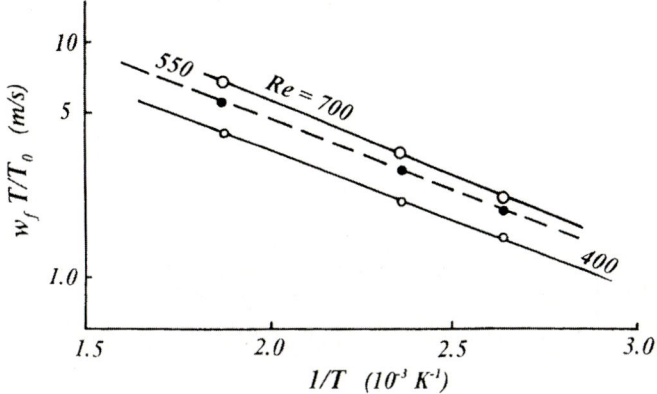

Fig. 10.8. Normalized flame propagation velocity $w_f T/T_0$ in a stoichiometric n-hexane–air mixture flow having different Reynolds number Re with respect to the reciprocal mixture temperature $1/T$

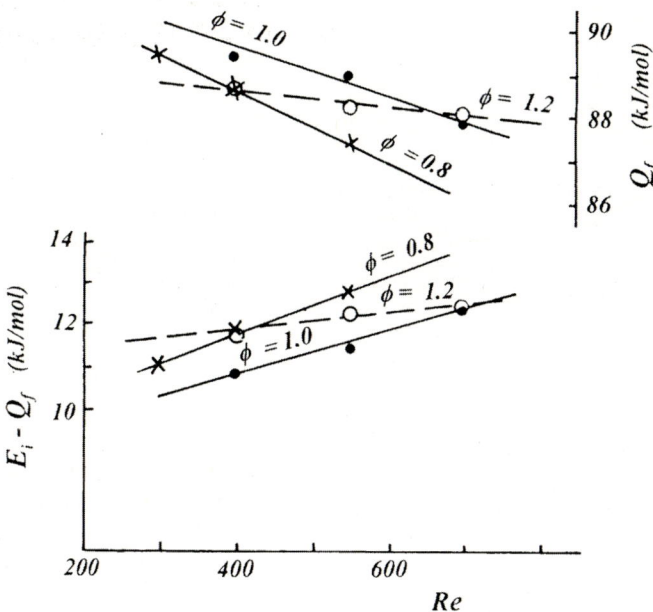

Fig. 10.9. Q_f and $E_i - Q_f$ in the flow of n-hexane–air mixtures having different equivalence ratio ϕ with respect to the Reynolds number Re of the mixture flow

mixture is already known from the experiments of spontaneous ignition using a shock tube, Q_f supplied from the flame to the fresh mixture is calculated. $(E_i - Q_f)$ and Q_f with respect to Re are illustrated in Fig. 10.9 with the equivalence ratio ϕ as parameter. The relation between A_F in (10.13) and Re is also shown in Fig. 10.10 with ϕ as parameter.

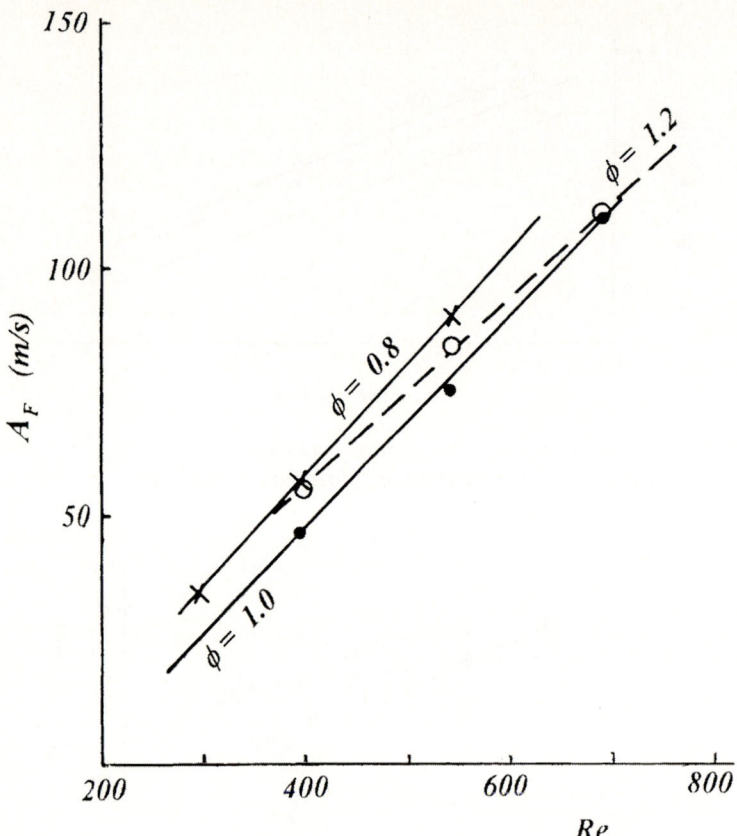

Fig. 10.10. Quantity factor A_F with respect to Reynolds number Re of n-hexane–air mixture having different equivalence ratios ϕ

In each equivalence ratio or mixture ratio, A_F increases proportionally to Re, namely, that means a quantity effect, which is proportional to the mixture quantity participating in the combustion reaction. The experimental results suggest that the fresh mixture quantity obtaining energy from the flame and participating in the combustion reaction increased proportionally to Re.

On the other hand Q_f supplied to the fresh mixture decreases with increase of Re. Just like with the spark ignition in a mixture flow, the energy to be supplied is dispersed and that per unit fresh mixture decreases with increase of the mixture quantity participating in the combustion.

The experimental results thus suggest that A_F as well as Q_f with respect to Re is expressed by a straight line in the whole region from the laminar area to turbulent one, any difference between laminar and turbulent flow cannot be recognized, though the form is different in both the flames.

10.2 Flame Propagation as an Irreversible Phenomenon

As the flame or combustion wave propagation has usually been investigated under room temperature, the flame propagation velocity seems to be proportional to $\sqrt{Re}$. We do not need to distinguish the turbulent combustion from the laminar one, as long as the definition of the propagation velocity of the flame or combustion wave is the same in both the flames. One has to consider only the quantity effect, A_F increases proportionally, while the energy Q_f supplied from the flames to the fresh mixture decreases inversely proportionally to the Reynolds number Re.

As the normalized combustion velocity w_b with respect to the reciprocal mixture temperature $1/T$ having Re as parameter in Fig. 10.11 or those having equivalence ratio ϕ as parameter in Fig. 10.12 shows the higher the mixture temperature, the higher the propagation velocity of the combustion wave. In order to have high-speed combustion, therefore, the mixture should be preheated and have a higher temperature.

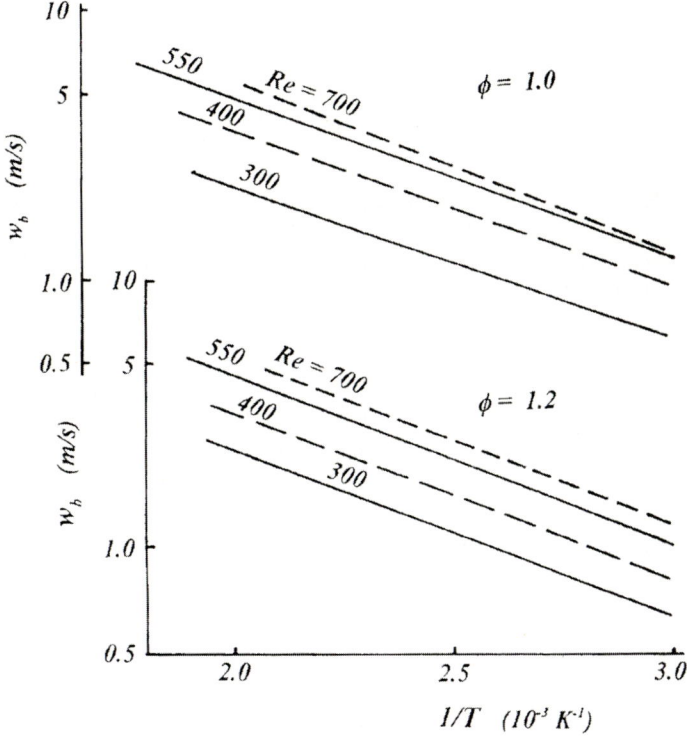

Fig. 10.11. Normalized flame propagation velocity w_b in n-hexane–air mixture having the equivalence ratios ϕ of 1.0 and 1.2 flowing with different Reynolds number Re with respect to the reciprocal initial mixture temperature $1/T$

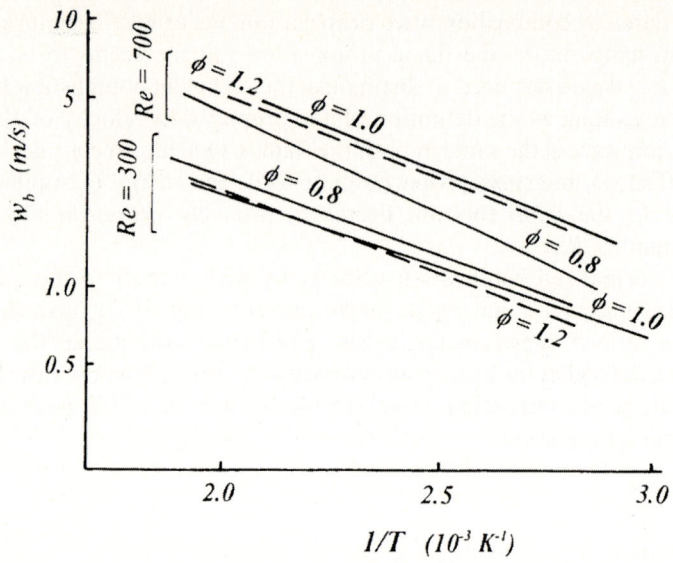

Fig. 10.12. Normalized flame propagation velocity w_b in n-hexane–air mixture of $\phi = 0.8$, 1.0, and 1.2 flowing with $Re = 300$ and 700 with respect to the reciprocal mixture temperature $1/T$

10.2.3 Transition from a Laminar Flame to a Turbulent Flame

Although there is no essential difference between laminar and turbulent flames, the laminar flame keeps the form stable and burns almost without any disturbance, while the turbulent flame burns unstably, always changing the form of the combustion surface.

Increasing the mixture quantity flowing through the tube, the flat flame front is transformed to a conical flame front, which is enlarged and separates first into two then into several conical forms, then grows to an unstable turbulent flame. The reason for it is explained as follows.

The whole ignition probability of the mixture of m mole is $m\mu$, if the mixture of 1 mole has an ignition probability of μ. As the mixture to burn must increase with increase of the flow velocity or Re, the area of the flame front which supplies the energy to the fresh mixture contacted with the flame front must increase, as long as the ignition probability $m\mu$ is not much changed. With increase of Re, the flame front has, thus, first two, then several conical forms and at last a saw tooth form. Thus, the thickness of the flame front increases, that is, m of the mixture to be ignited increase, while the energy to be supplied decreases with increase of Re, the ignition point fluctuates over the thick flame front and the ignition becomes unstable, consequently the flame front is unstable and a turbulent flame is formed.

10.3 Interaction between Combustion and Shock Waves

For the transition from a propagating flame to detonation waves, in reflected shock waves behind the detonation front, or for the combustion in resonance jet engines, we observe some phenomena caused by the interaction between combustion waves and shock or pressure waves, in which the combustion as well as the ionization in the flame is strongly stimulated. Some investigations of chemical reactions under weak shock or acoustic waves,[94] or those of oscillating combustions have been reported.

Considering that the combustion velocity and the ionization of gas in a flame have a remarkable relation with each other, we explain here some phenomena of combustion and ionization observed in flames propagating in a combustible mixture through a tube, when shock or pressure waves propagate through the same tube and pass by the flame or collide with it in the front.

10.3.1 Interaction Modes in the Experiments[95]

The experiments are carried out using a shock tube consisting of two stainless steel tubes for the low-pressure mixture, a stainless steel tube for the driver gas, and a plexiglas tube, as schematically illustrated in Fig. 10.13. The stainless steel tubes have the same inner-diameter of 51 mm, but the two low-pressure tubes have length of 2,056 and 1,000 mm, respectively, and the driver gas tube has a length of 1,000 mm, while the plexiglas tube set between both the tubes for the low-pressure mixture has an inner diameter of 30 mm and a length of 1,055 mm. The whole length of the shock tube is 5,100 mm. The driver gas tube is filled with compressed air under a pressure of 150 to 230 kPa, while in the other tubes a stoichiometric propane–air mixture (C_3H_8 + 25 air) is charged under a pressure of 50 kPa.

The high-pressure driver section and low-pressure section are separated by a polyester film of 16 to 25 μm in thickness.

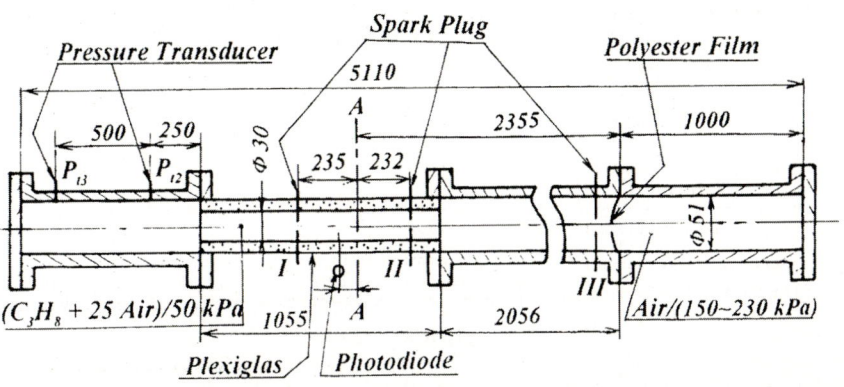

Fig. 10.13. Shock tube applied to the experiment

After the ignition of the mixture by an electrical spark at I or II, a flame propagates toward both the end plates of the shock tube with a velocity of about 23 m s^{-1}, while a shock wave propagates toward the other end of the shock tube through the stainless steel tube having a length of 2,056 mm and plexiglas tube, colliding with the propagating flame or overtaking it, as soon as the polyester film is broken by a spark at III. In Fig. 10.14 an example of the shock diagram is shown (time-distance diagram of the shock and flame propagation) in which the shock wave driven by the air of 150 kPa propagates against the flame and collides with it in the front near the measurement point.

At the measurement section A–A, shown in Fig. 10.15, a piezoelectric pressure transducer, a photodiode, and a pair of double probes are set for measuring the pressure, light emission from the flame, and ionization of the gas in the propagation flame behind shock waves.

Four series of experiment are carried out, as schematically shown in Fig. 10.16.

First mode. The flame ignited by the spark at I collides with the shock wave in the front at the measurement point, and they propagate against each other. As soon as the polyester film is broken by the spark at III initiated by the ignition at I through a delay circuit, a shock wave having a certain Mach number M_1 propagates toward the flame, so that the shock wave collides with the flame near the measurement point.

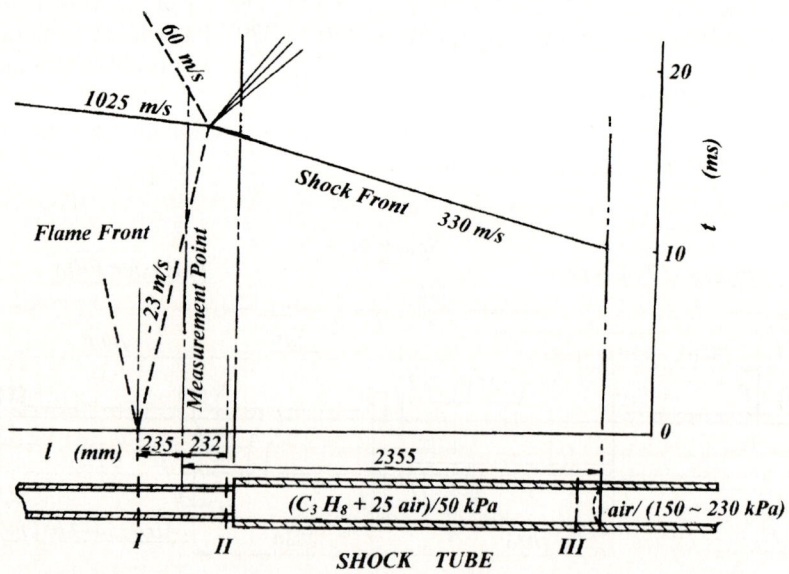

Fig. 10.14. Time-distance diagram of shock and combustion waves propagating in a stoichiometric propane–air mixture

10.3 Interaction between Combustion and Shock Waves

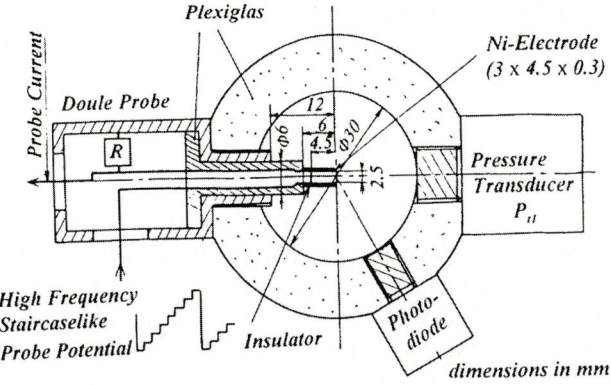

Fig. 10.15. Section A-A at the measurement point

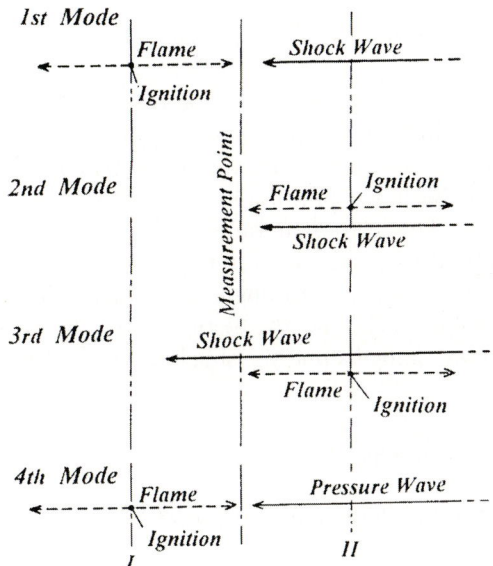

Fig. 10.16. Four modes of the experiment

After the collision, the flame is pushed back by the flow behind the shock wave. As shown in the shock diagram in Fig. 10.14, the flame, therefore, propagates after the collision backward with a velocity corresponding to the difference between the velocity of the gas particles and that of the flame propagation in the stagnant mixture.

Second mode. The shock wave propagating behind the flame overtakes the leading flame near the measurement point. The mixture is ignited at II and the flame propagating toward the left is overtaken by the shock wave propagating from the right side.

Third mode. The flame propagates behind the shock wave. After a shock wave propagating from right to left passes by the measuring point, the mixture is ignited at II and the flame propagates from there to the left, passing the measurement point.

Fourth mode. The flame front collides with a pressure wave. After the mixture is ignited at III and a pressure wave produced at the polyester film by the stagnation of the flow behind the flame from III propagates to the left, the mixture is ignited at I with a certain delay time, so that the flame collides with the pressure wave near the measurement point. The polyester film is not broken in this case.

The propagation velocity of the shock wave is measured by means of the piezoelectric pressure transducer at P_{t1}, P_{t2}, and P_{t3}, while that of the flame is measured by observing the passage instant of the flame front at the first photodiode on the A–A section and that of the second one at a distance of 46 mm from the first one, as shown in Figs. 10.13 and 10.15.

Remark 5. *Measurement of the shock propagation velocity*
From the instants of passage of the shock front at P_{t3} and P_{t2}, the instant t_4 of passage of the shock front at the end of the plexiglas tube can be estimated. From this instant t_4 and the passage instant t_1 of the shock front at P_{t1}, the propagation velocity of the shock wave can be estimated.

The adiabatic combustion temperature T_c in the flame is theoretically estimated to be about 2,300 K. The relation among the initial pressure P_d of the driving air, that of the mixture P_0, the Mach number M_1 of the shock propagation in the mixture before combustion through the plexiglas tube, the Mach number M_f of the shock behind the flame, the theoretical gasdynamic temperature T_s, and pressure P_s in the flame behind the shock waves are listed in the following Table 10.1.

10.3.2 Experimental Results

Pressure Variations in Combustion and Shock Waves

In the normal propagation flame without any shock or pressure wave, the pressure decreases so little that we cannot observe any change. The pressure almost retains its initial value of 50 kPa.

Table 10.1. Shock waves colliding with the flames

P_d (kPa)	M_1	M_f	P_s/P_0	P_s(kPa)	T_s/T_c	T_s (K)
150	1.33	1.05	1.12	0.56	1.02	2,350
180	1.55	1.09	1.20	0.60	1.04	2,390
210	1.87	1.13	1.32	0.66	1.06	2,440
230	2.18	1.17	1.42	0.71	1.07	2,460

During the collision with shock or pressure waves, the gas pressure P in the flame increases rapidly corresponding to the strength of the shock, although it is variable according to the mode. In Fig. 10.17 some examples of pressure history in each mode are shown together with the luminescence L of the flame, which is also stimulated by the shock wave.

Figure 10.18 represents the pressure variations measured in the flame with respect to the time t after the passage of the shock front at the measuring point when the shock waves propagate with different Mach numbers M_1 through the plexiglas tube against the flame propagating from the ignition point I to the right and colliding with it in its front (first mode).

The shock waves pass by the flame after the collision and reach the measuring point, while the flame is pushed back after the collision. We consequently observe at first the shock front (at $t = 0$), the tail of the pushed back flame ($t = 8$ ms), and finally the front of the flame ($t = 25$–40 ms). The highest peak pressure P_F in the flame behind the shock waves appears several microseconds after the passage of the flame tail at the measuring point in the reverse direction and increase with the Mach number M of the shock waves propagating in the mixture through the plexiglas tube, as shown in Fig. 10.19. P_{Ff} is the pressure ratio expressed with respect to the Mach number M_f of the incident shock into the mixture.

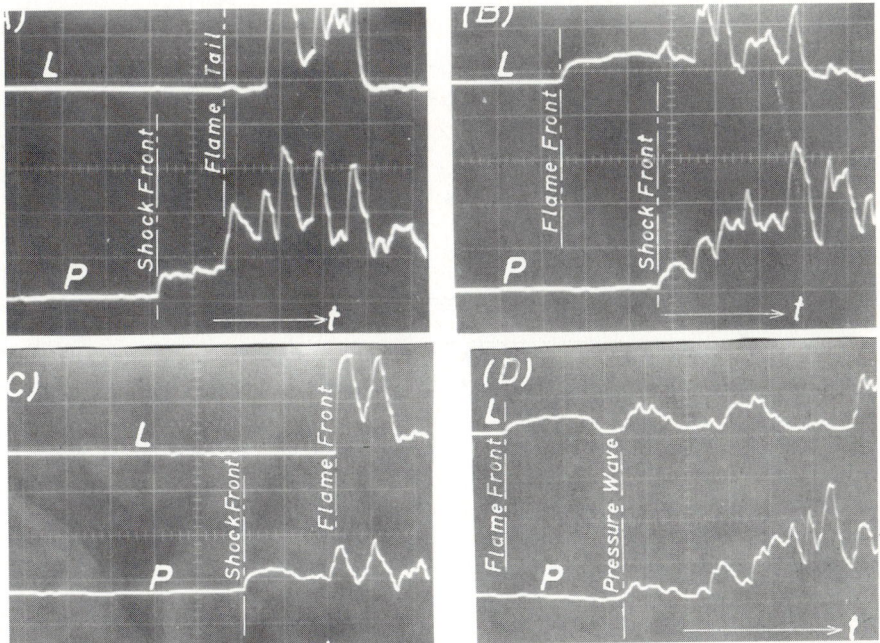

Fig. 10.17. Oscillograms of gas pressure P and luminescence L of the flame in different modes (**A**) First mode, (**B**) Second mode, (**C**) Third mode, and (**D**) Fourth mode Pressure: 100 kPa (div.)$^{-1}$, luminescence: an arbitrary scale, 5 ms (div.)$^{-1}$

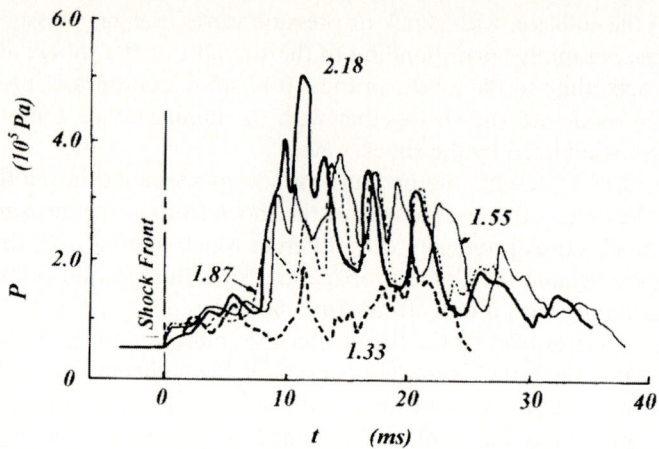

Fig. 10.18. Pressure variations P in the flame in the first mode with respect to the time t after the passage of the shock front at the measurement point. The number on each curve means the Mach number M_1 of the shock wave propagating in the mixture

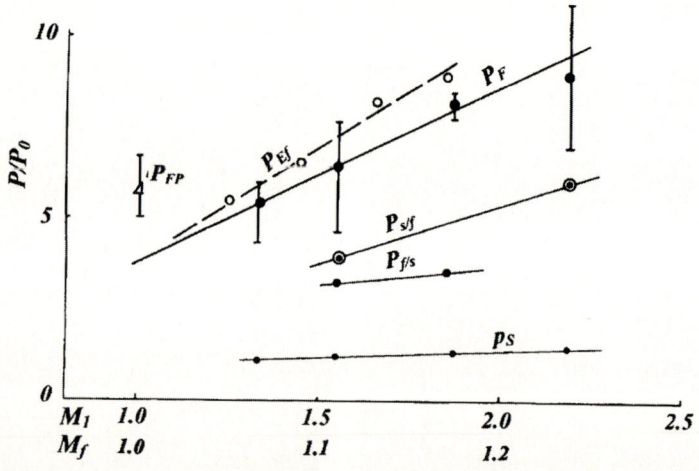

Fig. 10.19. P: the highest pressure in the flame, P_0: initial mixture pressure, M_1 is Mach number of the shock waves propagating in the mixture, and M_f that in the flame. P_F is the highest pressure in first mode vs. M_1 and P_{Ff} that vs. M_f, $P_{s/f}$ is the highest pressure in second mode, $P_{f/s}$ that in third mode, P_{FP} in fourth mode and P_s behind the shock waves propagating in the flame with respect to M_1

The pressure rise is also different according to the position where the shock wave and flame collide with each other, as shown in Fig. 10.20. In this case, the shock wave propagates behind the flame in the same direction (second mode). Figure 10.21 represents the ratio of the highest peak pressure $P_{s/f}$ in the flame

10.3 Interaction between Combustion and Shock Waves

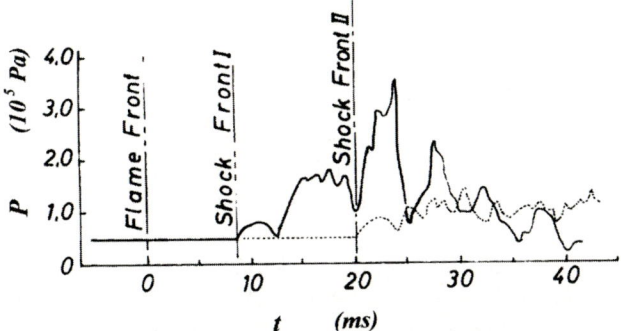

Fig. 10.20. Pressure variations P in the flame in comparison with the distance between the shock and flame front. t, time after the passage of the flame front at the measurement point. *Solid line* corresponds to shock I and *dotted line* to shock II

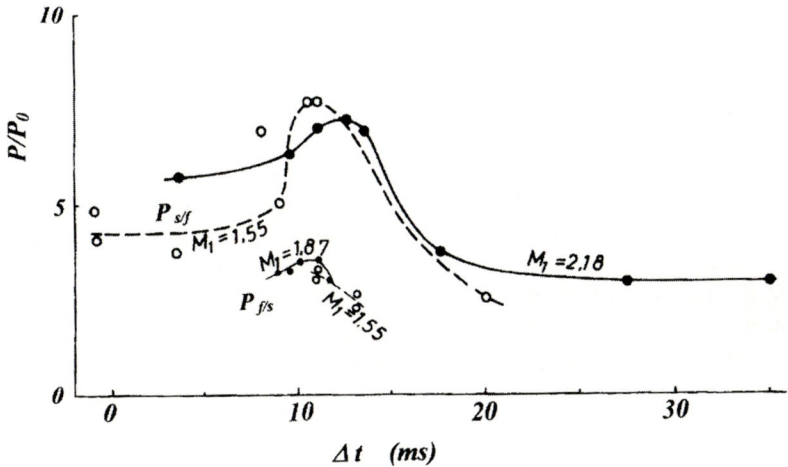

Fig. 10.21. The ratio of the highest peak pressure P to P_0 behind the flame or shock front in relation to the time difference Δt between the passage of flame front and that of the shock front at the measurement point. $P_{s/f}$: in second mode, $P_{f/s}$: in third mode

to the initial mixture pressure P_0 in the third mode, as a function of the time difference Δt between the passage of the flame front and that of the shock front at the measuring point. Δt is proportional to the distance between the flame and shock fronts.

As shown in Fig. 10.21, the highest pressure $P_{s/f}$ behind the shock waves in the second mode is almost constant when the shock waves propagate behind the flame with a delay time $\Delta t = -1.0 \sim +9.0$ ms, but in the region of $\Delta t = 10 \sim 14$ ms, it is 1.5~2.0 times higher, while in the region of $\Delta t > 15$ ms it is lower than that in $\Delta t < 9$ ms. Besides, the pressure P_{slf} in the region of $\Delta t < 9$ ms increases with the Mach number of the shock, while that in $\Delta t = 10\sim14$ ms is independent of it.

In the third mode where the flame propagates behind the shock wave in the same direction, we can measure the pressure only in a narrow range of Δt near 10 ms. We cannot recognize any influence of the shock strength.

Figure 10.19 also represents the ratio of the highest peak pressure P_{slf} behind the shock waves in the second mode to the initial pressure P_0 and that of P_{fls} in the flame in the third mode in the region of $\Delta t < 10$ ms as well as that of the peak pressure P_{FP} in the flame at the collision with pressure wave with respect to the Mach number M_1 and M_f of the shock waves. In each mode the highest peak pressure is much higher than the theoretical one behind the shock waves in the flame, and especially that in the first mode where the flame and shock waves propagate against each other is higher than that behind shock waves or in the flame when both the shock and combustion waves propagate in the same direction.

In the diagram of pressure variation we observe some pulsation in the flame after the collision with the shock waves. Considering the period between every two successive peak pressures, they must be caused by the reflection of the pressure wave from the shock tube end.

Ionization of Gases in the Flame

As already explained in Chap. 9, the combustion velocity increases almost proportionally to the ion density in the flame, it is very important to observe the ionization in the flame, with which we can recognize, if the combustion in the flame is really stimulated by shock or pressure waves, or not. The double probe method explained in Chap. 9 is applied to measure the ionization of gases in the propagating flame.

The probe is composed of two electrodes of nickel plates having the same size, that is, 0.2 mm in thickness, 3 mm in width, and 4.5 mm in length. Both the electrodes are set at the measuring position in the plexiglas tube, facing each other with a gap of 2.5 mm, so that each inner-side of 4.5×3 mm can effectively collect the charged particles, while the back side of each electrode is coated with an insulator as shown in Fig. 10.15.

An eight-step potential from -2.0 to 6.0 V having a frequency of 2 or 3 kHz, as shown in Fig. 10.22, is given to one of the electrodes, while the probe current from the other is measured by an oscilloscope. Figure 10.23 represents an example of the probe current observed in the normally propagating flame without any shock collision and Fig. 10.24 that at a collision with a shock together with light emission from the flame. Corresponding to the potential of each step an ion current should be observed. We can, thus, obtain in an oscillograph of the probe current many dotted lines corresponding to each probe potential. Connecting these dots for the same probe potential, we obtain a diagram composed of eight stripes as shown in Figs. 10.25 and 10.26. From such a diagram we can further obtain a probe characteristic, i.e., current vs. potential curve at an arbitrary instant after the passage of the flame or shock front at the measuring point as already explained in Sect. 9.2.3 of Chap. 9.

10.3 Interaction between Combustion and Shock Waves

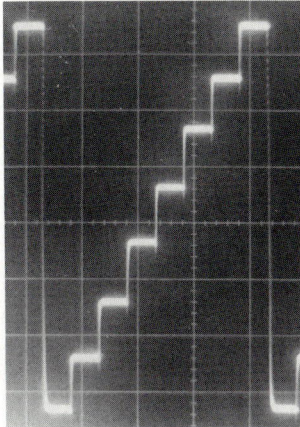

Fig. 10.22. Eight step probe potential 1.0 V (div.)$^{-1}$, 50 µs (div.)$^{-1}$

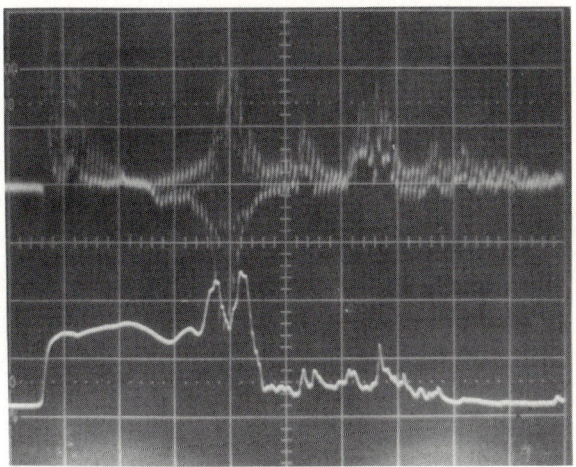

Fig. 10.23. Probe current (above, 0.1 µA (div.)$^{-1}$) corresponding to the probe potential shown in Fig. 10.22 and flame luminescence (below in arbitrary scale) in the normally propagating flame, 5 ms (div.)$^{-1}$

As also explained in Sects. 9.2.3 and 9.2.4 of Chap. 9, we obtain the electron temperature T_e, the ion mass m_i and ion density n_i from the probe characteristic curve, using the following equations:

$$T_e = \frac{e}{k} \frac{i_1 \cdot i_2}{(i_1 + i_2)\left(\frac{di}{dV}\right)_0} \tag{9.31}$$

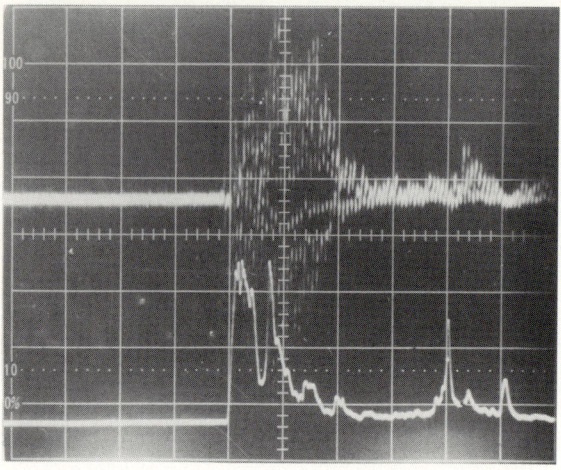

Fig. 10.24. Probe current (above, 0.5 µA (div.)$^{-1}$) and flame luminescence (below) in the flame at a collision with a shock wave, 5 ms (div.)$^{-1}$

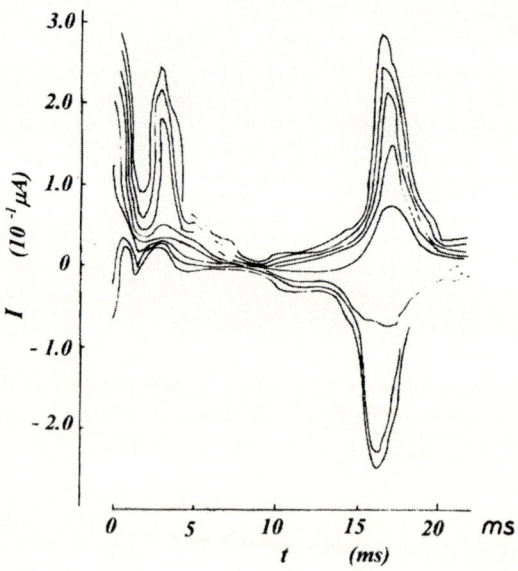

Fig. 10.25. Probe current of Fig. 10.23 expressed with stripes corresponding to each probe potential

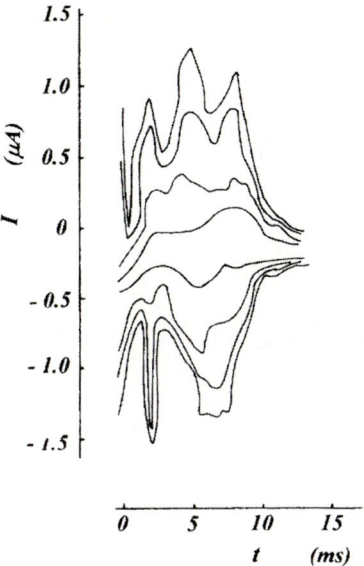

Fig. 10.26. Probe current of Fig. 10.24 expressed with stripes corresponding to each probe potential

$$V_f = \frac{kT_e}{2e} \ln\left(\frac{m_i T_e}{m_e T_i}\right) \quad (9.30)$$

$$n_i = \frac{J_i}{e\sqrt{\frac{kT_i}{2\pi m_i}}}, \quad (9.29)$$

where e is the elementary charge, k the Boltzmann's constant, i_1, i_2 the saturated probe currents, $(di/dV)_0 = \tan\alpha$ the slope of the characteristic curve at the turning point, V_f the wall potential, m_e, m_i the electron mass and ion mass, respectively, T_i the ion temperature which is here assumed to be equal to T_e and J_i the saturated ion current per unit area. In this case, however, the effective area of each electrode is the same, therefore $i_s = i_2 = i_s$ and

$$T_e = \frac{e}{k}\frac{i_s}{2\left(\frac{di}{dV}\right)_0}. \quad (9.32)$$

An example of the results calculated from the experimentally obtained probe characteristics is illustrated in Fig. 10.27.

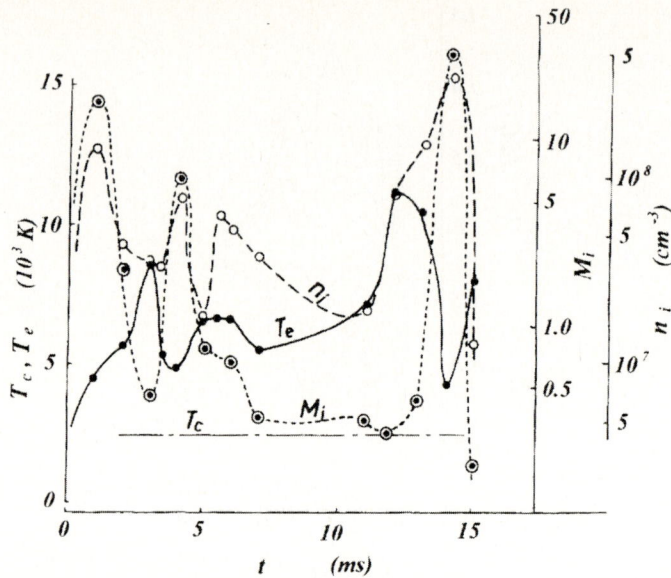

Fig. 10.27. Electron temperature T_e, ion mass M_i (in atomic weight), ion density n_i in the flame with respect to the time t after the arrival of the shock front at the measurement point in the first mode; T_c, theoretical adiabatic combustion temperature in the flame

Considering the cooling effect and disturbance occurred by the probe itself, the measured results are not exactly correct. Especially the measured electron temperature and ion density are much lower than the real values, as explained in Chap. 9, but they have nevertheless some meaning in their relative values. Let us make some discussion on the phenomena at least qualitatively by comparing the results.

Figures 10.28 and 10.29 represent the measured electron temperature T_e and ion density n_i, respectively, with respect to the time t after passage of the flame or shock front at the measuring point in the first mode, where the shock waves propagate against the flame with different Mach numbers and collide with the flame in its front. The number on each curve means the Mach number of the shock colliding with the flame. The curve having no number, therefore, means a normally propagating flame without any shock collision. As the flame is pushed back in this case, the flame tail passes by the measuring point after collision with the shock, flowing back toward the spark plug at I, while the flame front without any shock collision passes by the measuring point at $t = 0$. In the region, where $T_e = 0$ and $n_i = 0$, the electron temperature as well as the ion density cannot be measured, as the ion current is too small.

The variation tendencies with time in both the electron temperature and ion density at shock collision are similar to those in the flame without any shock collision, although their values are different. The electron temperature T_e has a

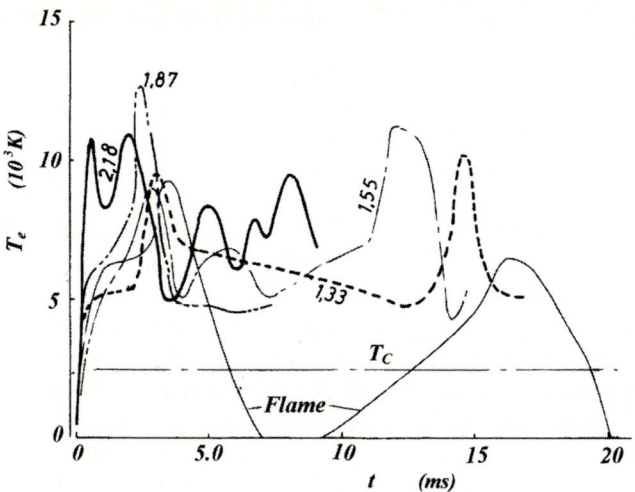

Fig. 10.28. Electron temperature T_e in the flame with respect to the time t after arrival of the shock front at the measuring point in first mode. Number on each curve means Mach number of the shock wave colliding with the flame. T_c, theoretical adiabatic combustion temperature

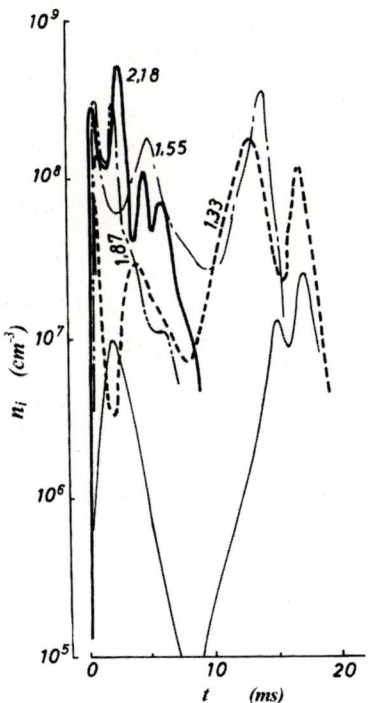

Fig. 10.29. Ion density n_i with respect to the time t under the same condition as in Fig. 10.28

value from 5,000 to 13,000 K, which is much higher than the theoretical adiabatic combustion temperature T_c of 2,300 K. Considering that the electron temperature is kept at a value much higher than the mean combustion temperature during the combustion reaction, the duration of combustion observed in these experiments is from 10 to 20 ms and not much different from the other.

The electron temperature T_e and especially ion density n_i are affected very much by the shock or pressure waves. Figure 10.30 represents the observed highest peak values of the electron temperature T_{ex} and ion density n_{ix} behind the flame or shock front with respect to the Mach number M_1 of the incident shock or pressure waves propagating in the mixture through the plexiglas tube. Both the peak values increase remarkably with Mach number.

Just like the pressure variation, the ionization of the gas in the flame is also variable according to the position of the shock front in relation to the flame front. In Fig. 10.31 the variation of ion density n_i and electron temperature T_e are illustrated with respect to the time after the passage of the flame front at the measuring point in the second mode where the shock waves propagate behind the flame in the same direction. In case A, where both the shock and flame fronts are overlapped, the highest electron temperature T_{ex} is about 10,000 K and ion density n_{ix} about 5×10^8 cm^{-3}, while in B, where the shock front passes by the measuring point 25 ms later than the flame front, $T_{ex} \approx 7,000$ K and $n_{ix} \approx 10^8$ cm^{-3}. The highest values of n_i and T_e in A are much higher than those in B.

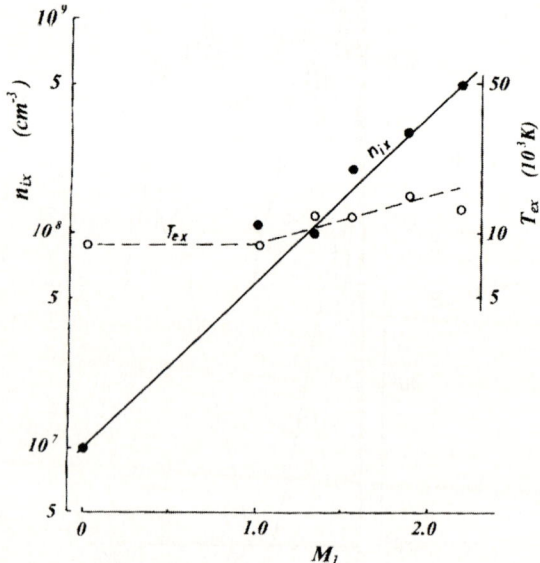

Fig. 10.30. The highest electron temperature T_{ex} and ion density n_{ix} with respect to Mach number M_1

10.3 Interaction between Combustion and Shock Waves

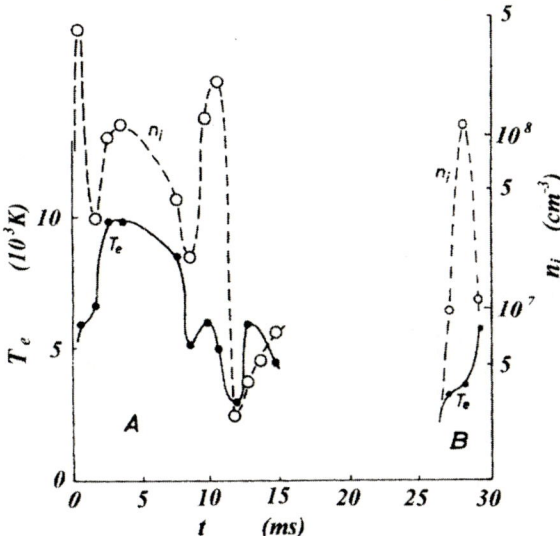

Fig. 10.31. Electron temperature T_e and ion density n_i with respect to the time t after arrival of the flame front at the measuring point. (**A**) both the shock and flame fronts overlap, (**B**) the shock front propagating behind the flame passes the measuring point 25 ms later than the flame front

According to the following Saha's equation:

$$r_i^2 = \frac{3.10^{21}}{n_0} \sqrt{T} \exp\left(-\frac{V_i}{T}\right), \tag{9.48}$$

where r_i is the ionization degree, n_0 the density of the neutral particles, T the gas temperature, and V_i the ionization potential, the logarithm of the square of ionization degree r_{ix} at the highest peak value in the ion density measured in the flame behind the shock waves with respect to the reciprocal gas temperature behind the flame front where the shock waves propagate is expressed in Fig. 10.32. All highest peak values of r_{ix}^2 observed in the first mode (marked with black circle) are expressed on a straight line having an ionization potential of 15.3V, while that in normally propagating flame without any shock collision is placed further below the line. Considering the ionization potential of 11–15 V for hydrogen, carbon, nitrogen, and oxygen, the here observed value of the ionization potential is a reasonable one.

In the other modes in which the shock waves propagate in the same direction of the flame, the same tendency as in the first mode is observed, although the measured values differ somewhat among each other.

As already mentioned, the results measured by the double probe method do not show the correct values, but much lower ones because of the cooling effect at the electrode surface. Comparing the results, however, we can discuss the

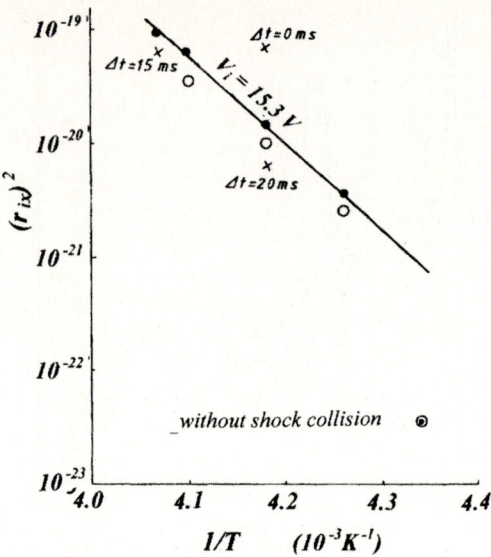

Fig. 10.32. r_{ix}^2 vs. $1/T$, r_{ix}: the highest ionization degree, T: theoretical gas temperature in the flame behind the shock waves. *Filled circle*: in the first mode, x: in the second mode, *empty circle*: in the third mode

phenomena at least qualitatively. The diagram in Fig. 10.32 suggests that the shock waves stimulate the ionization of the gases in the flame, i.e., a temperature increase in the flame more than the gasdynamic one occurs at the interaction with shock waves.

Interaction Between Combustion and Shock or Pressure Waves

The experimental results observed in a flame propagating in a propane–air mixture under an interaction with shock or pressure wave suggest:

1. An anomalously high-pressure increase is observed and its highest peak pressure increases with the Mach number of the incident shock waves three to nine times the initial pressure in the flame.
2. The ionization is stimulated by the shock or pressure waves corresponding to the Mach number of the shock so much that the highest electron temperature measured by the double probe method increases from 8,000 to 17,000 K and the ion density from an order of 10^7 to 10^9 cm^{-3} and this means an anomalous temperature increase in the flame at the collision with shock waves.
3. The increase in pressure and ion density in the flame by the shock waves are variable according to the position of the shock front in relation to the flame front and are even observed in the region where the combustion has ceased.

10.3 Interaction between Combustion and Shock Waves

For such an anomalously high-pressure increase, at first, a stimulation of combustion reaction by the shock or pressure waves should be considered. The pressure in the reaction zone at the flame front should rise rapidly by an almost isochoric combustion and propagate forward as well as behind the flame front. In the experiments described earlier, the highest pressure is observed not at the flame front but a few centimeters behind the flame front, where the combustion is almost completed. The stimulation of the normal combustion reaction by the shock heating cannot be recognized.

Next, some exothermic reaction other than the normal combustion taking place behind the flame front at the collision with shocks and producing new shock waves like detonation should be considered. From the ratio of observed pressure P_F behind the shock waves to the theoretical one P_s in the flame, the Mach number M_d of the new detonation waves and the reaction heat H_d for the self-sustained detonation waves are calculated as shown in Table 10.2. In comparison with them, the reaction heat of the normal combustion in a stoichiometric propane–air mixture is 2.63×10^3 kJ kg^{-1}.

As already explained in Sect. 9.3.3 of Chap. 9, the experimental results of the ionization of the gases in a flame investigated by a laser light scattering method which gives much more correct values than the double probe method suggest, that the gas in a flame is strongly ionized, partially with an ionization degree of 10^{-3}–10^{-2}, in which the charged particles having a temperature of 10,000–150,000 K are observed. This means that a very high energy state having a temperature of a few eV must appear once during the combustion reaction. Such a high energy state can only be realized by some exothermic reaction releasing much more heat than the normal combustion.

On the other hand, as described in Sect. 9.4 of Chap. 9, the gas behind shock waves is in a nonequilibrium state for a few hundred microseconds after the arrival of the shock front, where the ionization degree is about ten times and the temperature of free electrons as well as ions are several times higher than those in the equilibrium.

If a shock wave propagates into a flame, it excites the gas in the flame and increases the charged particles having a very high temperature of several eV to a great extent, perhaps on the order of ten times, especially during the nonequilibrium period of the shock wave. With such high-energy particles increased by the shock waves, the exothermic reactions which once take place during the

Table 10.2. Characteristics of detonation produced in the flame by shock collision

M_1	P_F/P_s	M_d	$H_d(10^3$ kJ kg$^{-1})$
1.33	4.80	2.10	3.17
1.55	5.25	2.19	3.40
1.87	6.21	2.38	4.04
2.18	6.30	2.40	4.12

combustion are stimulated so much that detonation waves are initiated in the flame at the collision with shock or pressure waves.

What exothermic reactions proceed to release so much energy is not yet clear, but if many molecules in the mixture are associated or combined together, it is possible to heat the gas to a temperature higher than 10,000 K.

The reason for the conclusions obtained from the experimental results observed in the combustion waves at the collision with shock or pressure waves are explained as follows:

The gas in the flame is in a nonequilibrium state where a strong ionization and the charged particles have a temperature higher than 10,000 K. The pressure or shock wave propagating into the flame stimulates the nonequilibrium state in the flame, increases the particles having anomalously high-temperature and initiates some exothermic reactions after normal combustion. Such exothermic reactions produce detonation waves in the flame, increasing the pressure and temperature.

10.4 Resonance Pulse Jet Engine (Schmidtrohr)[96–98]

Accompanying a combustion wave propagating in a combustible mixture along a tube, a pressure wave also propagates through the tube, reflects at the tube exit and returns as a rarefaction wave. A pressure wave, thus, goes and returns through the tube accompanied by a pulsating combustion. Such pulsating combustions are applied to some combustion systems for different purposes, for example, to heat water, disperse agricultural chemicals, and so on.

The Schmidtrohr invented and developed by a German engineer Schmidt is a jet engine for an airplane applying a resonance of the pressure waves oscillating along a tube to a spontaneous ignition of the mixture. Based on the idea of Schmidt a weapon V-1 (Vergeltungswaffe-1) was developed and used by the German Air Force at the end of the last Second World War, but as his idea was not perfectly understood, the performance was not so good as Schmidt planned and the valve life was only 20–30 min.

Although the Schmidtrohr is fit for the airplane engine because of its simple construction and light weight, it has not been practically used, as the noise is so loud. It is, however, still a very interesting apparatus applying shock waves and can be practically used for the plane or as a booster of rocket engine for the space. The mechanism and experimental results of the machine mainly reported by Schmidt is, therefore, explained here.

10.4.1 Construction and Action Mechanism[96, 99]

In Fig. 10.33 two Schmidtrohrs designed and constructed by Schmidt are schematically illustrated. The Schmidtrohr in this figure is the largest one having a combustion chamber of 500 mm in diameter and a thrust of about 5,400 N. The other one shown later is the smallest one of about 200 N in thrust, made of

10.4 Resonance Pulse Jet Engine (Schmidtrohr)

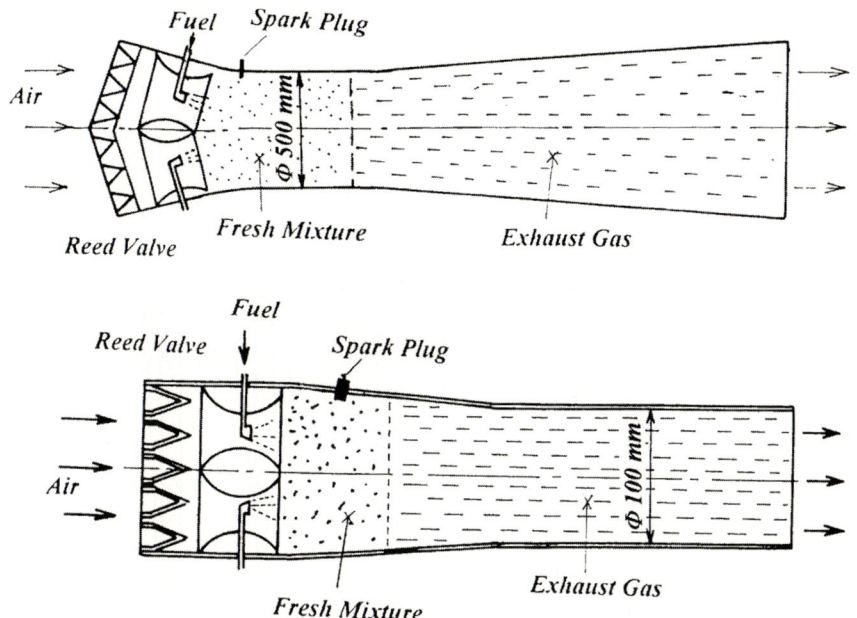

Fig. 10.33. Sketches of Schmidtrohr designed and constructed by Schmidt. Above: the largest one and below: the smallest one for the experiment

stainless steel plate having a thickness of 1.0 mm, has a whole length of 1,000 mm, a cylindrical air-intake part having an outer diameter of 150 mm, exhaust gas nozzle having an inner diameter of 100 mm and weight of about 8 kg.

In the air intake port many reed valves, carburetors having fuel-injector, a combustion chamber, and exhaust gas pipe are arranged one after another. When the pressure in the combustion chamber is lower than the atmosphere, the reed valve are opened and fresh air is introduced into the combustion chamber, while gasoline pressured by a fuel-pump is injected through the fuel-injector into the introduced air.

In order to start the engine, a mixture of air–gasoline introduced into the combustion chamber under a pressure higher than atmosphere is ignited by the spark plug set at the combustion chamber. With the combustion of the mixture a pressure wave propagates through the exhaust pipe and the gas flows out through the pipe exit. Then a rarefaction wave propagates from the pipe exit to the combustion chamber, where the gas is expanded by the rarefaction wave to a pressure lower than atmosphere, then the reed valves are opened and fresh air is introduced into the combustion chamber, where the fuel is injected and a combustible mixture is formed. The rarefaction wave reflected at the reed valve propagates to the pipe exit, decreasing its pressure more. If the rarefaction wave reaches the pipe exit, a shock wave produced by the pressure difference between the atmosphere and rarefaction wave propagates to the combustion

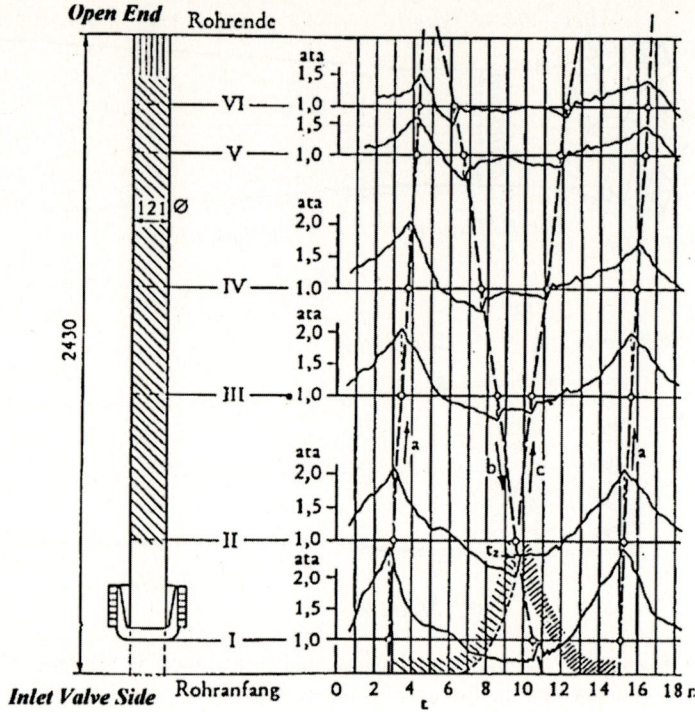

Fig. 10.34. Pressure variations at different positions of a tube having an inner diameter of 121 mm and length of 2,430 mm during a resonance combustion, presented by Schmidt

chamber and collides with the fresh mixture. By this collision with the shock wave the mixture is spontaneously ignited and the action is repeated to drive the engine steadily.

As the pressure, rarefaction, and shock waves propagate with a velocity near the sound velocity, the period between each ignition and combustion depends on the length of the Schmidtrohr and each wave is in a resonance state with each other.

In Fig. 10.34 a diagram of pressure variation measured at six different positions of a Schmidtrohr of 2,430 mm in length presented by Schmidt. In this diagram we can see how the combustion, pressure, and reflected rarefaction waves propagate.

10.4.2 Performance

In Fig. 10.35 an example of pressure diagram measured in the small Schmidtrohr schematically shown in Fig. 10.33 later is illustrated. In this case a thrust of the engine S_{ch} = 120 N, pressure difference ΔP = 230 kPa, and combustion frequency f = 135 kHz are observed. In Fig. 10.36 the pressure difference ΔP,

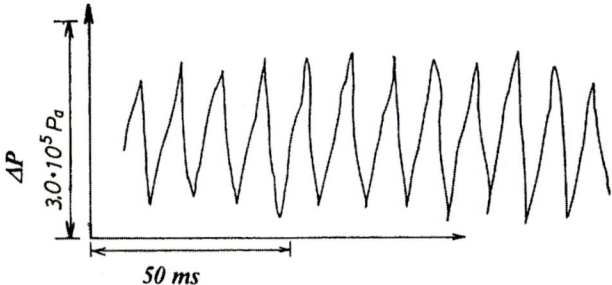

Fig. 10.35. Variation of combustion pressure observed in the combustion chamber of the Schmidtrohr shown in Fig. 10.33 later

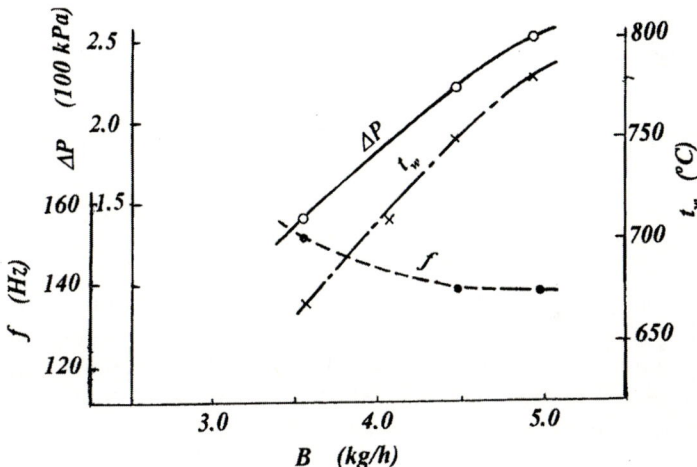

Fig. 10.36. Combustion pressure ΔP, frequency f of the periodical combustion and wall temperature t_w with respect to fuel supply B

frequency f, and wall temperature t_w of the engine are shown in relation to the supplied fuel.

A conventional performance curve, that is, the thrust S_{ch} and fuel consumption b are shown in Fig. 10.37. The fuel consumption is very large, as the fuel is injected always, even during the combustion and exhaust period, that is, only 1/4 of the whole supplied fuel is injected into the fresh air. Namely, if we could regulate the fuel injection only during the intake period of fresh air, we could decrease the fuel consumption to a value of about 1/4 of that shown in this diagram.

In the Science Museum in Munich the largest Schmidtrohr designed and constructed by Schmidt himself having an exhaust nozzle of 500 mm in diameter and thrust of 5,400 N is exhibited. The thrust is almost proportional to the cross-section area of the tube.

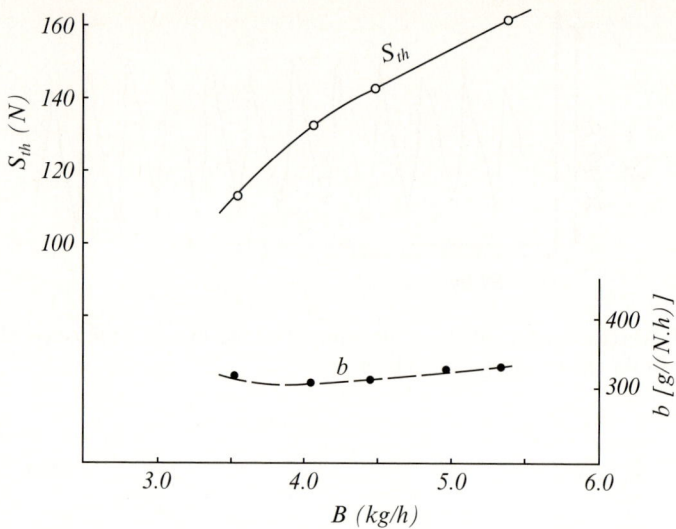

Fig. 10.37. Experimentally obtained performance curve of the Schmidtrohr shown in Fig. 10.33. S_{th}: thrust, B: fuel supply, b: rate of fuel consumption

10.4.3 Ignition and Combustion

A comparison of the combustion process among simple vibrating combustion, combustion in V-1 (Argus-Schmidtrohr) and that in Schmidtrohr reported by Schmidt is represented in Fig. 10.38. In the simple vibrating combustion system a continuous combustion proceeds, which is strengthened by weak pressure waves caused by the vibration of the air column and a pulsating combustion is observed. In the vibrating combustion, therefore, we observe a slow combustion velocity, little isochoric combustion accompanied by a small pressure rise and little mechanical work, while in V-1 a combustion having a higher isochoric combustion rate is observed, as the volumetric intake mixture efficiency reaches about 50%, and its thermal efficiency is higher than that of the simple vibrating combustion but less than that in Schmidtrohr driven by an almost 100% intake mixture efficiency.

These results are attributed to the construction of exhaust nozzle. As Schmidtrohr has a divergent nozzle, the rarefaction as well as shock waves produced at the exit of the nozzle is strengthened with the propagation of the waves in the nozzle reducing its cross-section area from the nozzle exit to the combustion chamber. Namely:

1. In the case of rarefaction wave, the pressure is lowered with propagation through the nozzle, and in the combustion chamber is much lower than that at the nozzle exit, when the rarefaction wave reaches there and the reed valves open. The fresh air is, thus introduced into the combustion chamber much more than that using a straight tube.[99]

10.4 Resonance Pulse Jet Engine (Schmidtrohr)   221

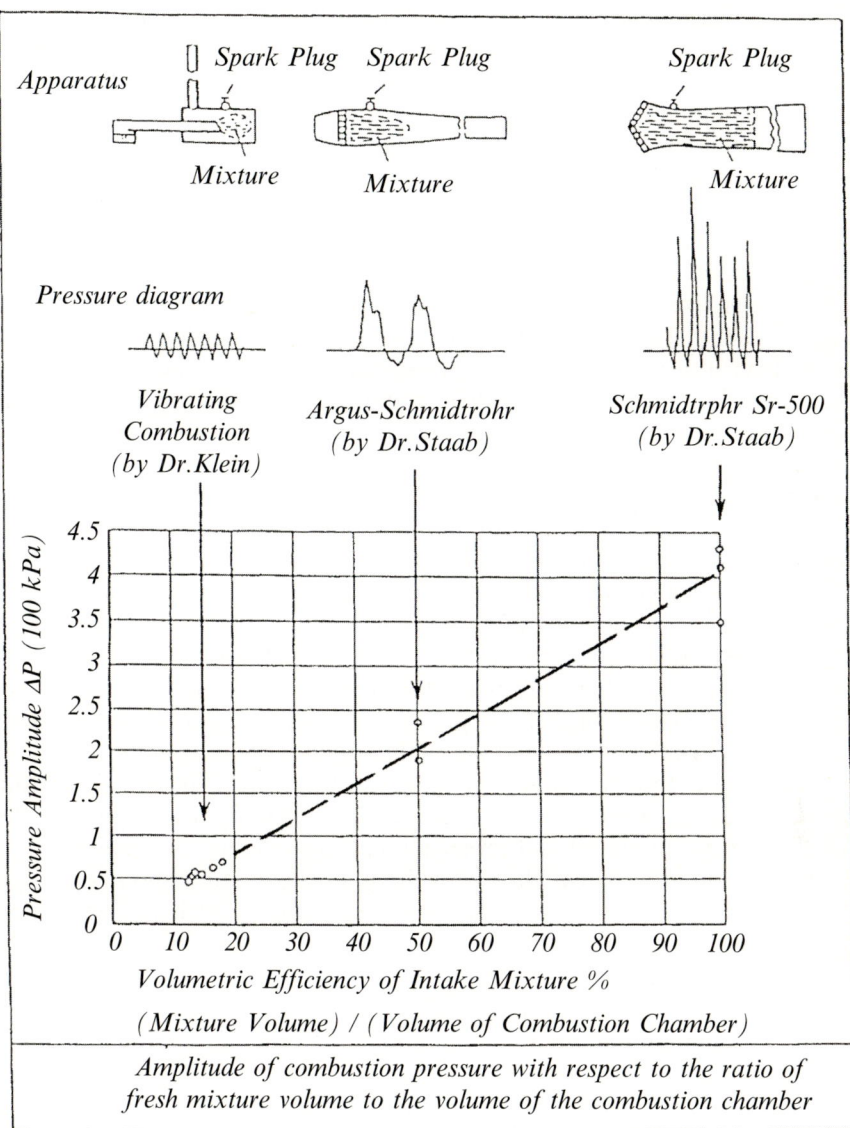

Amplitude of combustion pressure with respect to the ratio of fresh mixture volume to the volume of the combustion chamber

Fig. 10.38. Comparison of combustion aspect in different pulse combustion engines

2. In the case of shock wave, the pressure as well as the propagation velocity increases with propagation. The effect of collision with the mixture is much larger and ignites the fresh mixture stimulating its combustion much more than that using a straight tube.

In V-1 the rarefaction and pressure waves produced at the nozzle exit propagate through the nozzle increasing the cross-section area, increasing the pressure of rarefaction wave, and decreasing that of the pressure wave. The fresh air is, therefore, introduced into the V-1 much less than that into Schmidtrohr and the effect of shock wave on the ignition and combustion in V-1 is, therefore, much less than that in Schmidtrohr.

As the pressure waves as well as shock waves in V-1 or Schmidtrohr propagate with a velocity having a Mach number of 1.0 or a little higher than it, both the waves cannot heat the mixture so high that a spontaneous ignition can take place. The periodical ignition in V-1 or Schmidtrohr is, therefore, explained by an inflammation by the remained flame kept somewhere in the combustion chamber, just like the simple vibrating combustion. In Schmidtrohr, however, it is observed that the flame is extinguished after each combustion process.

As described in Sect. 10.4.2, the pressure in a flame increases several times, when a shock or pressure wave collides with it, accompanying another exothermic reaction than the normal combustion. The effect appears strongly behind the flame front where the combustion almost finishes more than that at the flame front. In Schmidtrohr a fresh mixture is introduced into the combustion chamber where the combustion just finishes, contacting with the tail gas of the combustion wave. If the shock or pressure wave propagating from the tube exit to the combustion chamber collides with the combustion gas in the tail of combustion wave, the pressure as well as the temperature of the gas increases several times by the interaction between both the nonequilibrium state behind the combustion and shock or pressure waves, and an exothermic reaction takes place, which propagates further into the fresh mixture and ignites it. A resonance combustion is, thus, repeated and the Schmidtrohr is driven steadily.

The stronger the shock wave colliding with the combustion wave, the more intensive the exothermic reaction proceeds. Especially in Schmidtrohr strong shock waves are produced and propagate into the combustion chamber, producing high-pressure by colliding with the gas behind combustion waves and fresh mixture. Schmidt thought the combustion having an isochoric combustion rate near 100%, but considering the pressure increase, a detonation wave should be produced.

11

Gaseous Detonation Waves

Gaseous detonation waves were first observed and reported in 1881 by Berthelot, Vieille, Mallard, and Le Chatelier. The definition of detonation waves is combustion waves accompanied by shock waves. Usually such phenomena are observed in a tube, through which a combustion wave propagates in a premixed combustible gas. Detonation waves propagate with a velocity higher than the sound velocity, between 1,000 and 3,000 m s^{-1}, while a normal flame propagates with a velocity from several centimeters to several meters per second. In these cases, however, we observe two different detonation waves, namely

1. A shock wave formed by the gas flow produced by the combustion propagates in the front of the combustion wave
2. The combustion wave following the shock wave is formed by spontaneous ignition behind the shock wave and the flow behind the combustion wave sustains the shock wave, which propagates fairly faster than that in the case of (1) described above.

Strictly speaking, the definition of the detonation waves should be the second case. It is, however, practically not easy to distinguish both the types. For it, it is often necessary to investigate the fine structure of the detonation waves.

In this chapter the classical theories of detonation waves are first briefly explained and then the theory constructed as an irreversible phenomenon is concretely explained, referring to many experimental results.

11.1 The Classical Theories of Detonation Waves[100–102]

The theories developed by Zeldovich, von Neumann, Doering, Sloukhin, and others are first briefly explained.

11.1.1 Macroscopic Structure of Detonation Waves

In the detonation wave a combustion wave follows a shock wave. Observing from a stationary point a detonation wave propagating in a combustible mixture through a tube, we have the one-dimensional diagram of pressure P and temperature T in relation to the time t after passage of the detonation front, as shown in Fig. 11.1.

First a shock wave having a pressure P_2 passes the observation point in a combustible mixture of an initial pressure P_1 and temperature T_1, then an ignition takes place at P_3 behind the shock wave. The gas expands with the combustion following the ignition which finishes at a pressure P_4. The mixture heated by the compression to T_2 behind the shock wave is further heated by the combustion from T_3 to T_4, where the combustion finishes and the gas expands. Such a diagram is called Neumann–Zeldovich–Doering model, or simply NZD-model.

In order to apply the gas dynamic theory to the detonation waves, the model shown above is further simplified, as shown in Fig. 11.2, in which we assume that the ignition and combustion reaction finish behind the shock wave at once, the initial mixture state before the detonation propagation is expressed by 1, that in the reaction zone behind shock wave by 2, that after the combustion reaction by 3, the detonation propagation velocity is denoted by D and the gas flows with a velocity of w in the region 3 after combustion.

Assuming a standing detonation wave just like the calculation of shock waves, we obtain the following three conservation equations:

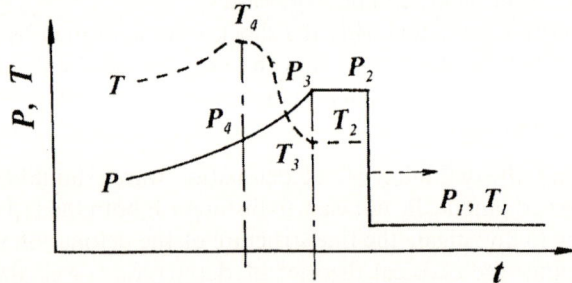

Fig. 11.1. *NZD*-model of detonation waves

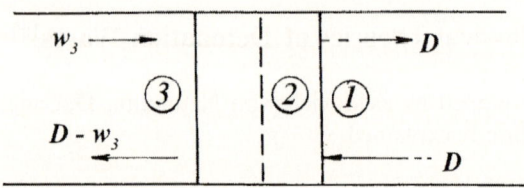

Fig. 11.2. Schematic expression of a detonation wave

11.1 The Classical Theories of Detonation Waves

$$\rho_1 D = \rho_3(D - w_3), \tag{11.1}$$

$$\rho_1 D^2 + P_1 = \rho_3(D - w_3)^2 + P_3, \tag{11.2}$$

$$\frac{1}{2}D^2 + E_1 + \frac{P_1}{\rho_1} + \Delta Q = \frac{1}{2}(D - w_3)^2 + E_3 + \frac{P_3}{\rho_3}, \tag{11.3}$$

where ΔQ is reaction heat per unit mixture mass. Applying the relation of a perfect gas

$$P = \rho RT \tag{11.4}$$

and sound velocity

$$a^2 = \gamma \frac{P}{\rho}. \tag{11.5}$$

under the assumption that the ratio of specific heats γ = constant, the following relation is obtained,

$$\frac{P_3}{P_1} = -\frac{\frac{\gamma+1}{\gamma-1} - \frac{\rho_1}{\rho_2} - \frac{2\gamma \Delta Q}{a_1^2}}{\frac{\gamma+1}{\gamma-1}\frac{\rho_1}{\rho_2} - 1}. \tag{11.6}$$

For shock waves without any combustion we obtain the following equation, as $\Delta Q = 0$

$$\frac{P_2}{P_1} = \frac{\frac{\gamma+1}{\gamma-1} - \frac{\rho_1}{\rho_2}}{\frac{\gamma+1}{\gamma-1}\frac{\rho_1}{\rho_2} - 1}. \tag{11.7}$$

From (11.6) the Hugoniot curve H_D for the detonation wave corresponding to an initial gas state of pressure P_1 and density ρ_1 is obtained in a diagram of pressure ratio P/P_1 in relation to density ratio ρ_1/ρ, while that of H_s for the shock wave follows from (11.7), as illustrated in Fig. 11.3:

In this diagram the initial state of the mixture is expressed by the point A ($P/P_1 = 1.0$, $\rho_1/\rho = 1.0$), on the Hugoniot curve H_s for the shock wave, while the state after the combustion falls on a point of the Hugoniot curve H_D for the detonation wave. The propagation velocity D of the detonation wave, consequently of the combustion wave is expressed by the following equation just like that of a shock wave:

$$D = \left(\frac{P_3 - P_1}{\rho_3 - \rho_1}\frac{\rho_3}{\rho_1}\right)^{1/2} \tag{11.8}$$

or

$$D - w_3 = \left(\frac{P_3 - P_1}{\rho_3 - \rho_1}\frac{\rho_1}{\rho_3}\right)^{1/2}. \tag{11.9}$$

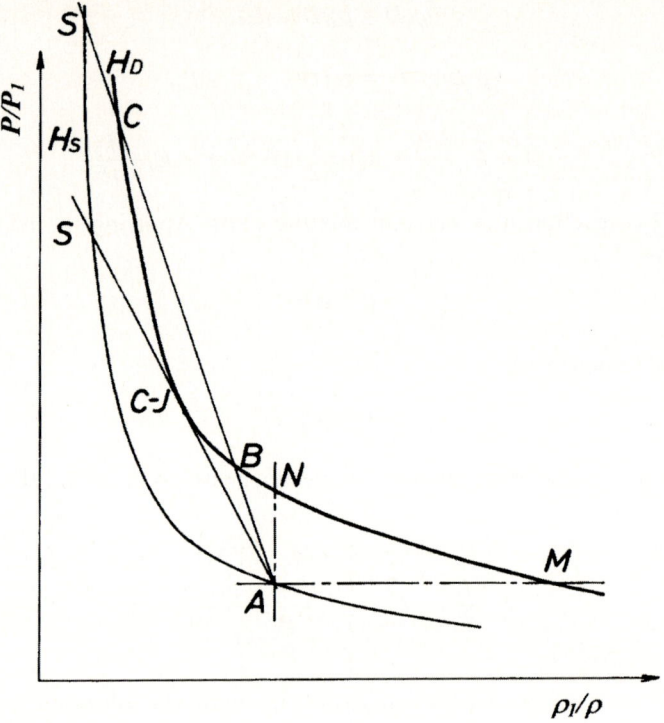

Fig. 11.3. Pressure (P)–density (ρ) diagram of a detonation wave. P_1, ρ_1: initial pressure and density of the mixture, respectively

The propagation velocity D of a detonation wave, thus, is proportional to $(\tan \alpha)^{1/2}$, where α is the slope angle of a straight line drawn from A to a point on H_D curve. As the propagation velocity corresponding to the slope of a straight line drawn from A to a point between N and M on H_D curve is imaginary, a detonation or combustion wave propagating with that velocity can never appear.

The state on H_D curve below M appears in a deflagration, which is a combustion wave without any shock wave, while the state above N appears after the combustion behind a detonation wave.

The straight line from the point A corresponding to a detonation wave intersects the H_D curve at two points B and C. In the diagram of Fig. 11.3, S means the shock front of the detonation wave. Behind S the ignition takes place, from which combustion waves propagate and the state of the gas behind the combustion waves reaches point C. The straight line intersects H_D at B as well, but as the entropy at C is larger than at B, only state C is observed.

Each straight line from A intersecting the H_D-curve can express pressure and density of the gas behind detonation waves. Usually, however, detonation waves propagating in a homogeneous mixture in a straight tube travels with a certain constant velocity after a period following the initiation. One says, the propagation

11.1 The Classical Theories of Detonation Waves

velocity of such a detonation wave is expressed by the tangent line from A to the H_D-curve, that is, the detonation wave propagates with its lowest velocity.

The point on the H_D-curve contacting with the straight line from A is called Chapman–Jouguet point, or simply C–J point. As at the C–J point the slope of the straight line from A is equal to the tangent of the H_D-curve, the following relation is obtained:

$$\frac{\frac{P_3}{P_1}-1}{1-\frac{\rho_1}{\rho_3}} = \gamma \frac{\frac{P_3}{P_1}}{\frac{\rho_1}{\rho_3}}. \tag{11.10}$$

At the same time the right-hand term of this equation means the tangential slope of an isentrope passing C–J point as well. Therefore

$$\frac{P_3-P_1}{\rho_3-\rho_1} = \gamma \frac{P_3}{P_1}, \tag{11.11}$$

from which the following equation is introduced

$$\frac{P_3-P_1}{\rho_3-\rho_1}\frac{\rho_1}{\rho_3} = \gamma \frac{P_3}{\rho_3}. \tag{11.12}$$

From (11.10)–(11.12) we further obtain the following equation:

$$(D-w_3)^2 = \gamma \frac{P_3}{\rho_3} = a_3^2. \tag{11.13}$$

This equation means that the gas behind the detonation wave flows with the sound velocity of the combustion gas.

In the NZD model behind the detonation wave the ignition, consequently the combustion takes place and the gas after the combustion expands, decreasing the pressure. Namely a rarefaction wave follows the detonation wave. In order to keep steady detonation propagation, the rarefaction wave has to follow the detonation wave neither overtaking the detonation wave nor leaving from it, that is, both the detonation and the rarefaction waves have to propagate with the same velocity. In the diagram of Fig. 11.3, the straight tangent line from A to the H_D-curve at the C–J point also a tangent to the isentropic of the combustion gas. On the straight line the detonation wave and rarefaction wave of the combustion gas propagate with the same velocity, so the detonation wave propagates with a steady velocity. Such detonation waves are called self-sustained detonation. As we have to take for the released heat the value at an equilibrium state after dissociation in the combustion, the real Hugoniot curve H_D locates much lower than that without any consideration of dissociation in the combustion.

According to (11.12) and (11.13), the relation of pressure P_{C-J} and density ρ_{C-J} of the gas at the C–J point is expressed by the following equation:

$$\frac{P_{C-J}}{P_1} = 1 + \frac{D^2}{\frac{P_1}{\rho_1}}\left(1-\frac{\rho_1}{\rho_{C-J}}\right). \tag{11.14}$$

Substituting the relation $M_D^2 = D^2/(\gamma P_1/\rho_1)$ into the (11.14), where M_D is Mach number of detonation propagation, the following equation is obtained:

$$\gamma M_D^2 = \frac{\dfrac{P_{C-J}}{P_1} - 1}{1 - \dfrac{\rho_1}{\rho_{C-J}}}. \tag{11.15}$$

Further we obtain

$$\frac{\rho_1}{\rho_{C-J}} = \frac{\gamma M_D^2 + 1}{(\gamma + 1) M_D^2}, \tag{11.16}$$

$$\frac{w_3}{D} = 1 - \frac{\gamma M_D^2 + 1}{(\gamma + 1) M_D^2}, \tag{11.17}$$

$$M_D^2 = \left\{ 1 + (\gamma^2 - 1) \frac{\Delta Q}{a_1^2} \right\} \pm \left[\left\{ (\gamma^2 - 1) \frac{\Delta Q}{a_1^2} \right\}^2 + 2(\gamma^2 - 1) \frac{\Delta Q}{a_1^2} \right]^{1/2}. \tag{11.18}$$

In the (11.18) having the plus sign of ± means the Mach number of the propagation velocity of detonation waves under C–J condition, while that having the minus sign the propagation velocity of the deflagration waves, which is a straight line tangent to the H_D-curve below point M from point A.

As the propagation velocity of detonation waves practically observed has usually a Mach number larger than 5, the Mach number M_D under C–J condition is approximately expressed as follows:

$$M_D^2 \approx 2(\gamma^2 - 1) \frac{Q}{a_1^2}. \tag{11.19}$$

The one-dimensional macroscopic structure of detonation waves is explained above according to the gas dynamic theory. The explanation is quite reasonable, but only for the idealized state. Considering the real states and phenomena, the detonation wave may be explained otherwise.

For example, it is questionable, if the C–J point really exists. Increasing the flow resistance of the tube, through which the detonation wave travels, the propagation velocity decreases to a value less than the C–J velocity according to a line lower than that from A to the C–J point, while the sound velocity of the combustion gas is unchanged, as the gas temperature after the combustion is the same. This suggests that the theory explained above does not fit to the phenomenon.

11.1.2 Microscopic Structure of Detonation Waves[101, 103, 104]

A detonation wave propagating in a combustible mixture through a channel having a rectangular cross-section marks a net like pattern, also called a cellular pattern, on a soot film coated on the inside of the bottom plate of the channel, as the photograph in Fig. 11.4 shows:

11.1 The Classical Theories of Detonation Waves

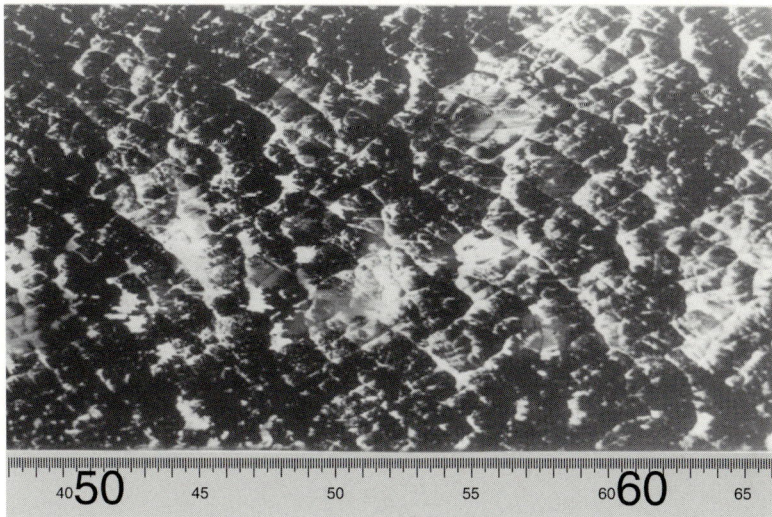

Fig. 11.4. Cellular pattern of detonation wave marked on a soot film. Detonation propagates in a stoichiometric propane–oxygen mixture from left to right (mm). Initial mixture pressure: 18.7 kPa

In a detonation wave propagating through a transparent channel, we also observe a luminous pattern just like the cellular pattern, in which the gas on each luminous line has such a high temperature that it can emit light. Such cellular patterns look geometrically regular. The reason for it has been attributed to an interaction between different shock waves formed behind detonation waves, namely

1. At the transition from deflagration to detonation, or at the initiation of detonation by other methods, some shock waves are produced. In several points behind the shock waves an ignition takes place, but not regularly, because of heterogeneity of the mixture or other reasons. As Fig. 11.5 schematically illustrates, the ignition takes place at several points behind shock waves. From each ignition point a new shock wave propagates forward and interacts with the initial shock wave, producing shock waves propagating in three different directions
2. This phenomenon is just like the configuration observed at Mach reflection in which an oblique shock wave collides with a wall, as schematically illustrated in Fig. 11.6. From the point O where the incident shock front OA and another shock wave, the so-called Mach stem, almost perpendicular to the wall collide with each other a reflected shock wave OR is produced. Such a point O is called triple point, which is also produced in many positions at the front of a detonation by the interaction of shock waves propagating in three different directions. These triple points propagate in oblique direction to the detonation propagation and collide with each other in several points.

Fig. 11.5. Scheme of shock waves produced by an irregular combustion

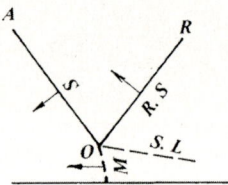

Fig. 11.6. Scheme of Mach reflection

3. In the mixture where two triple points collide with each other the pressure and temperature rise so high that a spontaneous ignition, i.e., an explosion takes place. At the detonation front, thus, explosions take place in many points consequently combustion waves propagate and sustain the detonation further,
4. On the other hand, behind the shock wave *OM* and reflected shock wave *OR* at the triple points the flow passing through *OM* with the supersonic velocity and the other flow passing through *OR* with subsonic velocity are observed. As both the flows have different velocities, a considerable amount of shear is produced in the borderline between both the flows, which creates a concentrated strong vortex. The vortex plays a role of stylus and marks the trace of the triple point motion on a soot coated the plate. Thus, cellular patterns shown in the photograph in Fig. 11.4 can be obtained on candle soot coated plate of a channel through which a detonation wave propagate.
5. Besides the hypothesis explained above Soloukhin proposed transverse waves TW propagating with sound velocity perpendicularly to the direction of the detonation. The transverse waves collide with the main shock waves at the detonation front and form many triple points having a geometrical regularity, as illustrated in Fig. 11.7. In each triple point a strong explosion takes place and forms a combustion zone RZ[105–107]

Soloukhin tried to prove his hypothesis by measuring the pressure variation around the triple point. Scaling up the cellular pattern much more in the detonation wave by adding argon to the mixture in which the detonation propagates,

11.1 The Classical Theories of Detonation Waves

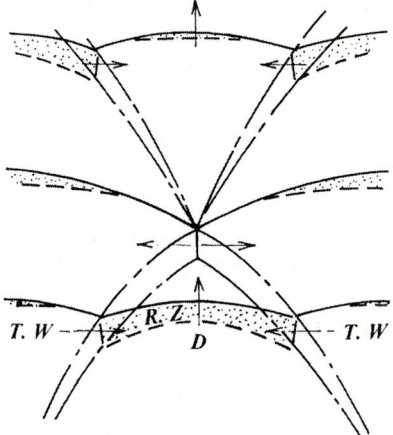

Fig. 11.7. Triple shock model proposed by Soloukhin. D, detonation wave; TW, transverse wave; RZ, reaction zone

he observed at the knot of the net a pressure much higher than that at the C–J point, while a much lower pressure exists between the knots.

As the explosion takes place in each collision of triple points, the triple shocks must be produced one after another and collide with each other continuously to sustain the detonation waves. Namely, in order to have a detonation wave propagating in a mixture, a certain definite number of triple points per unit length at the front of the detonation wave must be produced.

The mechanism of the cellular pattern formation seems to be explained well by the theories introduced above and generally recognized. Nevertheless there are some problems.

1. According to the theories explained above the cellular pattern must have an exactly geometric regularity, but practically some irregularities are always observed in each cellular pattern of detonation waves.
2. If the vortex formed by the shear between the flows having different velocities behind the Mach stem at the triple shock, has a scraping power stronger than a supersonic flow? What vortex is produced in three-dimensional flow?

The theories explained above based on only gas dynamics without any consideration of combustion, even if the detonation is a combustion phenomenon.

For the self-sustained detonation waves there is the Chapman–Jouguet condition. On the other hand, a certain definite density of triple points behind the detonation front is necessary to have a self-sustained detonation. Can self-sustained detonation waves satisfy both the two conditions which are independent with each other? The combustion mechanism in detonation waves must be investigated more accurately.

11.1.3 Transition from Deflagration to Detonation Waves[101–103, 108]

In order to produce detonation waves, there are many methods, but the simplest and basic one is the transition from a combustion wave propagating in a combustible mixture through a tube to a detonation wave, that is, a transition from deflagration to detonation takes place.

If a combustible mixture filled in a tube is ignited at an end of the tube, a flame propagates through the tube to the other end, being accelerated, then is suddenly transformed to a detonation wave. For it there is an explanation based on gas dynamics.

As schematically illustrated in the time–distance (t–x) diagram in Fig. 11.8, the mixture is ignited at O-point and a flame propagates to the right direction. From the rapid combustion some pressure waves propagate in the mixture in the front of the combustion wave one after another to right. The pressure waves stimulate the combustion and consequently produce stronger pressure waves, eventually shock waves, while the combustion wave accelerated by the pressure waves propagates according to the curve O–F. The shock waves propagating in the mixture in the front of the combustion wave are stimulated stronger with propagation of the combustion wave and at last have such a high pressure and temperature that the ignition takes place in several points of the mixture behind the shock waves. Thus, a detonation wave is produced.

According to the hypothesis explained above, the position in the tube where the transition takes place must be able to be estimated, as far as the initial state of the mixture and dimensions of the tube are known, but in reality there is still no theory with which the transition position can quantitatively be estimated, as it fluctuates over a wide range.

As the detonation waves are produced by spontaneous ignition in the mixture behind shock waves, we can initiate the detonation, if we can produce such a strong shock wave in the mixture. In addition to the transition methods from

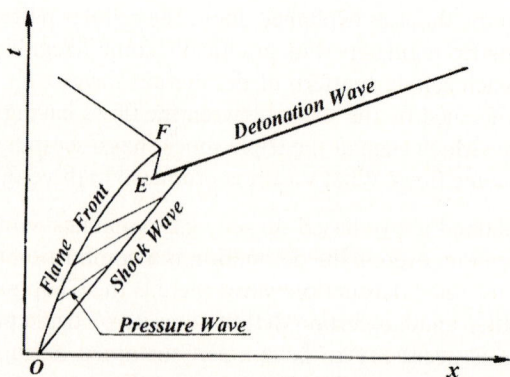

Fig. 11.8. Schematic time (t)–distance (x) diagram of transition from deflagration to detonation

deflagration, therefore, the detonation waves can be produced, applying shock waves from a shock tube or at the breakdown by focusing of a laser beam.

As described above, the detonation waves propagating in a gaseous mixture have been explained mainly by gas dynamic theories, but it must be necessary to analyze the phenomena, considering the ignition and combustion mechanism behind detonation waves as the detonation is also sustained by combustion phenomena.

11.2 Detonation Waves as Irreversible Phenomena

As described above, the detonation waves have mainly been explained, investigated, and discussed from the standpoint of gas dynamics. In order to investigate the detonation waves, however, the mechanism of ignition and combustion behind them should also be considered, as the detonation waves are composed of shock and combustion waves, especially the character of irreversible phenomena in the ignition and combustion must play important roles for production and propagation of detonation waves.

In this chapter the results different from those based on the gas dynamic theories, are presented, applying the stochastic ignition theory and considering the characteristic of combustion and shock waves at a nonequilibrium state.

First the microscopic structure of detonation waves and ignition behind the detonation front is explained, then the transition from deflagration to detonation, the Chapman–Jouguet condition for self-sustained detonation, the existence of the so-called Chapman–Jouguet velocity and nonequilibrium temperature behind detonation waves in each case, presenting many experimental results.

First of all, however, an experimental work on the interaction between converging shock and combustion waves is presented, since we can observe in this work some marking process of soot traces by shock and combustion waves and the process of shock production by combustion waves as well as the stimulation of the combustion by shock waves.

11.2.1 Interaction Between Converging Shock and Combustion Waves[109]

As described in Sect. 10.3., the combustion reaction is strongly stimulated by an interaction with shock waves. Such a phenomenon can be observed in every combustion wave propagating in a closed vessel. The so-called knock phenomenon is observed in a combustion chamber of spark ignition engines in which strong pressure waves propagate to and fro in the combustion chamber, repeating reflection from the chamber wall. This phenomenon may be due to a shock interaction, though it has been attributed to an autoignition in a part of unburned mixture in some corners of the combustion chamber.

With an intention to clarify such a phenomenon as knock in internal combustion engines and to find some configurations of predetonation, combustion waves propagating in a propane–oxygen mixture in an elliptical space are investigated by soot traces, schlieren photographs, and pressure measurement.

Combustion Chamber

As illustrated in Fig. 11.9, an elliptical space having a major axis of 300 mm, minor axis of 214 mm and a distance of 210 mm between the foci O_1 and O_2 together with a circular space having a radius of 76 mm from one of the elliptical foci O_1 is cut out from a plexiglas plates of 10 mm in thickness, the combustion chamber is prepared for the experiments. Both sides of the combustion chamber are further covered by steel plate of 10 mm in thickness having a circular window for the light path of the schlieren method and for the safety of the vessel chamber under the high combustion pressure.

In order to apply the schlieren method, avoiding the influence of buoyancy and sustaining the symmetry of the combustion and shock wave propagation, the combustion chamber should be installed, keeping the major axis O_2–O_1 perpendicular and the focus O_1 below the second one O_2.

A stoichiometric propane–oxygen mixture is filled in the combustion chamber under an initial pressure of 33.3 kPa and at room temperature of 20°C. The mixture is ignited by an electrical spark at the first focus O_1, from which a combustion wave propagates radially, keeping a circular form. A pressure wave initiated by the spark ignition at the first focus O_1 also propagates down to the

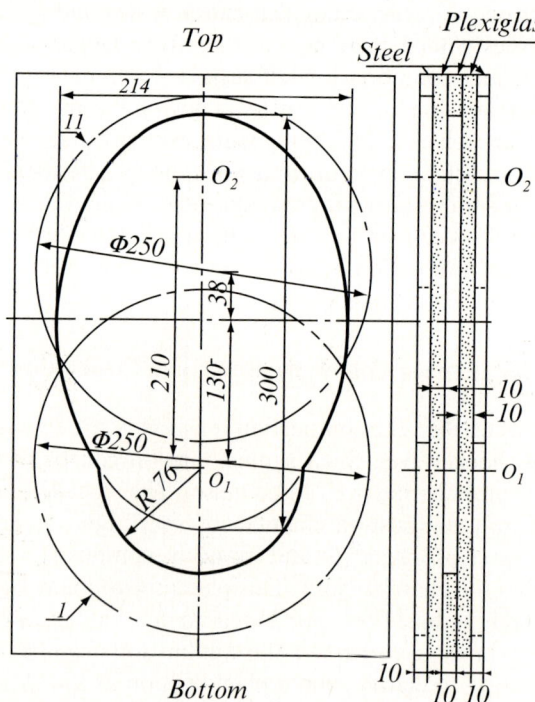

Fig. 11.9. Elliptical combustion chamber. I and II: Lower and upper exposure areas for schlieren photography, respectively

11.2 Detonation Waves as Irreversible Phenomena

lower direction, reflects from the circular circumference of the combustion chamber and converges to the focus O_1, being transformed to a shock wave. After focusing into O_1, the shock wave propagates further to the upper direction, reflects from the elliptical circumference, and converges to the other focus O_2. Near the second focus O_2 the mixture can be compressed and heated by the shock wave converging to O_2 and can cause a second ignition.

Soot Trace of Combustion Waves

After a combustion experiment a trace of combustion waves is observed near the second focus O_2 on the soot film made by a candle flame on the inner side of one plexiglas cover plate as the photograph in Fig. 11.10 shows. The trace must be made by a strong combustion reaction having an extremely high temperature, with which the carbon powder of the soot film is burned out. From the soot film trace we can recognize that a strong combustion reaction takes place near the second focus O_2.

Schlieren Photographs

In order to take schlieren photographs of the combustion waves, an optical system schematically illustrated in Fig. 11.11 is constructed, in which a light source, a condenser lens, pin hole, two concave mirrors on both sides of the combustion chamber, a knife edge, and a camera are arranged.

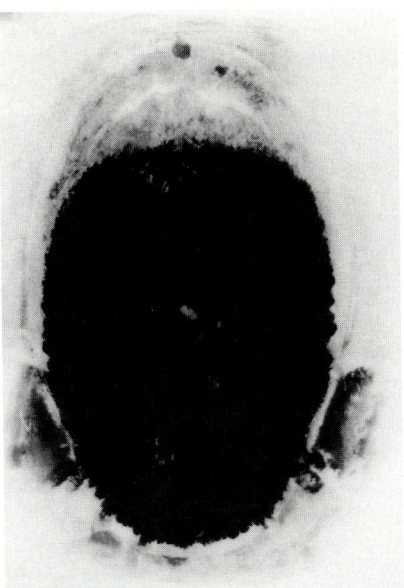

Fig. 11.10. Soot trace of combustion wave

236   11 Gaseous Detonation Waves

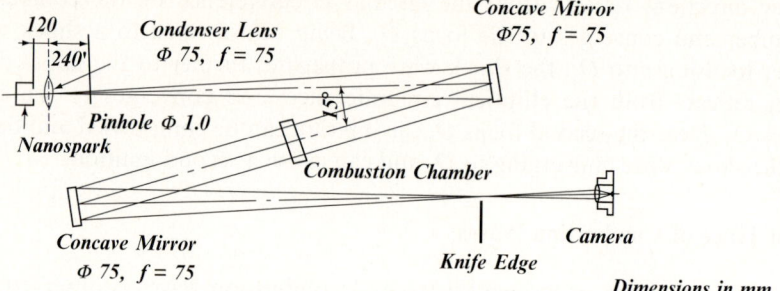

Fig. 11.11. Optical system of schlieren photography. Φ, diameter; f, focal length

As the light source an electrical spark between two electrodes having a gap from 3 to 4 mm in a nanospark[110] apparatus is used. The breakdown potential is 12–14 kV and spark duration is about 20 ns. The light beam from the light source focused by the condenser passes through a pinhole of 1.0 mm in diameter and is introduced to the first concave mirror.

The parallel light beam produced by reflection from the concave mirror passes through the combustion chamber, reflects again from the other concave mirror, is focused at the knife edge and taken into the camera.

The photographic area is divided into two parts as shown in Fig. 11.9 with I and II and the schlieren photograph of each part is taken separately from each other. The same experiment is repeated many times, so that schlieren photographs can be taken at different instants after the ignition. In Fig. 11.12 an example of schlieren photographs taken in the area I around the first focus O_1 where the mixture is ignited is shown, while in Fig. 11.13 two examples of those

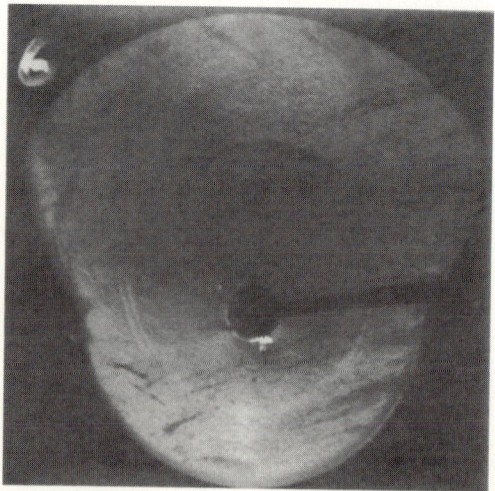

Fig. 11.12. Schlieren photograph in the lower area I of the combustion chamber at $t = 0.6$ ms after the spark

11.2 Detonation Waves as Irreversible Phenomena

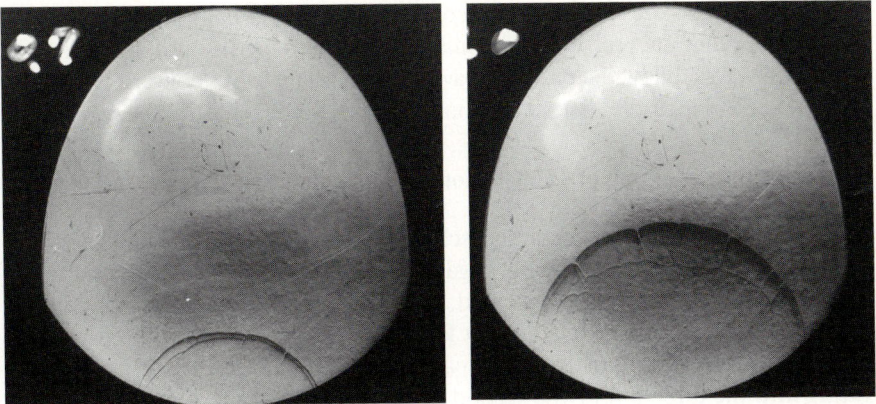

Fig. 11.13. Schlieren photographs in the upper area II of the combustion chamber at $t = 0.7$ ms (*left*) and 1.0 ms (*right*) after the spark

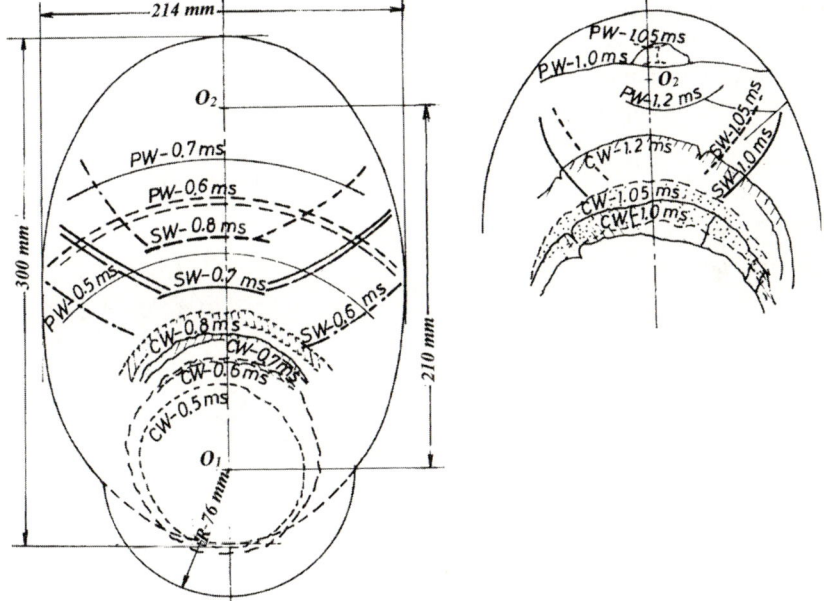

Fig. 11.14. Propagation of shock and combustion waves during $t = 0.5$–0.8 ms (*left*) and 1.0–1.2 ms (*right*). Time on each wave means the instant t after the spark at which the wave is observed

in the area II around the second focus O_2 taken at different instants after the spark are shown.

Composing such schlieren photographs taken at different instants t after the ignition, we obtain a propagation process of shock and combustion waves as shown in Fig. 11.14 during $t = 0.5$–0.8 ms (left) and during $t = 1.0$–1.2 ms (right). In these diagrams we recognize that before the combustion waves radially propagating from the ignition point the pressure waves initiated by the spark also propagate radially and shock waves follow the pressure waves. The pressure waves converge to a point near the second focus O_2, before the combustion waves reach it.

Time–Distance Diagram of Combustion, Pressure and Shock Waves

From the experimentally obtained schlieren photographs we obtain a time–distance diagram in which the propagation of combustion, pressure, and shock waves on the major axis are expressed, as illustrated in Fig. 11.15.

By the electrical spark at the first focus O_1 relatively rapid combustion and pressure waves are initiated and propagate radially. The pressure wave propagating into the direction of the second focus O_2 converges to O_2 after reflection from the elliptical circumference wall, while the pressure wave propagating to the opposite direction converges through the combustion waves to the first focus O_1 after reflection from the circular circumference, is transformed to a shock wave, which propagates radially to the elliptical circumference and then again converges to the second focus O_2 after reflection from the elliptical circumference following the first reflected pressure wave.

On the other hand the combustion wave first propagates radially, keeping a circular form, but then is stagnated by the reflected pressure waves near both the

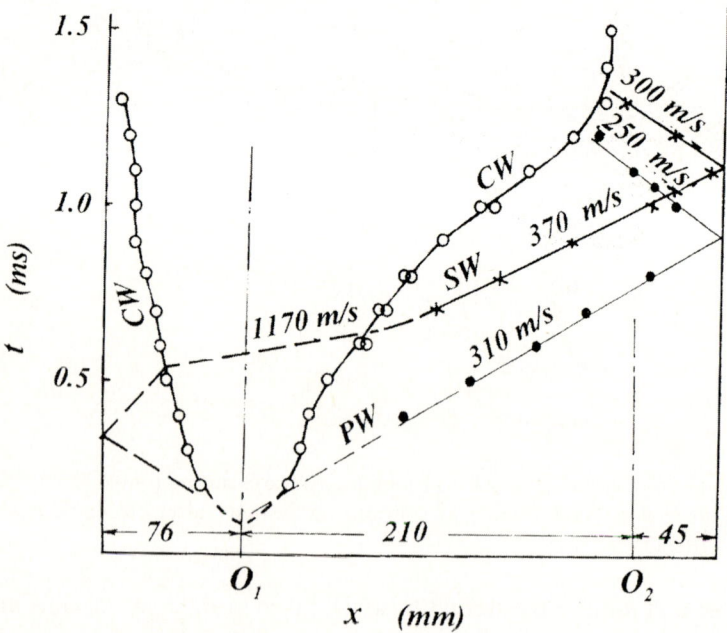

Fig. 11.15. Time–distance (t–x) diagram of the combustion (CW), pressure (PW), and shock (SW) waves. Propagation velocity of each wave is shown on each line

circular and elliptical circumferences. Behind the combustion waves observed in the schlieren photographs the combustion reaction is already over and the real flame should exist 30–40 mm in advance of it. If the flame is stagnated and stays there for a few milliseconds, the soot on the plexiglas plate is heated and oxidized, i.e., burns.

The trace on the soot film shown in Fig. 11.10 should be the trace where the soot thus burned. Besides, because of the interaction with the converged shock wave, a strong combustion reaction proceeds near the second focus O_2. The soot trace must be marked especially clearly by the strong combustion.

At several points on the major axis of the elliptical combustion chamber the pressure variation of the mixture gas is measured by a piezoelectric pressure transducer. The distance x from O_1 to each pressure measurement point is 35 mm, 75 mm, 135 mm, 140 mm, 160 mm, 210 mm, and 235 mm, respectively.

The measured results are illustrated in the time–distance diagram of shock and combustion waves, as illustrated in Fig. 11.16. The shock wave reflected from the elliptical circumference and converging to the second focus O_2 collides and interacts with the flame 30–40 mm in advance of the combustion wave, where a strong combustion reaction proceeds, producing a high pressure of 500 kPa, that is,

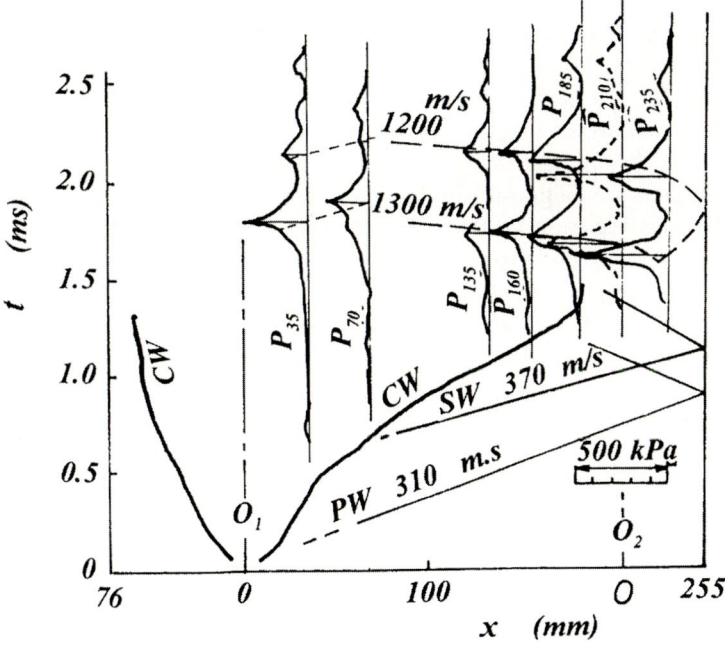

Fig. 11.16. Time–distance (t–x) diagram of the combustion (CW), pressure (PW), and shock (SW), waves along the major axis of the ellipse, together with the pressure histories (P) at different positions on the major axis of the ellipse. The propagation velocity is shown on each line and suffix number of P means the distance x (mm) each measurement point from O_1

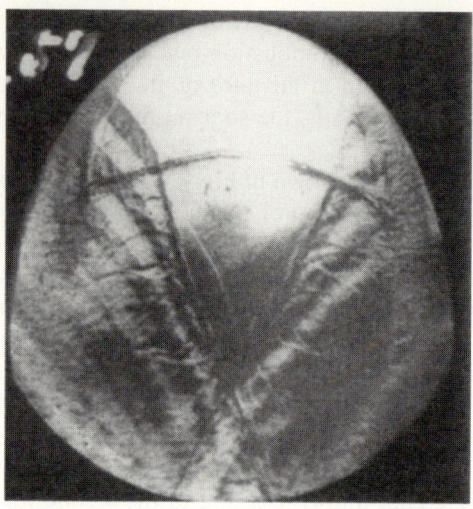

Fig. 11.17. Schlieren photograph in area II at $t = 1.57$ ms after spark

15 times higher than the initial one. The rapid combustion under the high pressure produces shock waves propagating in the direction to the first focus O_1 with a velocity of 1,300 ms^{-1}. This shock propagation velocity observed in the pressure diagram, however, does not show the real propagation velocity of the shock waves, but faster one, as the shock waves propagate obliquely to the major axis.

In the pressure diagrams obtained at $x = 35$ and 70 mm, we observe shock waves propagating from O_1 to O_2, which must be those reflecting from the circular circumference and propagating obliquely to the major axis, as the schlieren photograph in Fig. 11.17 shows. The real propagating velocity of the oblique shock wave is estimated from the angle to the major axis observed in the schlieren photograph in Fig. 11.17 to be about 1,200 m s^{-1}.

Stimulation of Combustion Reaction by Shock Collision

The reason of strong combustion reaction near the second focus O_2 stimulated by the collision with the shock waves is attributed to the nonequilibrium state at both the front of the combustion and shock waves, as already in Sect. 10.3 explained.

Behind combustion and shock waves there are nonequilibrium states in which the temperatures of ions and free electrons are much higher than those of the equilibrium state, as described in Sects. 9.3 and 9.4. By the interaction with shock waves the ionized high temperature particles in the combustion waves increase their density and stimulate the combustion reaction enormously.

In the elliptical combustion chamber the shock waves produced by the spark and combustion are strengthened by converging after the reflection from the elliptical circumference of the combustion chamber. Then the combustion reaction is stimulated by a collision with such strengthened shock waves, accompanying a high combustion pressure and new strong shock waves.

11.2.2 Cellular Structure Formation as a Stochastic Phenomenon

As described already, the cellular pattern of detonation waves marked on a soot film has been explained as a result of collision of detonation waves with transverse waves at the detonation front. Many shock waves initiated by some perturbations caused by irregular ignition collide with shock waves at the detonation front and form triple shocks, then a geometrically uniform pattern under the influence of transverse pressure waves. There are, however, no concrete, quantitative explanations of the perturbation or irregular ignitions which govern the phenomenon.

In this section, the phenomenon is interpreted as a stochastic phenomenon driven by the spontaneous ignition behind shock waves at the detonation front, investigating the cellular structure marked on soot films by detonation waves propagating in a stoichiometric propane–oxygen mixture with different velocities.

Shock-Produced Detonation Waves of Different Strengths[111]

In order to produce detonation waves of different strengths, i.e., of different propagation velocities, in a mixture of different densities, detonation waves are initiated and driven in a stoichiometric propane–oxygen mixture by shock waves produced in a shock tube of stainless steel. The procedure is schematically illustrated in Fig. 11.18.

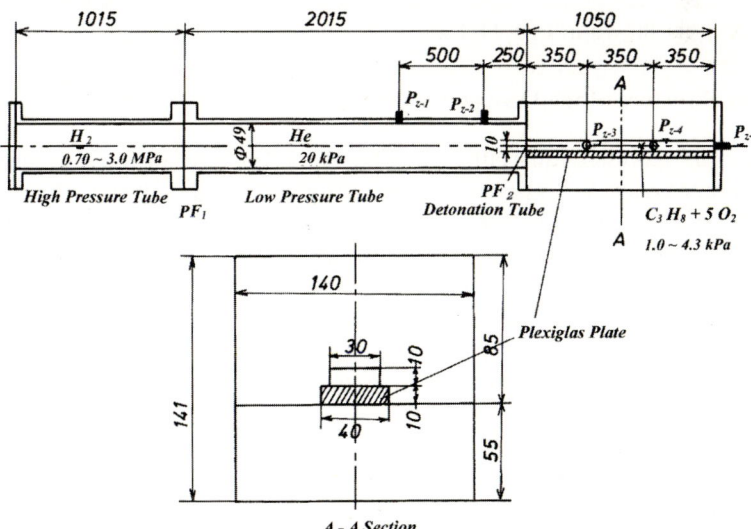

Fig. 11.18. Experimental apparatus for investigating the cellular structure of detonation waves using a shock tube. Dimensions in mm. PF_1, PF_2: polyester films; P_{z-1}, P_{z-2}, ..., P_{z-5}: piezoelectric pressure transducers

The shock tube is composed of three segments, 1,015 mm, 2,015 mm, and 1,050 mm in length, respectively. The first segment, high pressure tube and second one, low pressure tube, have the same inner-diameter of 49 mm, while the third one, detonation tube, has a rectangular cross-section of 10 mm × 30 mm. A 10-mm-thick plexiglas plate coated with candle soot is inserted below the rectangular space to record the cellular pattern.

A polyester film PF_1 having a thickness of 100–500 μm is set between the first and second segments, while PF_2 of 4 μm in thickness is set between the second and third segment.

As shock driver gas gaseous hydrogen is charged in the first segment at an arbitrary pressure between 700 kPa and 3.0 MPa, while helium gas in the second segment at 20 kPa and a stoichiometric propane–oxygen mixture in the third segment at an arbitrary initial pressure between 1.0 and 4.5 kPa. Breaking the first polyester film under different pressure ratios between the high and low pressure segment, shock waves of different propagation velocities are produced in the low-pressure segment and propagate into the third tube, where detonation waves are initiated by the shock waves. The detonation waves propagate with different velocities higher than that of the self-sustained detonation, corresponding to the initial pressure ratio. Thus, we have the mixture gas with different temperatures and densities behind shock waves at the detonation front. The cellular patterns are recorded on the soot film coating the surface of the plexiglas plate set in the third segment.

The propagation velocities of the incident shock and detonation waves are obtained by measuring the transit times between the piezoelectric pressure transducers P_{z1}, P_{z2}, …, P_{z5}, set at different positions of the shock and detonation tubes, as shown in Fig. 11.18.

Cellular Structure of the Detonation Waves

Figure 11.19 represents an example of Mach number M_s of the incident shock or detonation waves in the apparatus shown in Fig. 11.18 with respect to the distance L from the entrance of the rectangular tube where the detonations are produced. The propagation of the detonation waves is accelerated just after the initiation of the detonation very rapidly at first, but then decelerated slowly. The detonation tube in the third segment having a rectangular cross-section is so narrow that the gas flow behind the detonation front is decelerated by the wall friction with propagation through the tube. We can, thus, obtain cellular patterns of the detonation waves having different Mach numbers on the soot film. The temperature and density of the mixture behind the shock waves at the detonation front can be calculated from the propagation velocity and initial state of the mixture.

Figure 11.20 shows an example of the soot traces by the detonation waves at different positions where the detonation waves propagate with different velocities. Cellular patterns marked on the soot film under different Mach numbers are shown in these photographs. The formation of each apex point on which two lines intersect comes into question.

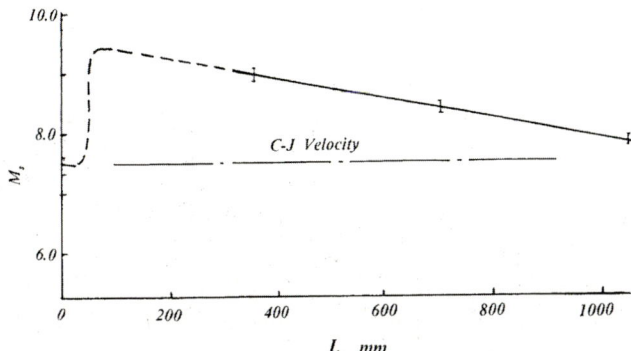

Fig. 11.19. Mach number M_s of shock or detonation propagation through the third channel shown in Fig. 11.18 in relation to the distance L from the channel entrance. Initial mixture pressure $P_0 = 23$ kPa

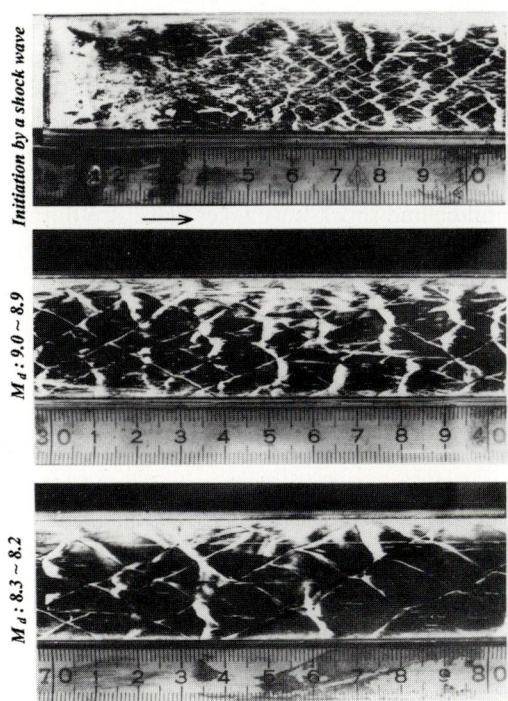

Fig. 11.20. Soot traces of the detonation waves obtained in the third tube. The scale under each picture indicates the distance from the channel entrance. M_D, Mach number of the each detonation propagation corresponding to the picture. $P_0 = 2.3$ kPa

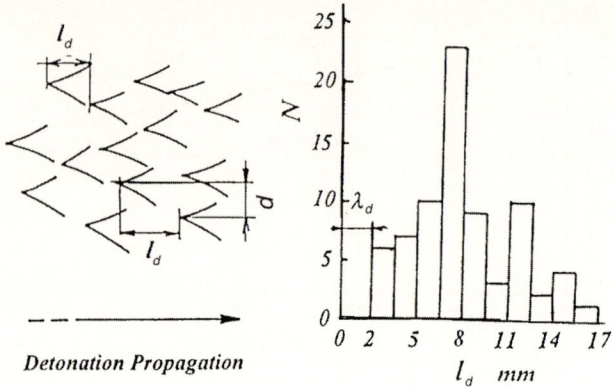

Fig. 11.21. Cellular pattern model of detonation waves (*left*) and histogram of the distance l_d between two successive apex points. *d:* Distance between two neighboring apex points perpendicular to the detonation propagation

As shown in left hand side of Fig. 11.21, the distance l_d between two successive apex points in the direction of the detonation propagation always have some fluctuation, as an example of histograms of l_d is shown in the right-hand side of Fig. 11.21. Applying (6.5) developed for calculating the probability of spontaneous ignitions behind shock waves in a shock tube to the histogram of apex point distance, the probability μ_d of apex point formation can be calculated, as both the phenomena have an analogical process, that is, the apex is formed during the shock propagation between two successive apex points, while the spontaneous ignition behind shock wave takes place also during the shock propagation from the end plate of the shock tube to the ignition point.

In order to apply (6.5') modified from (6.5),

$$\mu_d = 2V_d D \frac{\ln P(0) - \ln(l)}{Fl^2} \qquad (6.5')$$

to the apex formation in the cellular structure in detonation, the relation $P(l) = \int_l^\infty q(l)dl$, $l = l_d - \lambda_d$ where λ_d is the minimum distance of l_d, $q(l)$ the probability density of l_d obtained from the histogram of l_d, V_d the volume of 1 mol mixture behind the shock waves at the detonation front, D the propagation velocity of the detonation, and F the area where an apex point can appear and expressed by the product of two times the mean planar interval d_m between two neighboring apex points in the normal direction to the detonation propagation and half of it, that is $F = 2d_m \times d_m/2 = d_m^2$.

The diagram in Fig. 11.22 shows an example of the relation between $\ln P(l)$ and $(l_d - \lambda_d)^2$ from which the probability μ_d can be calculated.

Thus, the probability μ_d of apex point formation under different conditions, i.e., different temperatures and densities of the mixture corresponding to the propagation velocity of the detonation and the initial state is obtained. Fig. 11.23

11.2 Detonation Waves as Irreversible Phenomena

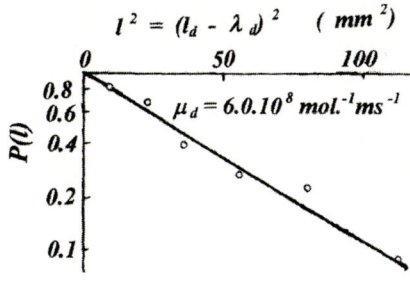

Fig. 11.22. ln $P(l)$ vs. $l^2 = (l_d - \lambda_d)^2$

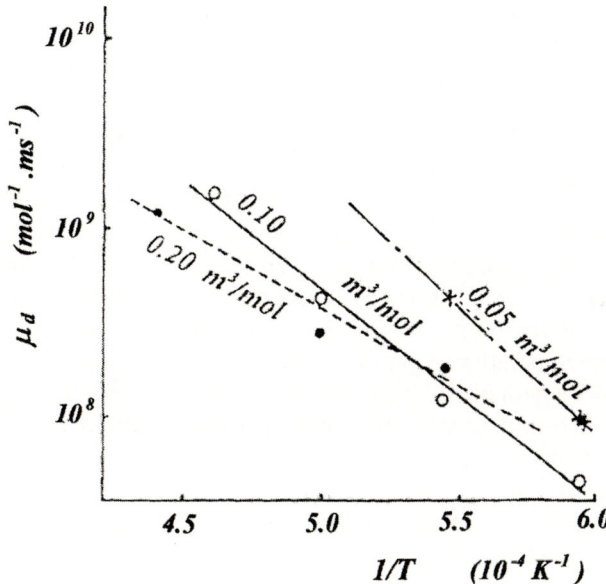

Fig. 11.23. Logarithm of apex point formation probability μ_d in the cellular structure of detonation waves propagating in a stoichiometric propane–oxygen mixture with respect to the reciprocal mixture temperature $1/T$ behind shock waves at the detonation front. Number on each line is the specific volume V_d of the mixture

illustrates the experimentally obtained μ_d of the apex point formation with respect to the reciprocal mixture temperature $1/T$ for different specific volumes V_d of the mixture behind shock waves at the detonation front.

The results are expressed with the following equation:

$$\mu_d = A_d \exp\left(-\frac{E_{ed}}{RT}\right), \tag{11.20}$$

where A_d is the frequency factor, E_{ed} the effective activation energy for apex point formation, R the gas constant, and T the mixture temperature at the detonation front. As described already in Chaps. 3 and 6, the frequency factor is expressed with (6.7), $A = b\rho^2$ and $E_e = (E_1 + E_2 - W)$ according to (3.15), in which W depends on the mixture density according to (6.8). Both A_d and E_{ed}, thus, depend on the density ρ or specific volume V_d of the mixture behind shock waves at the detonation front.

Cellular Pattern Traces on a Film of Incombustible Powder

The cellular structure of detonation waves is observed either optically or as a trace of soot-coated plate. As described already in Sect. 11.1.2, the reason for this has been attributed to a collision between shock and transverse waves behind the detonation front. We want to repeat it here briefly again. Since a discontinuity in Mach reflection made by the shock collision separates the region of different flow velocities, where the flow is supersonic on one side, but is subsonic on the other side, a considerable shear force is produced between both sides of the discontinuity. This creates a concentrated high temperature vortex and high pressure gas generated by the Mach stem which acts as a rotating needle producing the traces.

If this is correct, such a cellular pattern must be marked not only on the plate coated with soot, but also on that with other fine powders. In order to verify this hypothesis, we tried to produce such a detonation pattern on a plexiglas plate coated with an incombustible powder.

The experiment is carried out using an apparatus sketched in Fig. 11.24. A stainless steel tube of 62 mm in inner-diameter and 2,000 mm in length is filled with a stoichiometric propane–oxygen mixture under an initial pressure of 33.3 kPa and room temperature. A plexiglas plate having a thickness of 8 mm, a width of 38 mm and a length of 1,000 mm is set in the tube parallel to its axis. The plate surface is divided into several zones and each zone is alternatively coated with candle soot or calcium oxide (CaO) particles having a size of 2–30 μm.

The mixture is ignited by an electrical spark plug set at a tube end from which a flame, consequently a detonation wave propagates toward the other tube end along the plexiglas plate, marking the cellular structure on the plate surface.

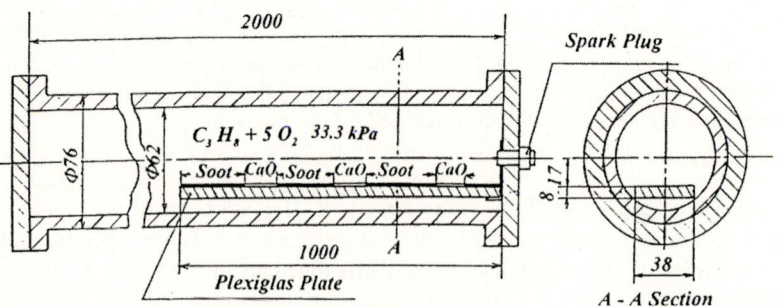

Fig. 11.24. Detonation tube for recording the cellular pattern on a film of incombustible CaO powder. Dimensions in mm

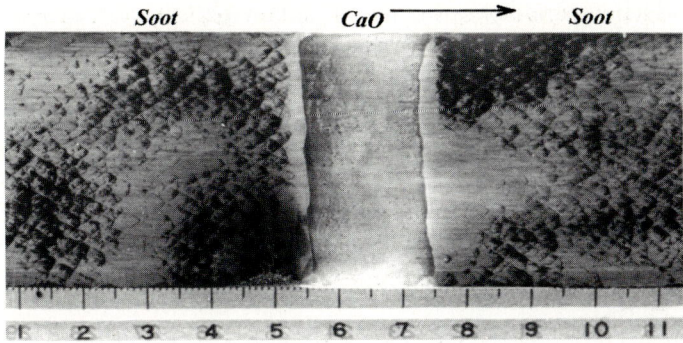

Fig. 11.25. Cellular pattern trace of detonation wave (scale in cm). The detonation wave propagates from the soot film zone over the CaO powder zone again onto soot film zone, as shown by the *arrow*

An example of the photographs of the trace marked on the plate by the detonation wave is shown in Fig. 11.25, in which the detonation wave propagates from a zone coated with candle soot over the next zone coated with CaO powder onto another zone coated with candle soot. The cellular structure of the detonation is marked on the soot film, but not on the CaO film.

Considering that the candle soot is combustible, while the CaO powder incombustible, the cellular pattern should be marked on the soot film not by the gas dynamic effect of shock collision but by the combustion of the soot. In section "Soot Trace of Combustion Waves" we also observe a trace on a soot film burned by a strong combustion.

Comparison of the Apex Point Formation with the Spontaneous Ignition

As far as the cellular structure of detonation wave is marked on a soot film by a combustion, we need to compare the cellular structure with ignition phenomenon. Therefore, the spontaneous ignition in the same mixture used for the detonation experiments, that is, in a stoichiometric propane–oxygen mixture is investigated applying a shock tube shown in Fig. 11.26.

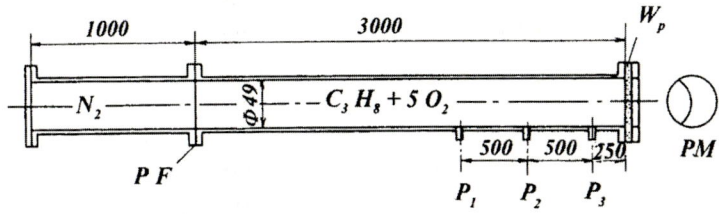

Fig. 11.26. Shock tube for investigation of spontaneous ignition behind reflected shock waves

The induction period of spontaneous ignition in the mixture behind reflected shock waves in the shock tube is again statistically investigated just like the ignition in hydrogen–oxygen mixtures described in Sect. 6.2. From the histogram of induction period of spontaneous ignition in the propane–oxygen mixture behind reflected shock waves we can obtain the probability μ_i of the ignition in one mole mixture per unit time.

The logarithm of the ignition probability $\ln \mu_i$ of the mixture having different specific volume V_i with respect to the reciprocal mixture temperature $1/T$ is illustrated in Fig. 11.27.

Just like in the propane–oxygen mixture the ignition probability μ_i is also expressed by the following equation:

$$\mu_i = A_i \exp\left(-\frac{E_1 + E_2 - W}{RT}\right)$$
$$= A_i \exp\left(-\frac{E_{ei}}{RT}\right), \qquad (11.21)$$

where A_i means the frequency factor, $E_{ei} = E_1 + E_2 - W$ the effective activation energy, and both A_i and E_{ei} depend on the density or specific volume of the mixture, as explained in Sect. 6.2.

Both probabilities of apex formation in cellular structure of detonation waves and of spontaneous ignition are expressed with the same formula. If $A_d = A_i$

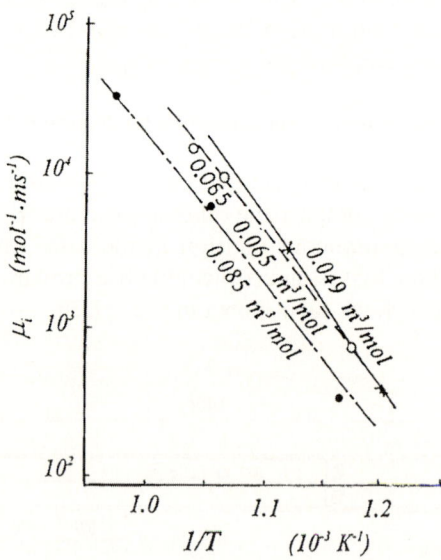

Fig. 11.27. Logarithm of ignition probability μ_i in stoichiometric propane–oxygen mixture with respect to the reciprocal mixture temperature $1/T$. Number on each line is specific volume V_i of the mixture

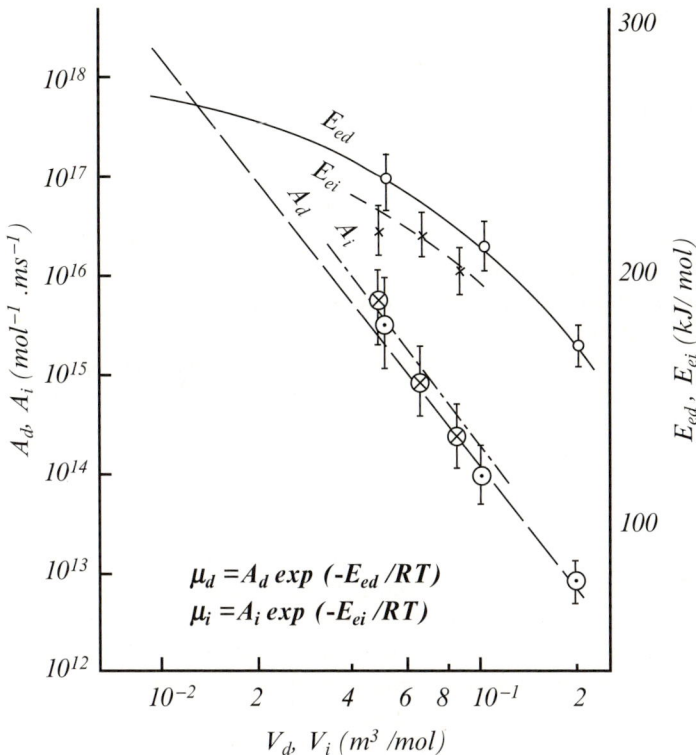

Fig. 11.28. Relations between ln A_d, ln A_i, E_{ed}, E_{ei}, ln V_d, and ln V_i obtained from the equations $\mu_d = A_d \exp(-E_{ed}/RT)$ and $\mu_i = A_i \exp(-E_{ei}/RT)$

and $E_{ed} = E_{ei}$, then $\mu_d = \mu_i$ and both the phenomena should be the same. Figure 11.28 illustrates the logarithmic relations of both the frequency factors A_d and A_i to the specific volume V_d and V_i, as well as that of the effective activation energies E_{ed} and E_{ei} to ln V_d and ln V_i. In this diagram we see that almost in the whole region of V_m, $A_d \approx A_i$ and $E_{ed} \approx E_{ei}$. From these results we can conclude, the apex point formation in the cellular structure of detonation waves to be the same phenomenon of the spontaneous ignition. Namely the apex point in the cellular structure of the detonation waves is the point of spontaneous ignition behind the shock waves at the detonation front.

Cellular Structure Formation

As the cellular pattern is marked on the soot film by combustion waves, the cellular structure of detonation should be constructed by the flames propagating from the ignition point at the apex point.

At several different positions behind shock waves at the detonation front a spontaneous ignition takes place with a certain probability corresponding to the mixture state. At each ignition point an almost isochoric combustion takes place, producing a blast wave. The blast wave having a circular flame in a two-dimensional space, spherical flame in a three-dimensional space, and high pressure zone radially or spherically propagates in its circumference, flowing downstream at first slowly, then accelerated by the flow behind the detonation front. Two envelope curves crossing at each ignition point, or an envelope surface having an apex at the ignition point are thus formed by the radially or spherically propagating flame which flows down in the gas flow behind the detonation front.

In the sector on the envelope curves or surface the gas flow against the soot film is much slower than that in the flame propagating downstream The flame is stagnated in the upward sector region much longer and soot is burned up, while that in the downstream region unburned, as schematically illustrated in Fig. 11.29. The soot in the sector region $OFCF'$ shown in No. 1 cell is burned up at different positions and the cellular patterns behind the shock waves at the detonation should be formed. In the photograph of cellular pattern marked on the soot film in Fig. 11.25 we observe such a cell having a burned region behind the apex point when detonation waves propagate through a tube having a rather large diameter.

In the cellular patterns shown in the photograph of Fig. 11.20, however, we observe only the cellular patterns composed of the envelope curves. In this case, the detonation waves propagate through a narrow channel having a distance 10 mm between two plates. Because of the interaction between the boundary layers and the blast wave propagating from the ignition point, the gas movement within the blast wave is prevented and the flame front stagnates on the envelope

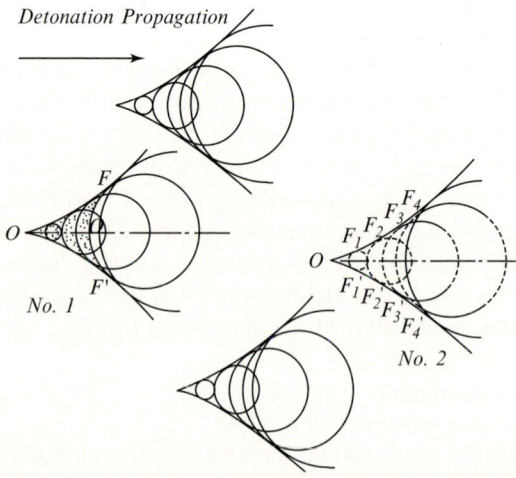

Fig. 11.29. Model of cellular structure formation in detonation waves

curves like those expressed with OF_1F_1', OF_2F_2', ... etc. in No. 2 cell in Fig. 11.28 which forms the cellular patterns.

Thus, the apex point constructing the cellular structure of detonation waves is not formed by the triple shock collision, but by a spontaneous ignition having some fluctuation in time and space. The cellular structure in detonation waves is constructed by such spontaneous ignition together with the combustion waves propagating from each ignition point spherically in the flow behind the shock waves at the detonation front.

The distribution of the apex point (spontaneous ignition point) in the direction perpendicular to the detonation propagation is also governed by the probability μ_d. Within the area having a width of the mean interval distance l_m between two successive apex points, for example, the average value d_m of the interval distance between two neighboring apex points in the normal direction to the detonation propagation has the following relation:

$$d_m = \frac{1}{\mu_d \frac{l_m d_m}{V_d} \frac{l_m}{D}}, \quad (11.22)$$

as the mean induction period of apex formation (ignition) is expressed by l_m/D and is equal to $l/(m\mu_d)$.

The hypothesis of the cellular structure formation in detonation waves by interaction between detonation front and transverse pressure waves may be applied to the detonation waves propagating through a narrow channel having a rectangular cross-section. The cellular structure is, however, observed not only in the detonation waves propagating through a narrow channel, but also through a large cylindrical tube having a diameter larger than 10 cm. What transverse pressure wave appears and how it propagates and interacts with the detonation front in such a three-dimensional space? What Mach reflection and vortex playing the role of needle takes place in such three-dimensional flow? The hypothesis might not be applied to such a three-dimensional cellular structure.

11.2.3 Interaction Between Shock and Detonation Waves[112]

In a frontal collision with a shock wave, detonation waves must be discontinuously propagate into a mixture having higher temperature and density behind the shock wave. The propagation velocity as well as the mixture state at the front of the detonation waves is suddenly changed in such a shock collision and a cellular structure different from that before the shock collision is observed on a soot film trace of the detonation waves.

Using a shock tube, the interaction between shock and detonation waves propagating in a stoichiometric propane–oxygen mixture is experimentally investigated. In this section the experiments and their results, especially the change of the propagation velocity and cellular structure in the detonation waves marked on a soot film are explained.

Experimental Method

The experiments are carried out using a shock tube connecting with a detonation tube having the same inner-diameter, but separated by a polyester film with each other as illustrated in Fig. 11.30. A stoichiometric propane–oxygen mixture is filled in the low-pressure tube of shock tube and detonation tube. Shock waves driven by high pressure He gas propagate to the right, while the detonation waves initiated by a spark set at the right end of the tube propagate to the left, and collide with each other in a certain position of the tube, as shown in the shock diagram in Fig. 11.30 above the sketch of the shock tube.

The propagation velocities of both the shock and detonation waves are measured by observing the transit times between piezoelectric pressure transducers set at several points of the tubes. Soot film traces of the detonation waves are marked on a plexiglas plate coated with candle soot.

Two shock waves having different propagation velocities, i.e., Mach number of 1.59 and 2.00 are applied to collide with detonation waves propagating in the propane–oxygen mixture approximately with Chapman–Jouguet velocity.

Propagation Velocity of Detonation Waves

The experimental results are shown in the following Table 11.1.

In this table u_1 is the propagation velocity of the shock waves, M_s its Mach number, P_1 the initial mixture pressure, w_2 the flow velocity behind the shock

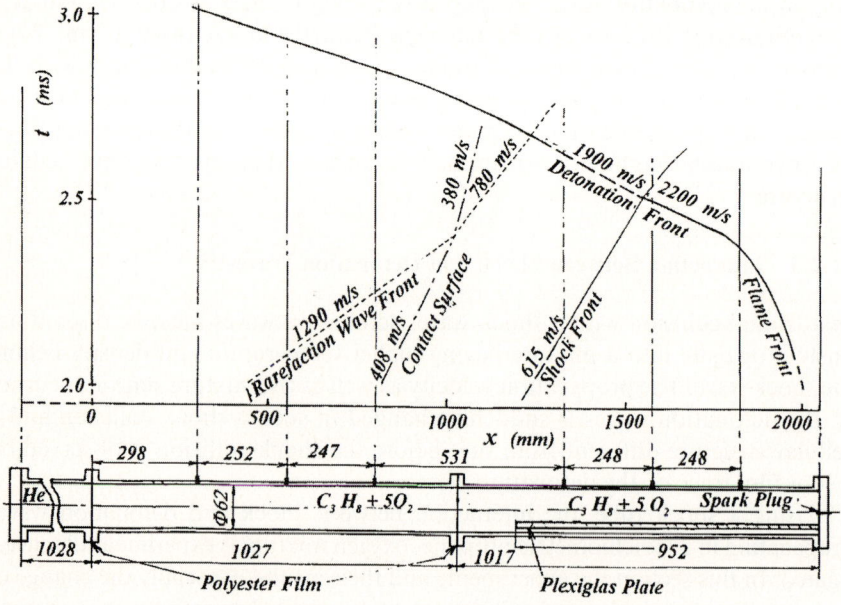

Fig. 11.30. Shock and detonation tube for the experiments (*below*) and time–distance diagram of the shock and detonation waves (*above*)

11.2 Detonation Waves as Irreversible Phenomena

Table 11.1. Observed results in detonation at shock collision

u_1 (m s^{-1})	M_s	P_1 (kPa)	w_2 (m s^{-1})	V_m (m^3 mol^{-1})	T_2 (K)	D_1 (m s^{-1})	M_{D1}	D_2 (m s^{-1})	D_3 (m s^{-1})	M_{D3}
484	1.59	40	258	29.9 × 10^{-3}	375	2,200	7.21	1,990	2,248	6.59
615	2.00	37	408	22.6 × 10^{-3}	439	2,200	7.21	1,900	2,308	6.34

waves, V_m and T_2 are the molar volume and temperature of the mixture behind the shock waves, D_1 and D_2 the propagation velocities of the detonation waves before and after the shock collision in the laboratory fixed co-ordinates, respectively, and D_3 the real propagation velocity of the detonation after the shock collision considering the flow velocity behind the colliding shock waves. M_{D1} and M_{D3} are the Mach numbers of D_1 and D_3. The measurement error should be estimated to be less than ±1.5%.

After the shock collision the detonation propagation velocity increases 2–5%, while the Mach number decreases 11–12%. According to the equation for the Chapman–Jouguet velocity described already in Sect. 11.1.1,

$$M_d^2 \approx 2(\gamma^2 - 1)\frac{Q}{a^2},$$

the Mach number of the detonation after the shock collision should decrease more, as the sound velocity behind the shock waves collided with the detonation increases 13–23%, while the propagation velocity should keep almost the same value. The experimental results suggest that the detonation propagation is accelerated by the shock collision. The detonation propagation velocity, thus, depends on the mixture temperature, while the theory of Chapman–Jouguet denies it.

Cellular Structure

Because of the sudden rise of temperature and density of the mixture in which the detonation waves propagate, we obtain a discontinuous cellular pattern on a soot film. An example of soot traces of the detonation waves at a shock collision is represented in Fig. 11.31. The cellular pattern after shock collision is much finer than that before the shock collision, namely, the density of apex points where two lines intersect with each other after the shock collision is much higher than that before it.

From each histogram of the distance l_d between two successive apex points before and after the shock collision shown in Fig. 11.32a, we obtain a diagram of ln $P(l)$ in relation to $l^2 = (l_d - \lambda_d)^2$ before and after the shock collision, respectively, as illustrated in Fig. 11.32b, from which we can further deduce the probability of the apex point formation in each case, as explained already in section "Cellular Structure of the Detonation Waves."

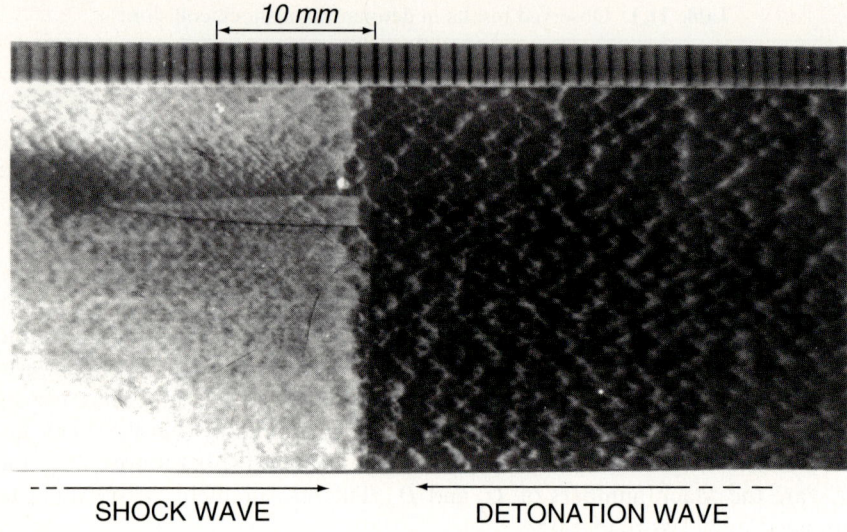

Fig. 11.31. Soot traces of detonation waves propagating in a ($C_3H_8 + 5O_2$) mixture at collision with a shock wave having a Mach number of 2.00

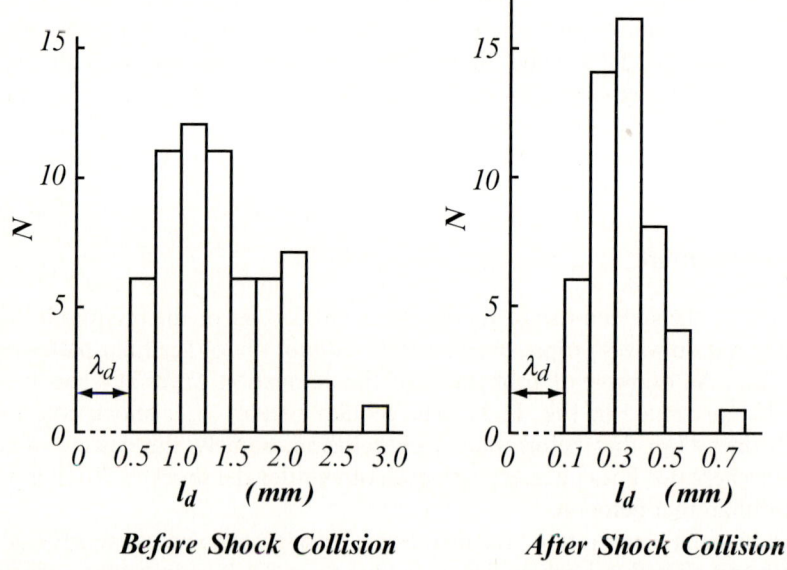

Fig. 11.32a. Histograms of distance l_d between two successive apex points in cellular patterns of detonation waves propagating in a ($C_3H_8 + 5O_2$) mixture before and after a collision with a shock wave having a Mach number of 2.00

11.2 Detonation Waves as Irreversible Phenomena

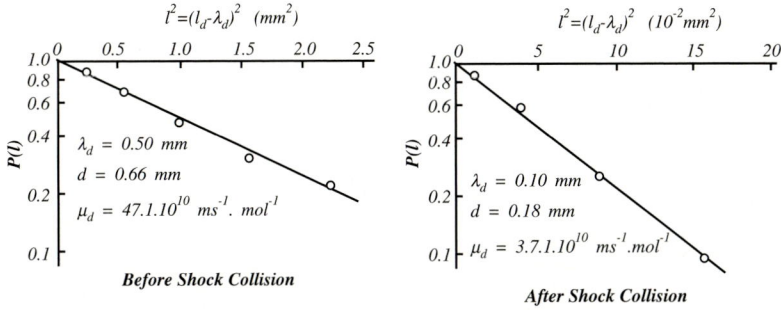

Fig. 11.32b. ln $P(l)$ with respect to $l^2 = (l_d - \lambda_d)^2$ corresponding to the histograms in Fig. 11.32a

Table 11.2. Experimental results at the shock collision with a detonation wave

		before shock collision				after shock collision			
M_s	P_1 (kPa)	D_1 (m s^{-1})	T_d (K)	V_d (m^3 mol^{-1})	μ_d ms^{-1} mol^{-1}	D_3 (m s^{-1})	T_d (K)	V_d (m^3 mol^{-1})	μ_d (ms^{-1} mol^{-1})
1.59	40	2,200	1,700	6.28 × 10^{-3}	5.9 × 10^{10}	2,248	1,780	2.93 × 10^{-3}	1.9 × 10^{12}
2.00	37	2,200	1,700	6.80 × 10^{-3}	4.7 × 10^{10}	2,308	1,905	2.27 × 10^{-3}	3.7 × 10^{12}

The results are listed in Table 11.2.

Applying these results to (11.20) described in section "Cellular Structure of the Detonation Waves":

$$\mu_d = A_d \exp\left(-\frac{E_{ed}}{RT}\right), \quad (11.20)$$

we can estimate the values of A_d and E_{ed}. If we express A_d and E_{ed} obtained above in the diagram of Fig. 11.28, we have an extended diagram of ln A_d, ln A_i, E_{ed} and E_{ei} in relation to ln V_m and ln V_d illustrated in Fig. 11.33.

Frequency factor A_d as well as the effective activation energy E_{ed} of the apex point formation probability μ_d are expressed on the extended line or curve of ln A_d or E_{ed} obtained on other experiments carried out in the same propane–oxygen mixture having lower density or less molar volume V_d. This suggests that the stochastic theory for the cellular formation in detonation waves proposed in Sect. 11.2.2 is a reasonable one.

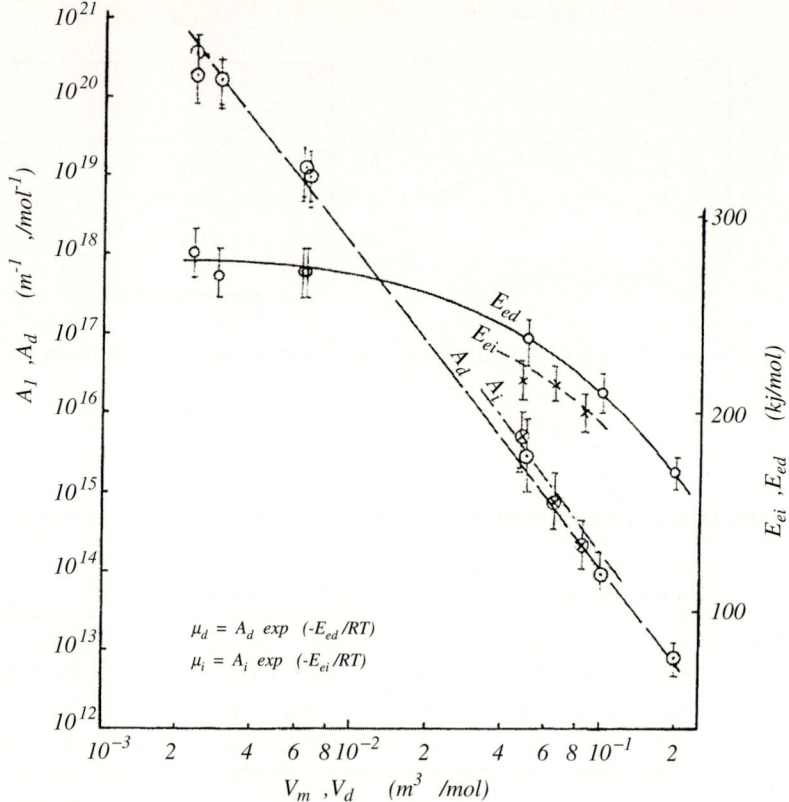

Fig. 11.33. Logarithmic relation of frequency factors A_i and A_d, effective activation energies E_{ei} and E_{ed} to the molar volumes V_m and V_d of the mixture ($C_3H_8 + 5O_2$)

11.3 Initiation of Detonation Waves

There are several different methods to initiate detonation waves, for example, transition from combustion or deflagration into detonation waves, initiation behind shock waves produced by strong electrical sparks, by focusing of a laser beam or by some explosives. The transition from deflagration into detonation waves is one of the easiest methods and has most been investigated, as described already in Sect. 11.1. In this case, however, the phenomenon has been treated as an one-dimensional one, while the transition actually proceeds in the two- or three-dimensional space and the phenomenon is much more complicated.[113–115]

In this chapter we try to investigate experimentally first a transition from combustion waves into a detonation in a two-dimensional space, then an initiation of detonation waves by shock waves. From the results we can extend our discussion to the three-dimensional phenomena.

11.3.1 Transition from Combustion to Detonation[116]

The transition from a propagating flame in a combustible mixture into a detonation wave has been explained as a self-ignition in the mixture behind the shock waves produced by the flame. As ignitions as well as detonation waves are irreversible and stochastic phenomena, the transition from combustion to detonation is expected to show some fluctuations in its physical properties. For example, the transition position as well as the induction period of transition should fluctuate corresponding to the mixture state. We consider that the investigation of these fluctuations should help us in clarifying the mechanism of the phenomena.

In this section, therefore, the induction distance from the ignition to the initiation of detonation is measured, observing soot film traces recorded by shock or explosions during the transition from the combustion propagating from the ignition point to detonation, which fluctuates over a fairly large range. By the pressure variations, propagation velocities of the combustion and shock waves and the probability of the detonation initiation obtained from the fluctuation the transition mechanism is discussed.

Experimental Method and Results

The experiments are carried out in a steel tube of 102 mm inner diameter, 1,000 mm length as shown in Fig. 11.34. A pair of plexiglas plates of 10 mm thickness, 100 mm width, and 1,000 mm length are set parallel to the axial direction of the tube. They are supported by steel plates kept 10 mm apart. The innersurfaces of both the plexiglas plates are coated with candle flame soot for recording the trace of shock waves or explosions during combustion and detonation. In one series of the experiments, a stoichiometric propane–oxygen mixture is introduced into the tube at 20.0 kPa and 22°C, while in the other series the initial pressure and temperature are regulated to be 19.5 kPa and 8.5°C,

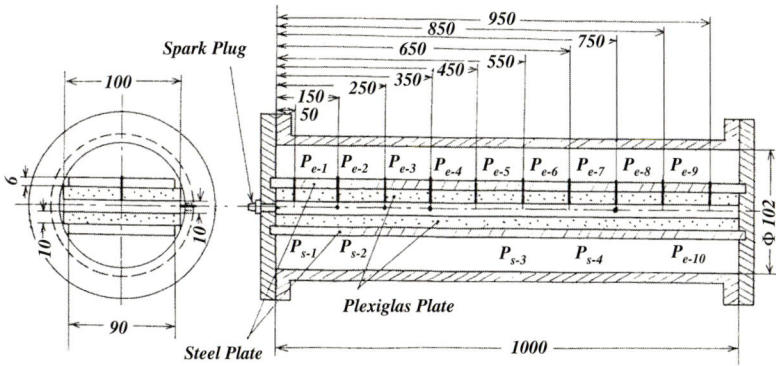

Fig. 11.34. Scheme of experimental apparatus for investigating the transition from combustion to detonation waves, dimensions in mm.

respectively, namely both the experiments are carried out in a mixture having the same specific volume or molar volume of 0.123 m^3 mol^{-1}, but different temperatures. The mixture is ignited by a spark plug set at the center of one end plate of the tube in such a way that the combustion waves in the mixture through the channel between the two plexiglas plates propagate toward the other tube end.

The pressure and the propagation velocity of the compression wave are measured as a function of distance from the spark plug. The pressure is monitored by piezoelectric pressure transducers at the different positions indicated by $P_{c-1}, P_{c-2}, \ldots, P_{c-10}$ on the center line of the channel and $P_{s-1}, P_{s-2}, \ldots, P_{s-4}$ at the channel side as also shown in Fig. 11.34 connected to an oscilloscope. The pressure varies at each measuring position in almost the same way as observed in detonation waves, namely the pressure first rises very quickly and decreases slowly, showing a peak pressure.

An example is illustrated in Fig. 11.35 of the measured results of the peak pressure and the propagation time of the compression wave produced by the combustion wave with respect to the distance from the ignition point at the end plate of the tube.

The pressures of the mixture calculated from the propagation velocity of the compression wave agree fairly well with the measured ones.

The results suggest that a shock wave produced by the combustion wave at a certain position between 100 and 200 mm from the spark plug propagates with a delay of several hundred microseconds after the spark ignition, keeping its pressure and propagation velocity almost constant over a distance of about 200 mm.

An example of the soot traces and the scheme of the shock formation process is given in Fig. 11.36. This shows that initially a flame propagates through the space between the plexiglas plates, followed by the production of

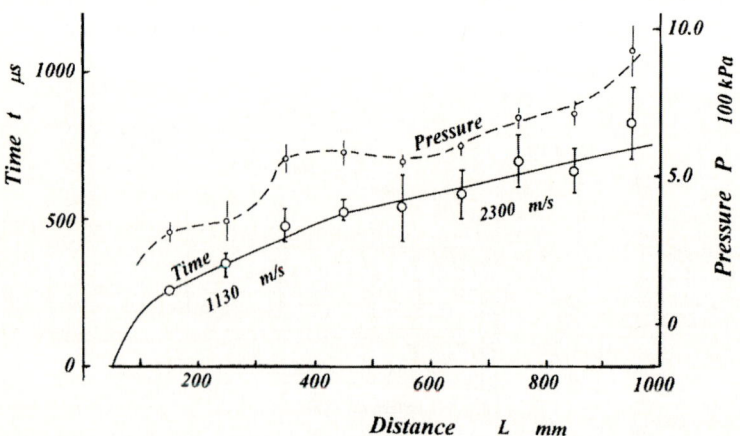

Fig. 11.35. Peak pressure and position of the front of the combustion wave propagating in a stoichiometric propane–oxygen mixture with respect to the distance from the spark plug. Initial mixture temperature is 8.5°C and pressure 20 kPa

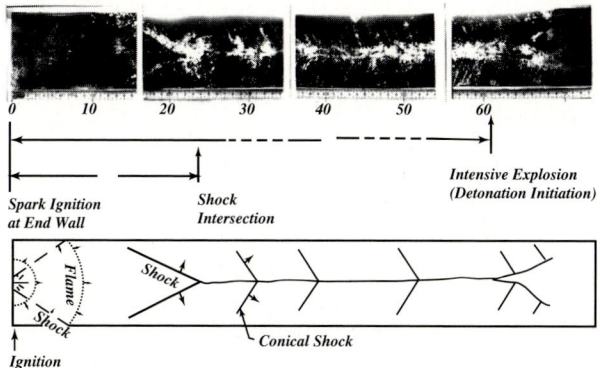

Fig. 11.36. Soot film trace at the transition from combustion to detonation in a stoichiometric propane–oxygen mixture having an initial temperature of 8.5°C and pressure of 20 kPa, and scheme of shock formation process (*below*). The scale shows the distance from the ignition point in cm

two almost symmetrical shock waves on both sides of the space. These shock waves intersect each other and propagate further, forming a conical shock wave. The measured pressure of the mixture and the propagation velocity of the compression wave indicate that a combustion wave also propagates behind the conical shock wave.

An intense explosion is observed after the conical shock wave had propagated for several decimeters. A cellular pattern usually observed in the detonation wave is formed following the explosion and spreads. Note that the cellular structure always appears in the detonation wave, while the combustion waves have no cellular structure. The point where first such a cellular pattern is observed, therefore, should indicate the beginning of the transition into detonation, i.e., the initiation of detonation.

Thus, the distance l_i from the spark plug to the point of intense explosion is recognized as the induction distance of the detonation initiation. The induction distance l_i fluctuates over a large range, as mentioned previously. Several experiments are carried out under different conditions, but each experiment, therefore, is repeated more than 50 times under the same conditions and the histogram of the induction distance of the detonation initiation in each experiment is obtained. Figure 11.37 shows an example of such a histogram.

Probability for Initiation of Detonation

Since the transition from combustion to detonation is caused by an ignition behind shock waves, it is expected that the probability of this transition can be obtained in the same way as that of the ignition behind shock waves. As described already in Chap. 6, the probability μ for the ignition behind shock waves in 1 mol mixture and in unit time is expressed by the following formula:

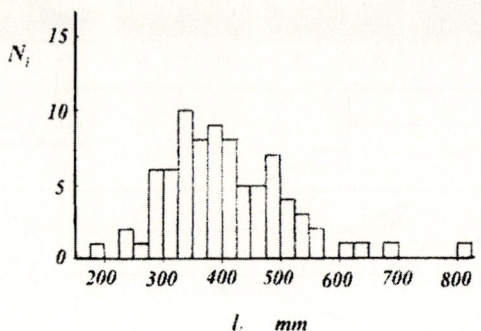

Fig. 11.37. Frequency of detonation initiation N_i with respect to distance l_i from the ignition point in a stoichiometric propane–oxygen mixture having an initial temperature of 22°C and pressure of 20 kPa

$$\mu = 2V_m D \frac{\ln P(0) - \ln P(l)}{Fl^2}, \tag{6.5}$$

where V_m is the molar volume of the mixture corresponding to the state behind the shock wave, D the propagation velocity of the shock, F the cross-section area of the space in which the shock propagates, and $P(l)$ the probability of ignition occurring at position beyond l.

Two examples of the relations of $\ln P(l_i)$ in the mixture of 22 and 8.5°C to the induction distance l_i of the detonation initiation is obtained from the histogram described above, as illustrated in Fig. 11.38. In each diagram there is a certain distance λ_i where no detonation initiation is recorded, because the

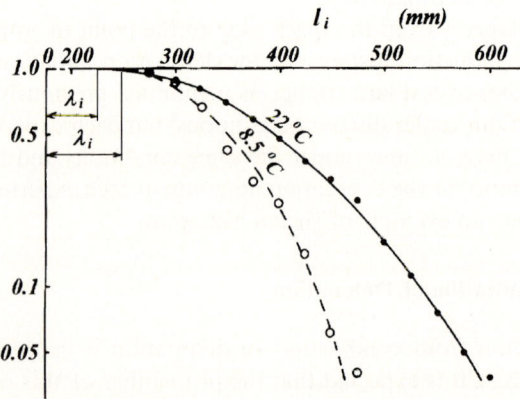

Fig. 11.38. $\ln P(l_i)$ with respect to distance l_i. The temperature on each curve means the initial mixture temperature and λ_i the distance where no detonation initiation is observed

11.3 Initiation of Detonation Waves 261

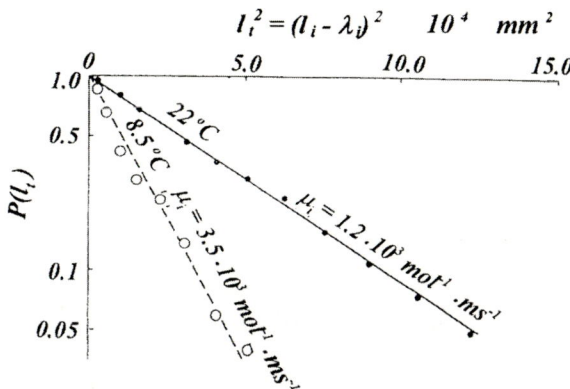

Fig. 11.39. ln $P(l_t)$ with respect to $l_t^2 = (l_i - \lambda_j)^2$ for detonation initiation. The temperature on the lines means the initial mixture temperature and μ_i probability for the detonation initiation

shock wave is produced only when the flame propagated a certain distance, two weak shocks produced in the corners intersect each other and form a conical shock. Consequently, $(l_i - \lambda_j)$ is used in place of l in (6.5).

Figure 11.39 represents ln $P(l_t)$ as a function of $l_t^2 = (l_i - \lambda_j)^2$. The probability for the initiation of detonation μ_i in 1 mol mixture and in unit time is obtained from the slope of ln $P(l_t)$ against l_t^2 according to (6.5). Thus, μ_i is estimated to be 1.2×10^3 per (ms mol) in the mixture having an initial temperature of 22°C and 3.5×10^3 per (ms mol) in the same mixture but at 8.5°C. The transition from combustion to detonation in a mixture with lower initial temperature proceeds more easily than in that of a higher initial temperature. In the mixture of lower initial temperature at 8.5°C the shock wave produces by the combustion propagates with a Mach number of 3.8, while that in the mixture of higher temperature 22°C with a Mach number of 3.3. The temperature behind the shock waves is estimated to be 860 K in the mixture of lower initial temperature, which is higher than 760 K estimated in the mixture of higher initial temperature 22°C.

The probability for the detonation initiation can be expressed by the following Arrhenius' formula analogous to the probability of ignition or that of apex formation in the cellular structure in detonation waves explained in Chap. 10:

$$\mu_i = A \exp\left(-\frac{E_e}{RT}\right), \quad (11.23)$$

where A is the frequency factor, E_e the effective activation energy, R the gas constant, and T the mixture temperature behind the shock wave. According to this equation E_e is estimated to be 60 kJ mol^{-1} from the experimentally obtained relation of ln μ_i to the reciprocal mixture temperature $1/T$ illustrated in Fig. 11.40. The effective activation energy E_{ei} for spontaneous ignition in the same stoichiometric propane–oxygen mixture of the same molar volume (0.123 m mol^{-1}) is estimated to be 270 kJ mol^{-1}, as shown in Fig. 11.33.

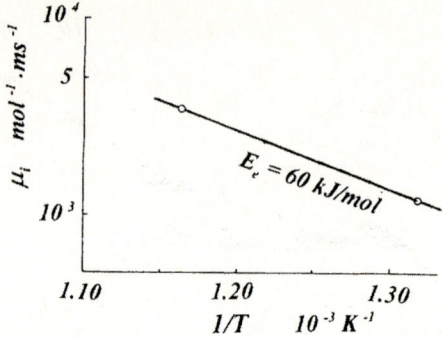

Fig. 11.40. Logarithm of the probability for the detonation initiation μ_i with respect to the reciprocal mixture temperature $1/T$. E_e is the effective activation energy

These results suggest that the transition from combustion to detonation is initiated not by a completely spontaneous ignition, but by an ignition to which some energy $(E_{ei} - E_e) = 210$ kJ mol^{-1} is supplied from the flame behind the shock wave. The next photograph in Fig. 11.41 shows a soot trace of a detonation initiation, in which the top of the flame propagating behind the shock wave is split into two tips. From each tip of the flame a detonation wave is initiated, spreading

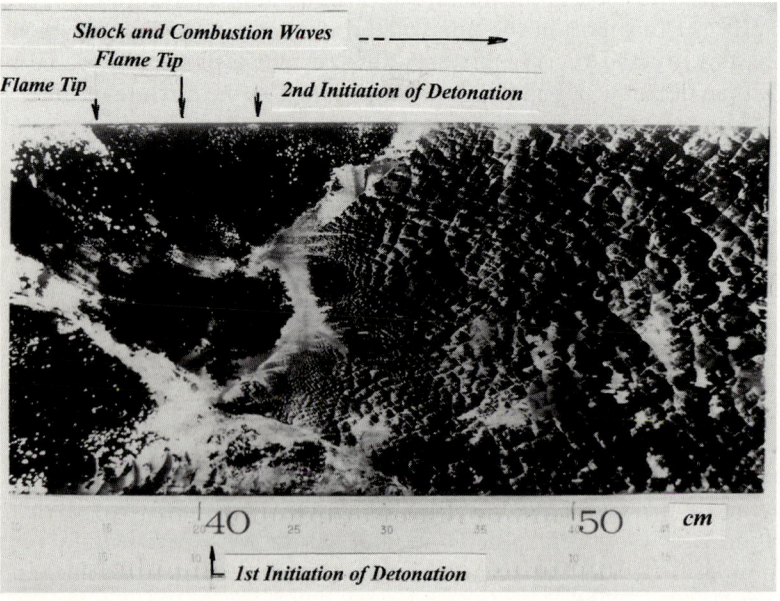

Fig. 11.41. Soot trace of a detonation wave initiated at two points in a stoichiometric propane–oxygen mixture

11.3 Initiation of Detonation Waves

its cellular structure region. This proves that the transition from combustion to detonation takes place supported by the energy supplied from the flame as explained above. We remember Sect. 9.3 in which a lot of high energy particles are observed also in the front of the flame (Fig. 9.28).

Development from the Initiation to the Self-sustained Detonation

Figure 11.42 shows one of the photographs of soot traces marked by detonation waves during the development from initiation to self-sustained detonation waves. From the initiation point the region of cellular pattern radially spreads in a sectorial form, enlarging the cell size, and develops to the self-sustained detonation.

The probability of the apex point formation in the cellular pattern is obtained from the histogram of the interval l_{ts} between two successive apex points, as explained already in Sect. 11.2.2. From such a photograph of cellular pattern we can obtain the histogram of l_{ts} on different radii r from the detonation initiation point. Figure 11.43 shows an example of such a histogram on a radius r = 59 mm.

From such histogram we can obtain a diagram of $\ln P(l_{ts})$ in relation to $l_{ts}^2 = (l_{ts} - \lambda_{ts})^2$, in which $\lambda_{ts} \approx 0$ in this case, as illustrated in Fig. 11.44, in which $P(l_{ts})$ is the probability of the apex point formation in a distance larger than l_{ts}, while λ_{ts} is the minimum l_{ts}. From these relations we further calculate the

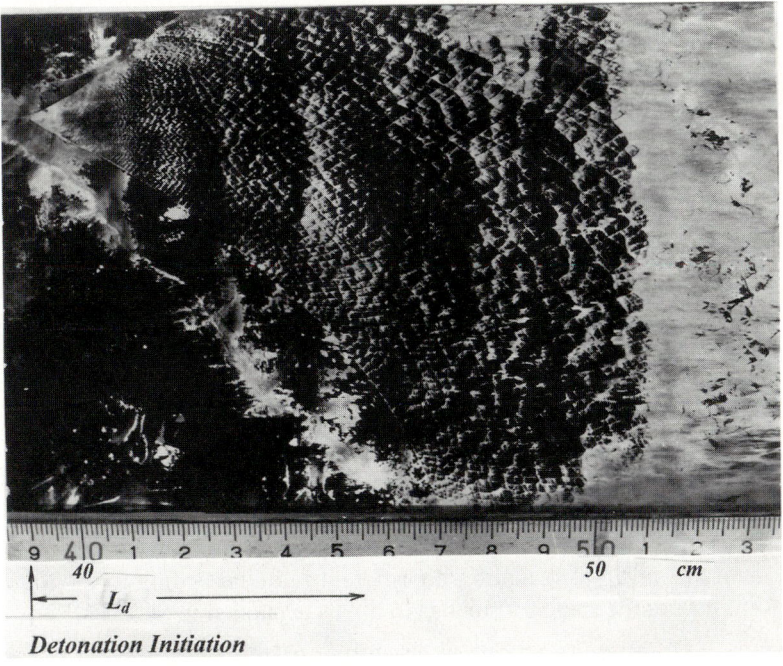

Fig. 11.42. Soot film of a developing detonation wave

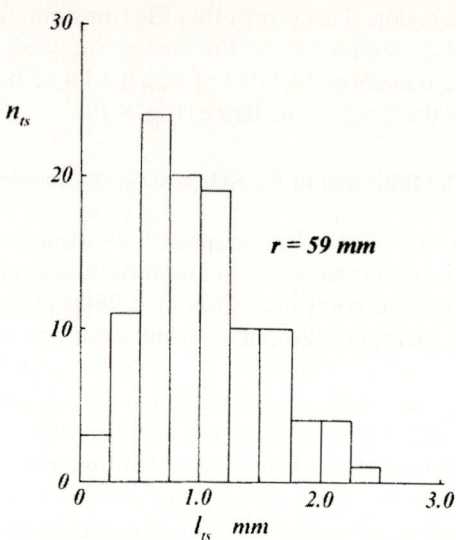

Fig. 11.43. Histogram of apex point at a distance $r = 59$ mm from the detonation initiation point. l_{ts} is the interval between two successive apex points. Initial mixture temperature: 22°C

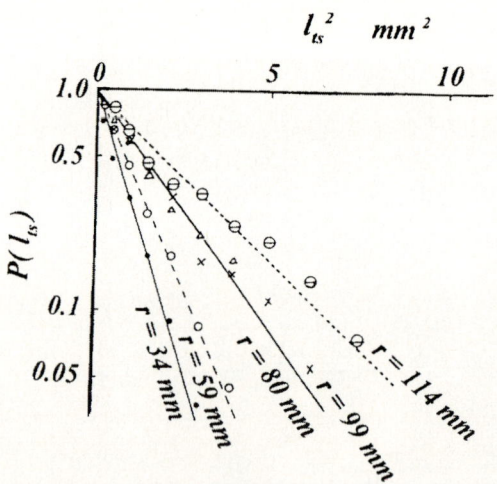

Fig. 11.44. ln $P(l_{ts})$ vs. $(l_{ts} - \lambda_{ts})^2$. The number on each line is the distance from the detonation initiation point in mm, $\lambda_{ts} \approx 0$

probability μ_d of apex formation on different radii, according to the following equation having the same formula as (6.5′), as explained in Sect. 11.2.2.

$$\mu_d = 2V_{ts} D \frac{\ln P(0) - P(l_{ts})}{Fl_{ts}^2}. \qquad (6.5')$$

11.3 Initiation of Detonation Waves

The probability μ_d is expressed by the following equation, as also described in Chap. 10:

$$\mu_d = A_d \exp\left(-\frac{E_{ed}}{RT}\right). \tag{11.20}$$

From μ_d we can estimate the mixture temperature behind the shock wave at the detonation front, as far as the values of A_d and E_{ed} are known. Both the values depend on the density or molar volume of the mixture and are obtained in the diagram in Fig. 11.33, while the mixture density can be estimated from the Mach number of the shock or detonation wave propagating in the mixture.

In Fig. 11.45, experimentally obtained μ_d and mixture temperature T behind shock waves at the detonation front estimated from the values of μ_d are illustrated with respect to the distance from the spark ignition point. Just downward from the detonation initiation point the mixture temperature is as high as seven times the initial mixture temperature and decreases with propagation and enlargement of the detonation front. The peak pressure P_c measured by the piezoelectric pressure transducers P_{c-1}, P_{c-2}, ..., P_{c-10} on the center line of the detonation channel shows a very high pressure at the initiation point having a magnitude about 10 times of the initial one, then decreases with the propagation and enlargement of the detonation just like the temperature variation.

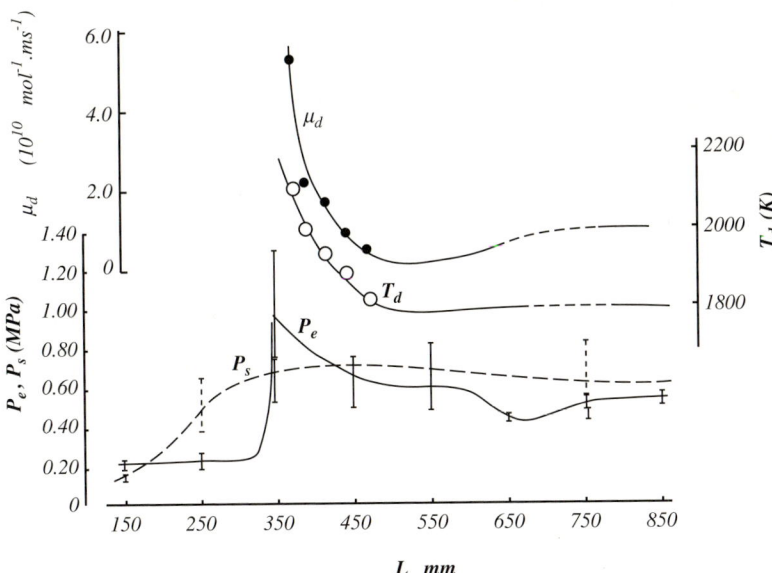

Fig. 11.45. Apex point formation probability μ_d, the mixture temperature T_d behind shock waves at the detonation front estimated from μ_d, the peak pressure P_c measured on the center line and that P_s measured on the side of the channel with respect to the distance L from the ignition point

Both the variations of the temperature and pressure have almost the same tendency and they suggest the same process of the mixture. From the results we can conclude, thus, that at the initiation point a very intense explosion caused by an isochoric combustion takes place, showing a very high pressure and temperature, then a cylindrical or spherical shock wave propagates radially in the mixture, decreasing the pressure and temperature by an adiabatic expansion. The detonation region having a cellular structure is formed and radially or spherically enlarged in the flow behind the main conical shock wave produced by the combustion before the detonation initiation, and consequently the cellular region has a conical or sectorial form, as shown in the photograph in Fig. 11.41.

The propagation velocity of the shock wave accompanied by a cellular pattern and radially propagating from the detonation initiation point can be estimated from the flow velocity behind the main conical shock wave and the slope of the developing detonation region having a sectorial form against the propagation direction of the conical shock. In Fig. 11.46 such propagation velocities W_n of the shock waves radially propagating from the detonation initiation point in both the mixtures of 8.5 and 22°C in initial mixture temperature are shown with respect to the distance L_d from the detonation initiation point. The radially propagating shock wave produced by the intense explosion is accelerated during the development of the detonation up to the Chapman–Jouguet velocity, decreasing the pressure and temperature.

Summarizing the experimental results, the process of the transition from combustion to detonation in a closed tube is concluded as follows:

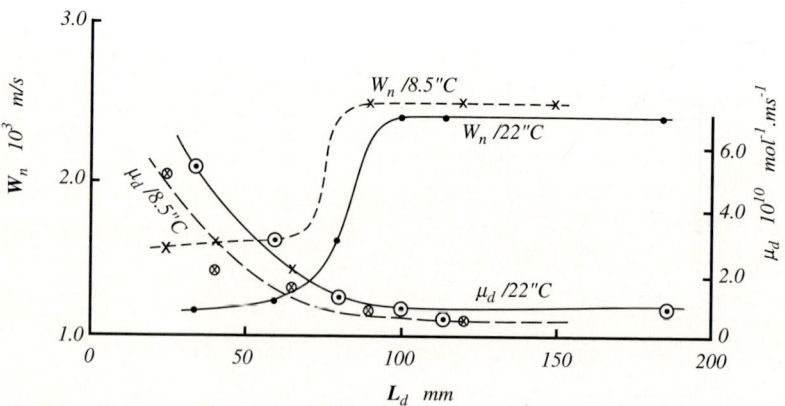

Fig. 11.46. Propagation velocity W_n of the radial shock wave from the detonation initiation point and the apex point formation probability μ_d in the developing detonation with respect to the distance L_d from the detonation initiation point. The initial mixture temperature is noted on each curve

11.3 Initiation of Detonation Waves

1. By an ignition of a detonable mixture at one end of the tube, a flame, consequently a flow behind it propagates in the mixture, producing a compression wave.
2. Reflecting from the side walls, the compression wave produces a conical shock wave which propagates further, being supported by the combustion wave behind it.
3. After the conical shock wave has propagated a certain induction distance, an intense explosion takes place at a position behind the conical shock wave, affected by the combustion wave, where a detonation wave is initiated.
4. A cylindrical or spherical shock wave propagates radially from it, trailing and overtaking the preceding shock wave and increasing its propagation velocity.
5. The cellular structure is formed behind the radial or spherical shock wave.
6. In the flow behind the main conical wave caused by the combustion wave the region of the detonation wave having a cellular structure is first enlarged in a sectorial or conical form, and then spreads in the whole space, accelerating its propagation to the Chapman–Jouguet velocity.

Pseudodetonation Wave

According to the definition of detonation waves, a combustion wave following a shock wave in a detonation wave drives the shock wave by its combustion energy. In a steady detonation wave the combustion wave takes place by spontaneous ignitions behind the shock wave which occur everywhere apparently at random, forming a cellular structure, while each state of the mixture and combustion gas appears on the Hugoniot-curve according to the Rankine–Hugoniot equations.

As described already in section "Development from the Initiation to the Self-sustained Detonation," a shock wave propagates in the mixture with a Mach number of 3–4 during the transition from the propagating flame to the detonation wave and a combustion wave follows the shock wave. According to the definition, this shock wave accompanied by a combustion wave should also be called a detonation wave. But such a process is quite different from the normal detonation, as

1. The combustion wave behind the shock wave takes place not by spontaneous ignition, but by flame propagation
2. The phenomenon proceeds not homogeneously and cannot be expressed by the one-dimensional Rankine–Hugoniot equations

We call such a propagating shock wave driven by a combustion wave a pseudodetonation wave. In real detonation waves sustained by the combustion developed from spontaneous ignition behind shock waves at the detonation front we always observe some cellular structure marked on a soot film, while in a pseudodetonation such a cellular structure is never observed. Such a pseudodetonation takes place in a mixture having a high temperature and density where the

268   11 Gaseous Detonation Waves

combustion reaction proceeds faster than the sound velocity showing an almost isochoric process.

At the transition from combustion to detonation wave in a two-dimensional space of a flat channel like that used in the experiment in this section, the shock waves produced by the combustion waves reflect from the side walls and converge on the center line of the channel forming a conical shock wave, as far as the mixture ignites at the center of the channel end wall. The mixture on the center line is compressed and heated by reflected and converging shock waves. The pseudodetonation driven by a combustion wave, therefore, propagates in the mixture on the center line having a high temperature and density than that in the other region. As the combustion is, however, a stochastic phenomenon, the top of the pseudodetonation propagates not correctly on the center line of the channel, but fluctuating according to the mixture conditions, sometimes splitting the top of the propagating flame, as described already.

Figure 11.47 shows a photograph of a soot trace marked at a transition from combustion to detonation, in which the pseudodetonation produced by the combustion propagates not on the center line of the channel, but shifted from it. The fluctuation of the detonation initiation position as well as the shock intersection position is expressed by the probability density as shown in Fig. 11.48. Both the detonation initiation and shock intersection position fluctuates over a large range in the wave propagation direction but also in the direction perpendicular to the propagation.

If the spark ignition position is shifted from the center to the side, the shock waves reflected from the side walls cannot converge well on the line and the transition to the detonation hardly takes place, even if pseudodetonation is observed, as the pseudodetonation is not strong enough to produce a strong explosion in which a detonation wave is initiated.

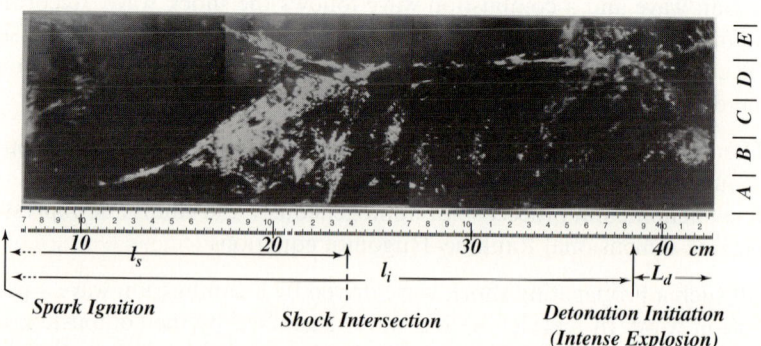

Fig. 11.47. Soot film trace at a transition from combustion to detonation in a stoichiometric propane–oxygen mixture in which a pseudodetonation wave propagates deviating from the center line of the channel. The initial mixture temperature is 22°C and pressure 20 kPa

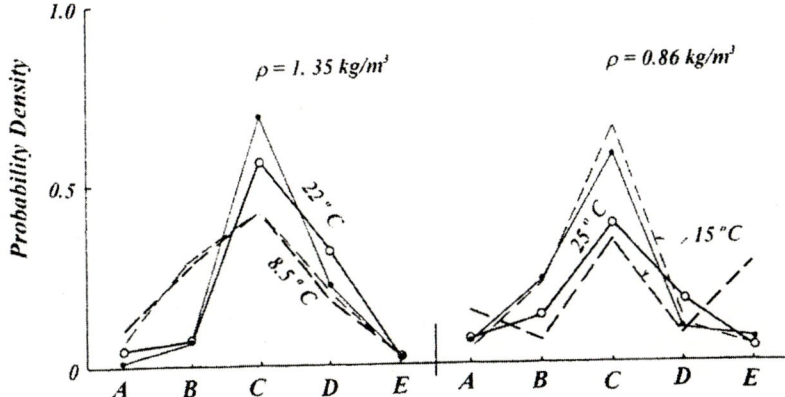

Fig. 11.48. Probability density of detonation initiation point (*thick lines*) and that of shock intersection point (*thin lines*) with respect to the segment A, B, C, D, and E divided in the direction perpendicular to the wave propagation, as shown in Fig. 11.48. ρ: the initial mixture density, temperature on each line

At the transition from combustion to detonation in a mixture in a tube having a circular cross-section, the shock waves produced by the combustion waves propagated from the ignition point at the center of the tube end reflect from the circular tube wall and converge to the axis of the tube, on which a pseudodetonation propagates, producing a strong explosion and initiates a detonation wave. The three-dimensional convergence produces a much stronger pseudodetonation, and therefore the induction distance of the detonation initiation is much shorter than that in two-dimensional channels.

The larger the diameter of the tube, the longer the distance to the shock intersection and formation of the conical shock, but the stronger the pseudodetonation and the shorter the induction distance of the detonation initiation. Therefore, there is an optimum diameter to have the minimum length of the induction distance of the detonation initiation corresponding to the mixture state.

Applying several ignition plugs we can produce a detonation wave within a smaller induction period and distance, since by this method some pseudodetonation waves are produced which intersect with each other, causing several strong explosions and initiating a detonation wave more easily.

As explained already in section "Cellular Structure Formation," a conical combustion wave propagates from each ignition point behind shock wave at the detonation front. In a combustible mixture having a state near the border line of the detonation limit, the spontaneous ignition takes place only at one or a few point behind the shock wave at the detonation front. If such a spontaneous ignition occurs near the wall, the combustion wave propagating from the point interacts with the main shock wave. At the interaction point a shock wave driven by the combustion wave, i.e., a pseudodetonation wave propagates, making a spiral trace on the soot film coating the inner side of the tube wall. Such a phenomenon is called a spinning detonation. Such spinning detonation waves have different aspects according to the mixture states and form of the tube or channel.

11.3.2 Transition from Shock to Detonation Waves

As described already in Sect. 11.3.1, at the transition from combustion to detonation waves, the detonation waves are initiated behind shock waves produced by the combustion waves, or as explained in Sect. 11.2, the detonation waves are sustained by the spontaneous ignition behind shock waves at the detonation front, the detonation waves can be directly initiated behind shock waves produced using a shock tube. An example of the shock-induced detonation is already briefly explained in section "Shock-Produced Detonation Waves of Different Strengths."

In this section the experimental method and results of the initiation of detonation behind shock waves produced in a shock tube are explained more detailed.

Experimental Method

The experimental method using a two-stage shock tube has been already mentioned in section "Shock-Produced Detonation Waves of Different Strengths." The principle of the method applying the shock tube shown in Fig. 11.18 is again explained as follows using the shock diagram in Fig. 11.49.

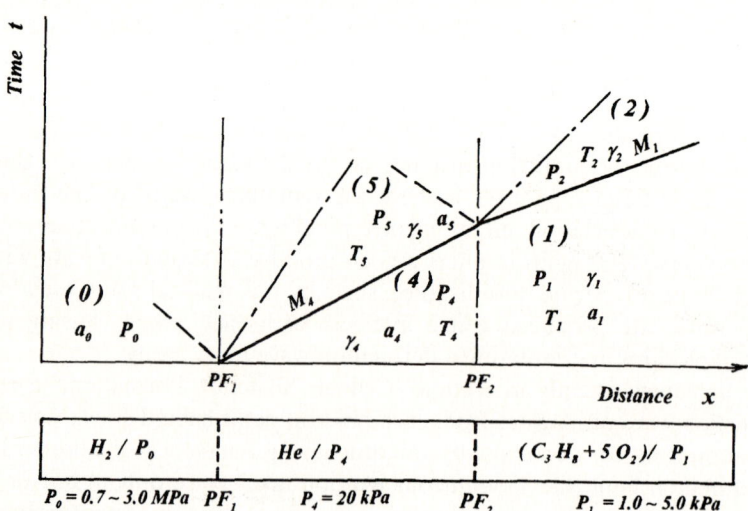

Fig. 11.49. Shock diagram in a two-stage shock tube. Driver gas: H_2 of a pressure $P_0 =$ 0.70–3.0 MPa, first driven gas: He of a pressure $P_4 = 20$ kPa, second driven gas: $C_3H_8 + 5O_2$ of a pressure $P_1 = 1.0$–5.0 kPa, PF_1, PF_2: diaphragms of polyester film, M_1, M_4: Mach number, γ: ratio of specific heats

11.3 Initiation of Detonation Waves

A shock tube driven by hydrogen gas having a pressure between 0.7 and 3.0 MPa under room temperature of about 20°C propagates in He keeping a pressure of 20 kPa. Assuming that the second diaphragm PF_2 is broken without any resistance and stagnation of the flow behind the shock wave as soon as the shock reaches the diaphragm, we can obtain the Mach number M_1 of the incident shock wave in the propane–oxygen mixture in the last stage of the shock tube, applying the elementary equations of the shock wave.

Namely from (5.13)

$$\frac{P_5}{P_4} = \frac{2\gamma M_4^2 - (\gamma_4 - 1)}{\gamma_4 + 1} \qquad (11.24)$$

and from (5.17)

$$\frac{P_5}{P_1} = \frac{2\gamma_1 M'^2 - (\gamma_1 - 1)}{\gamma_1 + 1} \frac{1}{\left(1 - \frac{\gamma_5 - 1}{\gamma_1 + 1} \frac{a_1}{a_5} \frac{M'^2 - 1}{M'}\right)^{2\gamma_1/(\gamma_5 - 1)}}, \qquad (11.25)$$

where $M' = M_1 - M_4$, P is the pressure, T the temperature, γ the ratio of specific heats, and a the sound velocity of each gas in the shock tube, the subscript 0 means the initial state of the driver gas H_2, 1 that of the stoichiometric propane–oxygen mixture in the last stage, 4 that of He gas, 2 the state of the gas behind shock waves propagating in the propane–oxygen mixture, 5 that propagating in He, M_1 Mach number of the shock waves incident into the propane–oxygen mixture, while M_4 that propagating in He. From these relations, we can estimate the Mach number M_1 of the shock wave incident into the mixture in the last stage of the shock tube, i.e., detonation tube, where detonation waves should be initiated, as some examples shown in Fig. 11.50.

The propagation of the shock waves in He gas as well as that of the detonation waves are observed by the piezoelectric pressure transducers $P_{z-1}, P_{z-2}, \ldots, P_{z-5}$ set on the shock tube, while the cellular structure of the detonation is recorded on the soot film coated on the surface of the plexiglas set in the detonation tube, as illustrated in Fig. 11.18.

From the results observed by the pressure transducers and the cellular pattern marked on the soot films, we obtain the Mach number of the shock and detonation waves propagating in the propane–oxygen mixture, as described later in section "Propagation Velocity of Shock-induced Detonation Waves."

Soot Traces of Shock-Induced Detonation Waves

Several photographs of soot traces marked on the plexiglas plate by the detonation waves initiated by shock waves and propagating in the stoichiometric propane–oxygen mixture in the detonation tube under different conditions are represented in Figs. 11.51–53.

In Fig. 11.51, (1) shows an example of soot traces marked by detonation waves initiated by a shock wave incident into the mixture having an initial pres-

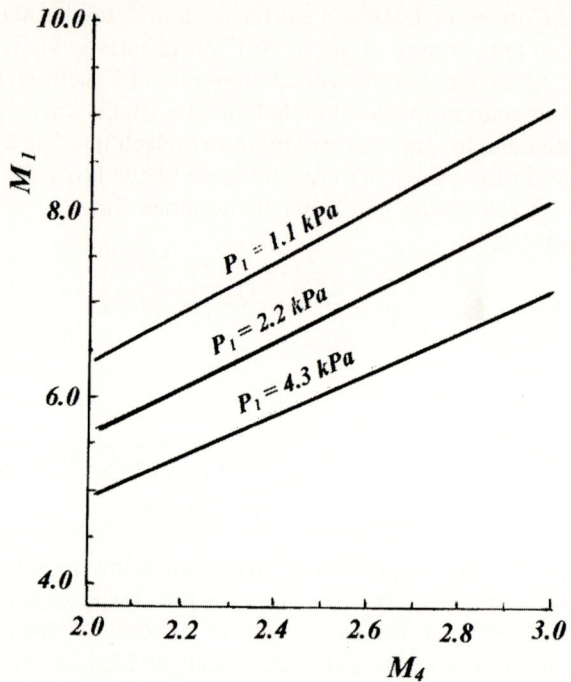

Fig. 11.50. Mach number M_1 of a shock wave propagating in $(C_3H_8 + 5O_2)$ mixture in the second stage with respect to that M_4 of a shock wave propagating in He in the first stage of the driven side of the two-stage shock tube. Initial mixture pressure is on each line

sure of 2.2 kPa with a Mach number $M_1 = 7.5 \pm 0.2$ which is almost equal to that M_{C-J} of the Chapman–Jouguet velocity. The cellular pattern appears first at a distance $L = 3.0$ cm from the entrance of the detonation tube, while the detonation wave is estimated from the cellular pattern to be established near $L = 5.0$ cm quite smoothly without any explosion like that always observed at the transition from combustion to detonation waves.

In (2) of Fig. 11.51 showing a soot trace of a detonation initiated by a shock wave of $M_1 = 7.6$, almost equal to M_{C-J}, but into the mixture having a lower pressure $P = 1.1$ kPa, we observe a little longer induction distance to the establishment of the detonation wave at $L = 6.0 - 7.0$ cm, as the mixture density is lower and the probability of spontaneous ignition in the mixture behind the shock waves is less. The cellular pattern is also coarser because of the same reason.

In the case of the transition to detonation from a shock having a Mach number M_1 much higher than M_{C-J} of the Chapman–Jouguet velocity, the cellular pattern first appears after a shorter induction distance, observed at $L = 2.0$–3.0 cm and the cellular structure is much finer because of the higher mixture temperature behind the shock wave, as shown in (3) of Fig. 11.51.

In Fig. 11.52 two examples of the transition to detonation waves by a shock wave having a Mach number lower than M_{C-J}. In (4) the detonation initiated by a

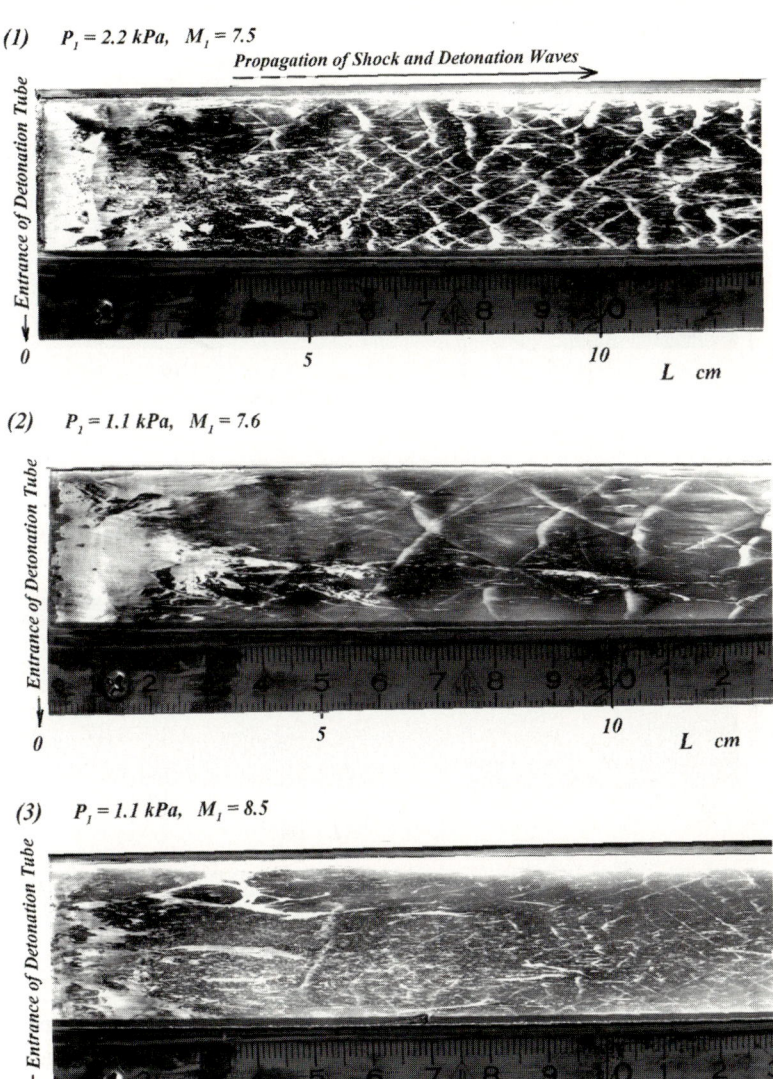

Fig. 11.51. Soot traces of a transition from shock to detonation wave. $M_1 > M_{C-J}$. P_1, initial pressure of $(C_3H_8 + 5O_2)$ mixture; M_1, Mach number of the shock incident into the mixture; L, distance from the entrance of the detonation tube

shock of $M_1 = 6.6$, the cellular pattern near $L = 4.0$ cm following a strong explosion, while in (5) the detonation is initiated by a shock wave having a lower Mach number $M_1 = 5.3$, in which the cellular pattern is observed at $L = 2.0$–3.0 cm, but very irregular because of the lower mixture temperature behind the shock waves.

(4) $P_1 = 4.3$ kPa, $M_1 = 6.6$

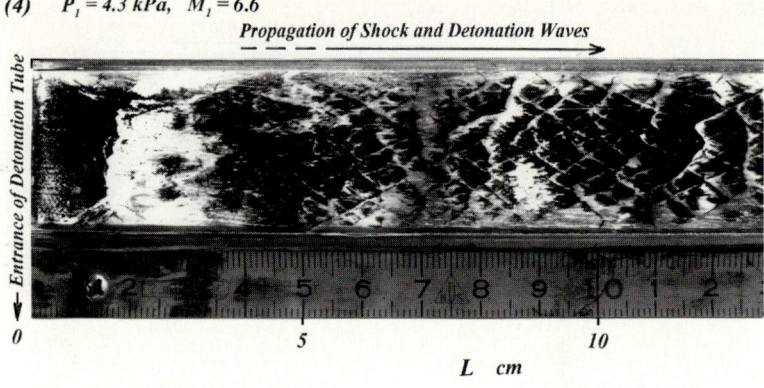

(5) $P_1 = 4.3$ kPa, $M_1 = 5.3$

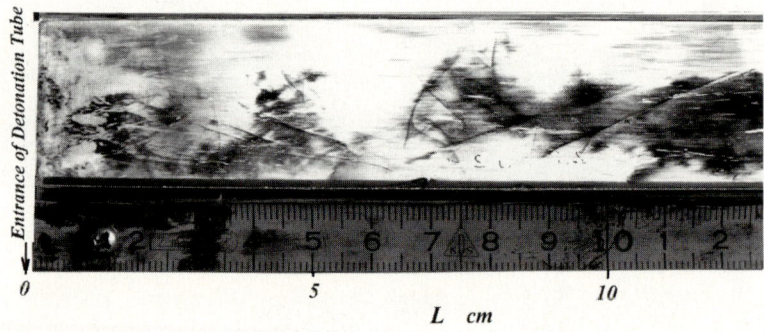

Fig. 11.52. Soot traces at transition from shock to detonation waves. $M_1 < M_{C-J}$. P_1, initial mixture pressure; M_1, Mach number of the shock wave incident into the mixture; L, distance from the entrance of the detonation tube

At the transition to detonation from a shock wave of low Mach number, the induction distance to the detonation initiation fluctuates over a large range. In (6) of Fig. 11.53 an example of soot traces of detonation waves initiated by a shock having a Mach number $M_1 = 5.8$ in a stoichiometric propane–oxygen mixture of an initial pressure $P = 4.3$ kPa which shows an induction distance of about 13 cm, while in (7) that initiated by a shock wave of $M_1 = 5.7$ in the same mixture having an initial pressure of 22.2 kPa, which shows a much longer induction distance of about 54 cm.

From any shock wave incident into a stoichiometric propane–oxygen mixture having an initial pressure lower than 1.1 kPa, no transition to detonation waves is observed, as far as the Mach number M_1 of the incident shock wave is less than 5.0.

Shock waves propagating with a Mach number higher than a certain value into a combustible mixture having a high enough density are smoothly transformed to detonation waves through a cellular structure, i.e., spontaneous ignition

11.3 Initiation of Detonation Waves

(6) $P_1 = 4.3$ kPa, $M_1 = 5.8$

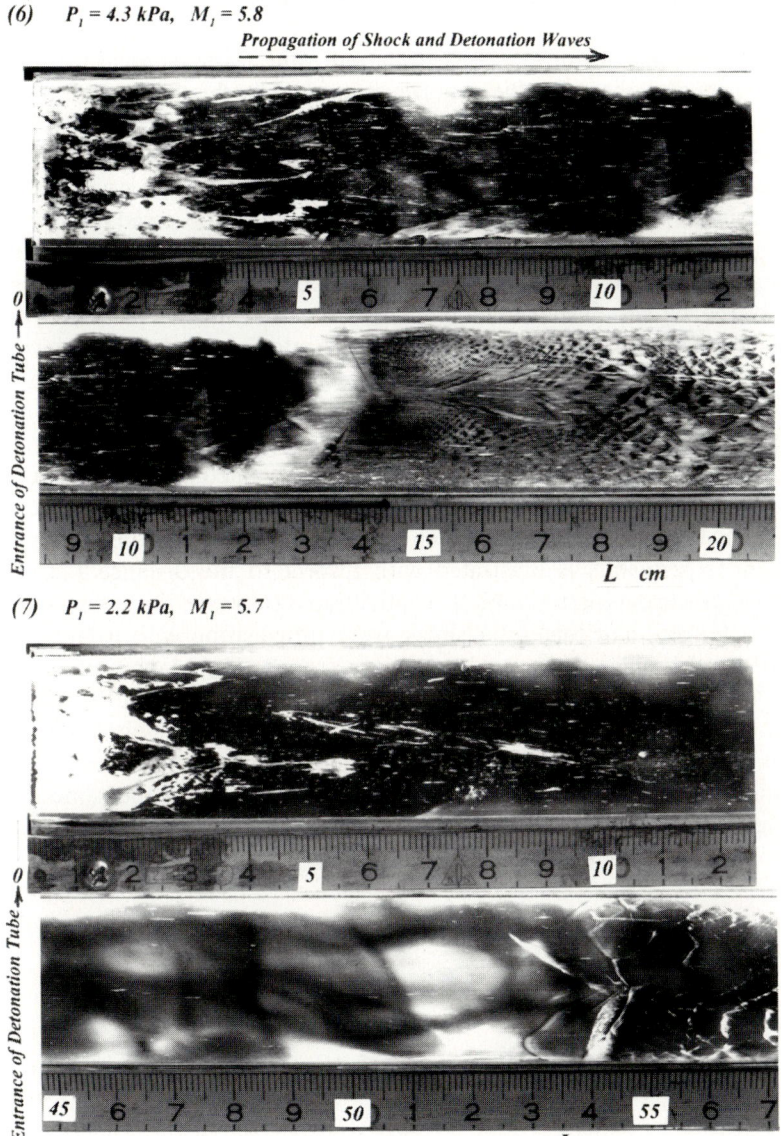

Fig. 11.53. Soot traces at transition from shock to detonation waves in the case of a long induction distance. P_1, initial mixture pressure, M_1, Mach number of the incident shock wave; L, distance from the entrance of the detonation tube

behind the shock waves. If the incident shock wave of a lower Mach number into the mixture having lower density, the transition to the detonation occurs after a long and fluctuating induction distance, sometimes following a strong explosion, or does not take place at all.

Propagation Velocity of Shock-Induced Detonation Waves

From the passage instants of the detonation front at the piezoelectric pressure transducers and observing the soot traces marked by the detonation, the propagation velocity of the detonation waves corresponding to each position in the detonation tube can be estimated.

In Fig. 11.54, Mach numbers M_s of shock and detonation waves transformed from a shock wave having a Mach number of about 7.5 observed in three different experiments are illustrated in relation to the distance L from the entrance of the detonation tube. The incident shock has almost the same Mach number as M_{C-J} of the self-sustained detonation under Chapman–Jouguet condition, but the detonation waves initiated by the shock propagate first a much higher velocity, that is, so-called overdriven detonation waves are formed, which decelerate, however, their propagation velocity with their propagation, approaching the Chapman–Jouguet velocity. The induction distance to the detonation formation fluctuates only a little, while the propagation velocity fluctuates over a large range.

In Fig. 11.55 the Mach number M_s of the detonation waves initiated by a shock wave propagating with a Mach number less than M_{C-J} observed in three different experiments is illustrated with respect to the distance L from the entrance of the detonation tube. The induction distance to the detonation initiation is longer than that by a shock wave propagating with a larger Mach number and besides fluctuates over a large range. The Mach number of the detonation propagation first jumps up to a value near M_{C-J}, but then decreases to a value lower than M_{C-J}, that is, slow down detonation are observed.

In Fig. 11.56, the Mach number of the detonation waves initiated by almost the same shock wave as in Fig. 11.55, but propagating in the same mixture having a lower pressure of 1.1 kPa observed in four different experiments are

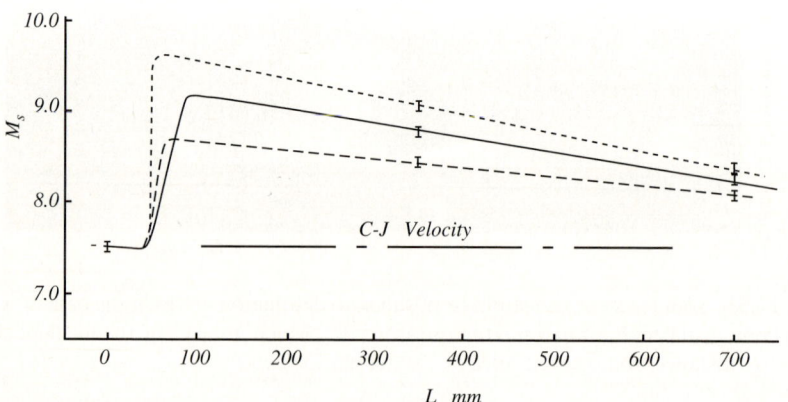

Fig. 11.54. Mach number M_s of the incident shock and detonation waves propagating in $(C_3H_8 + 5O_2)$ in relation to the distance L from the entrance of the detonation tube. The incident shock waves have almost the same Mach number M_{C-J} of the detonation wave under Chapman–Jouguet condition

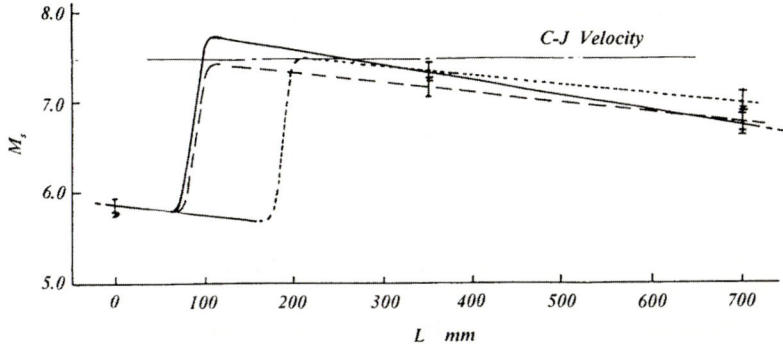

Fig. 11.55. Mach number M_s of the shock and detonation waves propagating in (C_3H_8 + $5O_2$) mixture with respect to the distance L from the detonation tube entrance. $M_1 < M_{C-J}$

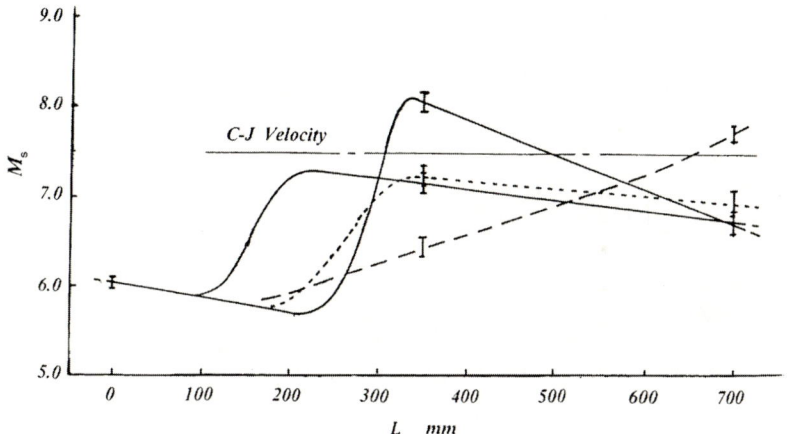

Fig. 11.56. Mach number M_s of the shock and detonation waves propagating in ($C_3H_8 + 5O_2$) with respect to the distance L. The detonation waves are formed hardly or much delayed

illustrated with respect to the distance L. The induction distances as well as the propagation velocities fluctuate much more than those in the previous case of Fig. 11.55. Besides, with most of them the propagation velocity is always less than M_{C-J}, namely slow down detonation waves are also formed here.

At the transition from shock to detonation waves the propagation velocity of the detonation waves always jumps up to a Mach number by 1.0–2.0 higher than that of the incident shock wave, then the detonation waves further propagate in the detonation tube, decreasing their propagation velocity, as the flow behind the detonation front is decelerated by the friction and heat loss on the tube surface. Corresponding to the Mach number of the incident shock and the mixture density, thus, overdriven, Chapman–Jouguet or slow down detonation waves are produced, which are decelerated with their propagation.

The Process of the Transition in a Pressure–Density Diagram

The diagram of Fig. 11.57 shows several processes of transition from shock to detonation waves propagating in a stoichiometric propane–oxygen mixture in a pressure–density diagram in which the pressure P_0 of the shock driver gas H_2 is set at 2.94 MPa.

O_1 is the initial state of the mixture having a pressure P_1 and density ρ_1, while P_2 and ρ_2 represent the pressure and density behind the shock or detonation waves, H_S and H_D are the Hugoniot curves for shock and detonation waves, respectively, $O_1 S_n$ are the Rayleigh lines of the shock waves incident into the mixture, while $O_1 D_{ml}$ are those of the detonation waves at the transition. The pressure on each line means the initial pressure of the mixture.

$O_1 S_1$, $O_1 S_2$, and $O_1 S_3$ can be obtained from the Mach number of each incident shock wave, while $O_1 D_{11}$, $O_1 D_{21}$ and $O_1 D_{31}$ stem from those of each

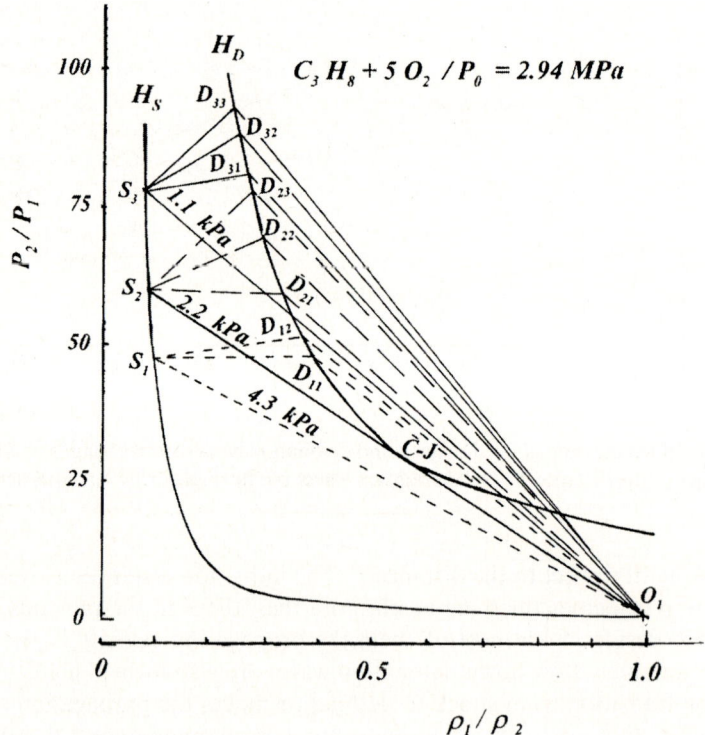

Fig. 11.57. Shock and detonation waves at the transition in a pressure–density diagram P_0: the initial driver gas pressure; P_1, ρ_1 at O_1: initial pressure and density of $(C_3H_8 + 5O_2)$; P_2, ρ_2: those behind the shock and detonation waves; H_S, H_D: Hugoniot-curves of shock and detonation waves; $O_1 S_n$, $O_1 D_{ml}$: Rayleigh lines of shock and detonation waves, respectively. Pressure on each Rayleigh line means the initial mixture one

detonation wave transformed from the shock wave. The Rayleigh-line touching the H_D-curve is that under the Chapman–Jouguet condition and express Chapman–Jouguet (C–J) velocity. All shock-induced detonation waves propagate with a velocity higher than C–J velocity, even those initiated by shock waves with a Mach number lower than M_{C-J}. The points S_n and D_m are connected by straight lines, as the real process is unknown. Just after the transition to the detonation, the states of the combustion gas are expressed by D_{m1} on the H_D-curve, but as the propagation velocity decreases with the propagation, the state of the gas slides down along the H_D-curve to the C–J point.

The process from S_n to D_{m1} is still not clear, but in most cases some pressure increase is observed against the classical theory for the equilibrium state described in Sect. 11.1.1. Considering that an approximately isochoric combustion is observed at the transition from combustion to detonation waves, we should also consider a rapid combustion accompanied by a pressure increase at the transition from shock to detonation waves.

The diagram in Fig. 11.58 shows the same relations in the same mixture, in which the pressure of the shock driver gas H_2 is set at 1.47 MPa. The Mach numbers of the incident shocks are less than those in Fig. 11.57, but the results suggest the same tendencies.

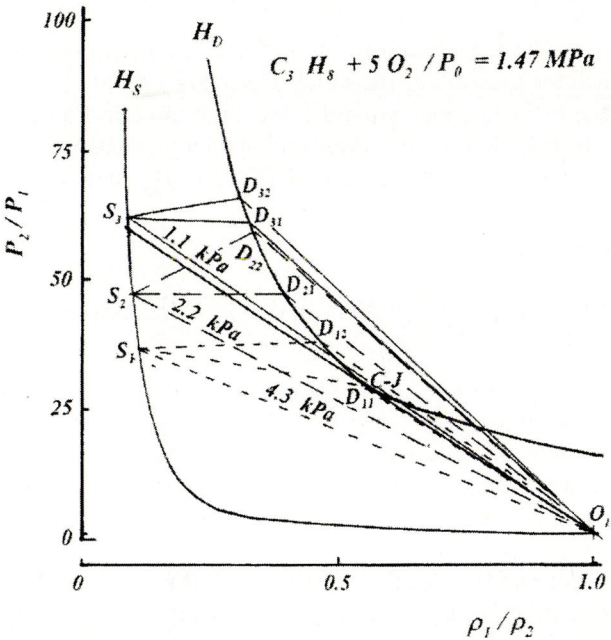

Fig. 11.58. Shock and detonation waves at the transition to detonation waves in pressure–density diagram. $P_0 = 1.47$ MPa

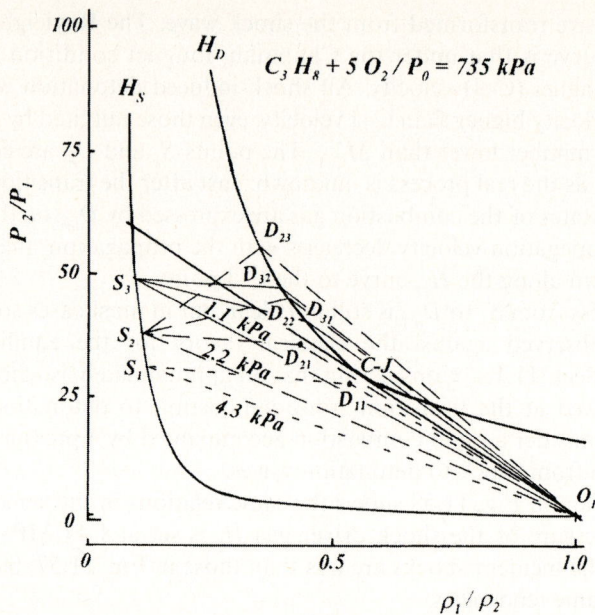

Fig. 11.59. Shock and detonation waves at the transition to detonation wave in a pressure–density diagram. Driver gas pressure $P_0 = 735$ kPa

In Fig. 11.59 the same relations of the shock-induced detonation are illustrated, in which the pressure of the shock driver gas is set at 735 kPa. The Mach numbers of the incident shock are much less than those in Fig. 11.57 as well as in Fig. 11.58. Several detonation waves initiated by weak shocks propagate with a velocity less than the C–J velocity, as O_1D_{11} or O_1D_{21} suggests.

Concluding Remarks

In the experiments of the transition from shock to detonation waves the following three remarkable phenomena are observed:

1. From a strong shock wave propagating with a velocity having a high Mach number into a high density combustible mixture a smooth transition to detonation waves proceeds through a fine cellular structure without any explosion, while the transition from a weak shock wave incident into a low density mixture takes place through a coarse cellular pattern, sometimes after a strong explosion.
2. At the transition from shock to detonation waves a jump of Mach number of the detonation propagation by 1.0–2.0 is always observed, after which the propagation velocity decreases, approaching a certain value expressed by an asymptote. Comparing such a propagation velocity shown by the asymptote, an overdriven detonation is, thus, always observed at the transition.

3. The detonation wave transformed from a shock wave can propagate with different velocities, sometimes higher than the Chapman–Jouguet velocity, sometimes with the same one as that under Chapman–Jouguet condition or sometimes with a lower one than the C–J velocity, that is, the Chapman–Jouguet velocity never means a critical velocity for the detonation wave.

These three phenomena can be explained as follows:

1. At the transition from a weak shock to a detonation wave in a low density mixture, the probability of spontaneous ignition is so small that only a few ignition take place and a coarse cellular structure is formed behind the shock wave.
2. At the transition from shock to detonation wave the spontaneous ignition of the first stage behind the shock wave takes place in the flow velocity w_S behind the shock wave. From the ignition points the combustion waves propagate with a velocity w_D, forming a detonation wave. In the early stage, therefore, the detonation wave is driven by the flow having a velocity composed of both the flow velocity behind the incident shock and that behind the combustion wave, namely with a velocity $w_3 = w_S + w_D$, while the flow velocity behind the shock is later stagnated by the counterflow behind the combustion waves, that is $w_S = 0$ and the detonation wave is driven by the flow with the velocity w_D. The detonation wave driven by the flow with a velocity $w_3 = w_S + w_D$ propagates faster than that driven only by the flow with w_D, as described already in Sects. 5.1 or 11.1. At the initiation of detonation waves, therefore, an overdriven detonation is often observed.
3. The detonation wave is sustained by the spontaneous ignition behind shock waves at the detonation front. As far as the detonation wave can propagate such fast to produce a strong shock wave at its front in which many spontaneous ignitions take place in a short time to drive the detonation wave, the detonation is sustained. The self-sustained detonation wave should propagate faster than such a minimum velocity, but must not always with the Chapman–Jouguet velocity or faster than it.

11.4 Propagation of Detonation Waves

A detonation wave is composed of a shock wave at its front and a combustion zone behind the shock wave. According to the Rankine–Hugoniot equations the shock wave as well as the combustion zone are expressed by the Hugoniot-curves and Rayleigh-lines on the pressure–density diagram, as explained already in Chap. 10. In the steady propagation detonation waves the propagation velocity of the shock front must be equal to that of the combustion zone. Both the Rayleigh lines of the shock wave and combustion zone, therefore, coincide with each other.

In the self-sustained detonation wave, the so-called Chapman–Jouguet condition should be satisfied, namely, the Rayleigh-line in this case is the tangent from the initial mixture state to the Hugoniot-curve of the detonation wave in the pressure–density diagram. The point on the Hugoniot-curve touching the Rayleigh-line is called Chapman–Jouguet point (C–J point) and the propagation velocity of the detonation Chapman–Jouguet velocity (C–J velocity), which means the minimum propagation velocity.[101, 117]

The Mach number M_D of the detonation propagation under the Chapman–Jouguet conditions is approximately expressed by the following equation as described already in Sect. 11.1.1,

$$M_D^2 \approx 2(\gamma^2 - 1)\frac{Q}{a_1^2}, \tag{11.19}$$

Or the propagation velocity D_{C-J} by the equation

$$D_{C-J} \approx 2(\gamma^2 - 1)Q. \tag{11.19'}$$

The higher the initial temperature, the smaller the value of the ratio γ of specific heats and reaction heat Q because of dissociation after the combustion. According to (11.19'), therefore, the propagation velocity D_{C-J} must decrease, if the initial mixture temperature rises.

The classical detonation theory, thus, leads us to conclude that

1. The propagation velocity of the detonation wave must decrease with increase of the initial mixture temperature
2. The detonation propagation velocity lower than the Chapman–Jouguet velocity is never realized in any self-sustained detonation wave

In the experiments to investigate the interaction between shock and detonation waves described in Sect. 11.2.3, an acceleration of detonation in a mixture heated by a shock wave is observed. In the experiments of the transition from shock to detonation waves described in Sect. 11.3, several detonation waves having a propagation velocity lower than the Chapman–Jouguet velocity are also observed. These results suggest that the Chapman–Jouguet condition is questionable.

In this chapter the mechanism of the detonation propagation is discussed observing experimentally the temperature dependence of the propagation velocity of detonation waves.

11.4.1 Detonation Propagating in Mixtures having Different Temperatures[118]

According to (11.9'), (11.14), and (11.16), we can obtain the Chapman–Jouguet velocity D_{C-J}, the ratio of pressure P_{C-J}/P_0 at C–J point and that of temperature. T_{C-J}/T_0 in relation to the initial mixture pressure P_0 and temperature T_0. In Fig. 11.60 D_{C-J} in Fig. 11.61 P_{C-J}/P_0 and T_{C-J}/T_0 in mixtures of $(C_3H_8 + 5O_2)$ and $(C_3H_8 + 15O_2)$ having two different initial pressure of 100 and 56 kPa are illustrated with respect to the mixture temperature T_0.

These theoretical results suggest that the C–J velocity as well as the pressure P_{C-J} and temperature T_{C-J} of the mixture at the C–J point decrease with increase of the initial mixture temperature. At the collision of a detonation wave with a shock wave, however, we once observe in Sect. 11.2.3, that the propagation of the

11.4 Propagation of Detonation Waves

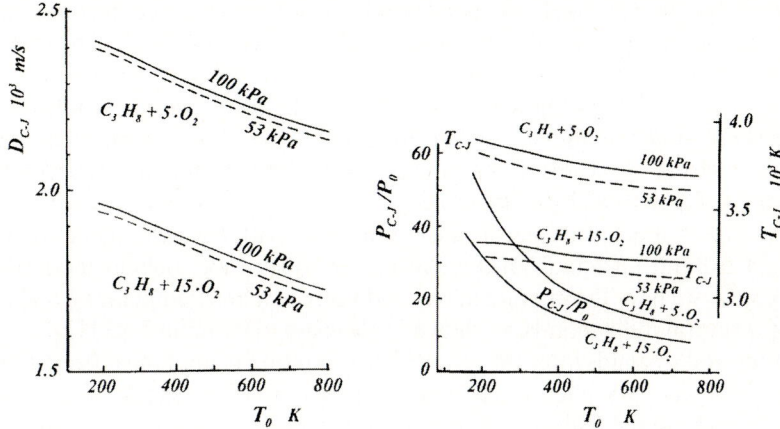

Fig. 11.60. Detonation propagation velocity D_{C-J} (*left*) under Chapman–Jouguet condition, pressure ratio P_{C-J}/P_0 and temperature T_{C-J} (*right*) in the mixtures of $(C_3H_8 + 5O_2)$ and $(C_3H_8 + 15O_2)$ with respect to the initial mixture temperature T_0. P_{C-J} and T_{C-J}: pressure and temperature at the Chapman–Jouguet point, respectively. Initial mixture pressure on each line

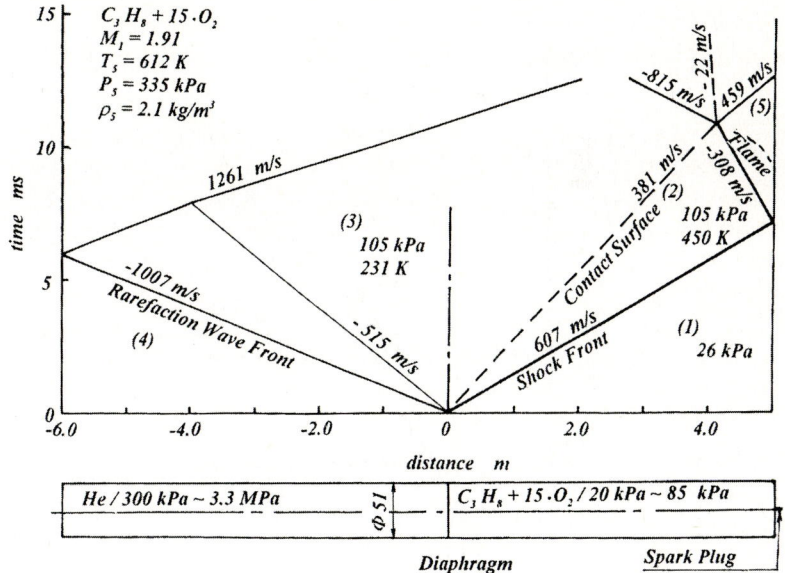

Fig. 11.61. Shock tube (*below*) used for the experiments and wave diagram (*above*) experimentally obtained under the conditions shown *above left*

detonation is accelerated in the mixture behind the shock wave where the temperature is higher than that before the collision. We have, therefore, to confirm the influence of the mixture temperature on the detonation propagation. To do so we have tried to carry out some experiments using shock tubes. In Fig. 11.61

below an example of the shock tube applied to the experiments and above that of wave diagram experimentally obtained are illustrated. The shock tube of stainless steel has an inner-diameter of 51 mm and is composed of a 6 m long driver gas tube and a 5–7 m long low pressure tube. The driver gas He is filled in the driver tube under a pressure from 300 kPa to 3.3 MPa, while a propane–oxygen mixture of ($C_3H_8 + 15O_2$) is filled in the low pressure tube under a pressure from 20 to 85 kPa. Breaking the diaphragm between the driver gas and the mixture, a shock wave propagates to the right with a Mach number between 1.3 and 2.0 and reflects from the end of the shock tube. Behind the reflected shock wave we have the mixture at rest and keeping a high temperature and pressure for several milliseconds, as shown in the wave diagram in Fig. 11.61.

A few milliseconds after the reflection the mixture behind the reflected shock is ignited by a spark plug mounted at the end wall of the shock tube. A flame propagates from the ignition point to the left transforming to a detonation wave. The flame or detonation propagation is observed by several photodiodes set at the shock tube end part keeping a distance of 200 mm with each other.

Figure 11.62 represents an example of the time–distance (t–L) diagram of the flame fronts (combustion waves) propagating in a mixture of ($C_3H_8 + 15O_2$) under different conditions, where t represents the time after the ignition and L the distance from the shock tube end wall. From such diagrams we can obtain

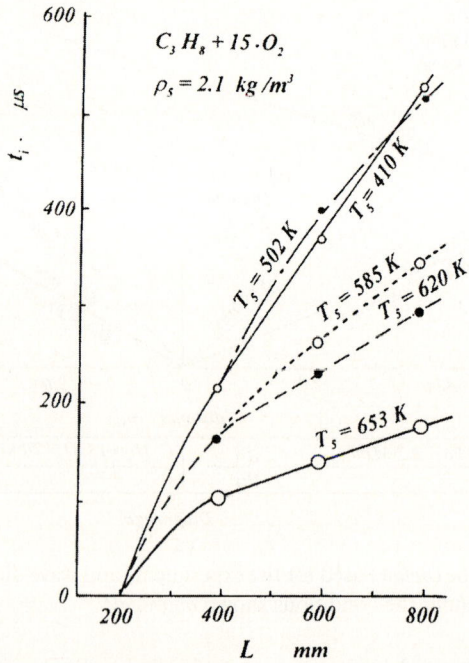

Fig. 11.62. Time–distance diagram of combustion waves propagating in a propane–oxygen mixture behind reflected shock waves

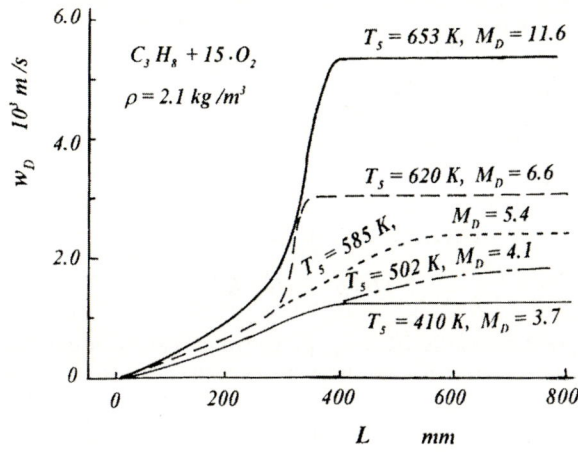

Fig. 11.63. Propagation velocity w_D of the combustion or detonation waves in a propane–oxygen mixture behind reflected shock waves

diagrams of the propagation velocities w_D of the combustion wave with respect to the distance L, as illustrated in Fig. 11.63.

The results suggest the following:

1. The propagation velocity of combustion wave increases with the rise of the mixture temperature before the ignition
2. After passing the point near $L = 200$ mm, the flame propagation is accelerated to a velocity higher than the sound velocity. A shock wave must be formed in the front of the flame
3. In the region of $L > 600$ mm the flame propagation reaches a constant velocity from 1,000 to 5,400 m s^{-1}, with which a self-sustained detonation wave propagates. In the mixture having a temperature before the ignition higher than 580 K, the self-sustained detonation wave propagates with a velocity higher than the C–J velocity.

The experimental results, thus, suggest, the higher the mixture temperature, the higher the propagation velocity of the detonation wave, namely, the detonation propagation velocity shows a quite inverse tendency to the results calculated according to the hypothesis of Chapman–Jouguet. As the experimental results of the transition from shock to detonation waves described in Sect. 11.3.2 also suggest, we can now conclude that there is no Chapman–Jouguet velocity in detonation waves.

The Chapman–Jouguet hypothesis has been formed according to the theory of gas dynamics without any consideration of the combustion mechanism. Considering that the flame and the shock wave formed in its front propagate together with the same velocity, the temperature T_D behind the shock waves can be calculated from the propagation velocity w_D of the combustion or detonation wave according to the shock wave theory explained in Chap. 5.

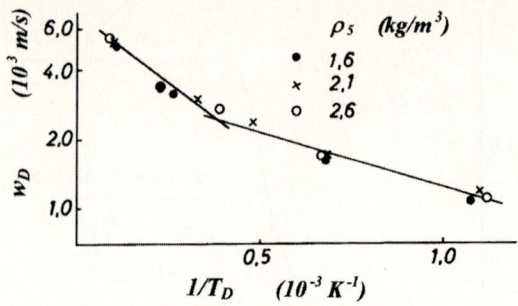

Fig. 11.64. ln w_D with respect to $1/T_D$. w_D, propagation velocity of combustion or detonation waves; T_D, temperature behind shock waves at the detonation front

In Fig. 11.64 the logarithms of the propagation velocity w_D of combustion or detonation wave are illustrated with respect to the reciprocal mixture temperature $1/T_D$ behind the shock wave, in which we have two straight lines having different slopes and crossing with each other at $T_D = 2{,}600$ K. Calculating from the Arrhenius' equation:

$$w_D = \exp\left(-\frac{E_D}{RT_D}\right), \qquad (11.26)$$

we obtain

$$E_D \approx 30 \text{ kJ mol}^{-1} \text{ where } T_D > 2{,}600 \text{ K}$$
$$E_D \approx 10 \text{ kJ mol}^{-1} \text{ where } T_D < 2{,}600 \text{ K}.$$

This critical temperature 2,600 K must be different according to the mixture state. In Fig. 11.65 the pressure variations in the mixture measured by a piezoelectric pressure transducer at $L = 200$, 400, and 600 mm during the

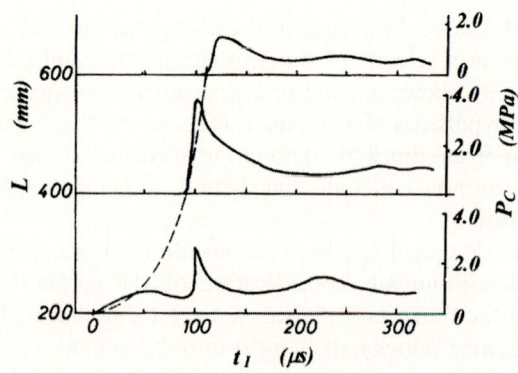

Fig. 11.65. Pressure variation measured at the distance $L = 200$ mm, 400 mm, and 600 mm from the shock tube end wall or ignition point

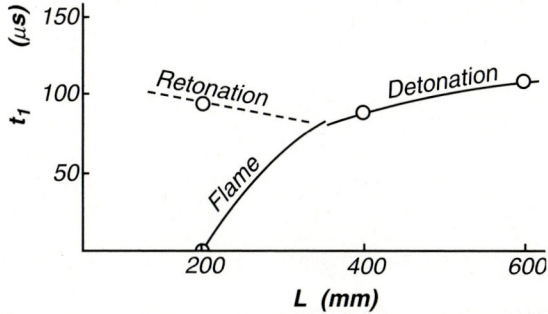

Fig. 11.66. Propagation of flame, detonation, and retonation fronts estimated from Fig. 11.65

flame propagation in a mixture of $(C_3H_8 + 15O_2)$ having an initial temperature of 664 K and density of 2.03 kg m^{-3} are presented. These pressure variations mean first after the ignition the flame propagates under acceleration in the mixture where a strong explosion takes place near $L = 350$ mm, the detonation wave is produced and propagates, while a retonation propagates from the explosion point to the opposite direction of the detonation, as illustrated in Fig. 11.66.

Together with the measurement of the propagation velocity, the microscopic structure of the propagating flame and detonation wave is recorded by a soot film coated on the surface of a polyester film sticked on the inner-surface of the shock tube where the flame propagates. In Fig. 11.67 (I–IV) photographs of four parts of soot trace at the transition from a flame to detonation wave in the propane–oxygen mixture having a temperature of 590 K and density of 1.53 kg m^{-3} are presented. The polyester film is partially destroyed, but we can recognize the trace.

In the first picture (*I*) meandering trace (marked with *Pseudo-D*) of the pseudodetonation is observed from the ignition point to L–200 mm, then in the next picture (*II*) a strong explosion, from which a sectorial cellular pattern (marked with *Cellular-P*) spreads. In the next *(III)* and *(IV)* we observe the cellular pattern developed further. This process suggests the same one illustrated in Fig. 11.66 and that observed at the transition from a combustion to detonation wave described already in Sect. 11.3.1. In this case, the temperature behind the shock wave in the front of the flame is estimated to be about 2,900 K. A soot trace (Fig. 11.67(V)) of a combustion wave propagating in the same propane– oxygen mixture of 499 K is presented in Fig. 11.67(V). In this picture we do not observe any cellular pattern, though the flame propagates with a Mach number of 3.5–4.5, that is, we observe no detonation wave, but only a pseudodetonation wave. The temperature behind the shock in the front of the flame is estimated to be about 1,800 K.

Considering the soot traces of the propagating flame at the transition from the combustion to detonation waves, the Arrhenius' relations between w_D and $1/T$ shown in Fig. 11.64 suggest that in the propane–oxygen mixture having a temperature higher than 2,600 K real detonation waves propagate in which the spontaneous ignition behind the shock waves at the detonation front lead the combustion, while in that of a temperature lower than 2,600 K only pseudodetonation waves driven by combustion propagate.

288   11 Gaseous Detonation Waves

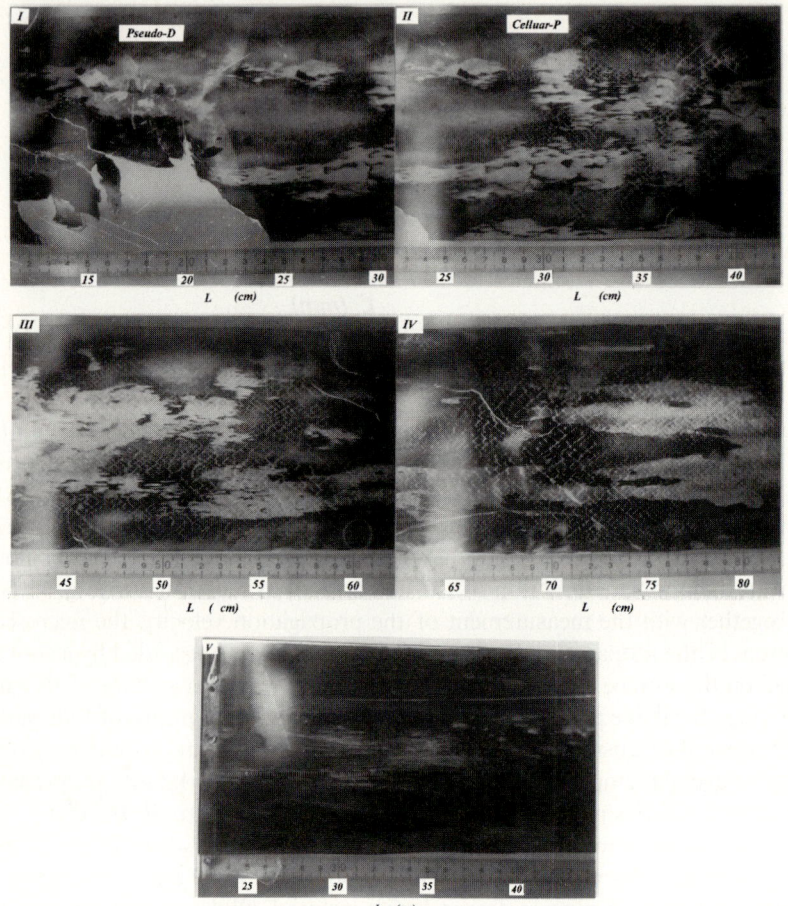

Fig. 11.67(I–IV). Soot trace pattern at the transition from combustion to detonation wave in a mixture of $(C_3H_8 + 15O_2)$ having a temperature of 589 K and density of 1.53 kg m^{-3} behind the reflected shock wave, L: distance from the tube end and ignition point. **(V)** A soot trace of a combustion wave propagating in $(C_3H_8 + 15.O_2)$ of 499 K and 2.3 kg m^3

In a combustible mixture filled in a tube having a closed end the gas heated by the combustion flow follows the combustion wave ignited at the closed tube end wall, driving the mixture in the front of the combustion wave. If the combustion releases heat enough, the heated gas flows so fast as to form a shock wave in the front of the combustion wave and pseudodetonation is produced. If the gas flow is further accelerated, a stronger shock wave having a higher pressure and temperature is formed, behind which many spontaneous ignitions take place and drive the mixture more strongly, that is, a detonation wave is formed. In a mixture having a higher temperature, the combustion in the mixture proceeds with a higher velocity, the gas behind the combustion wave flows faster and produces a stronger detonation wave having higher temperature and pressure, propagating faster. In order to control the propagation velocity of the detonation, therefore, the flow velocity behind the combustion zone should be controlled. In order to accelerate the detonation propagation in a mixture, the mixture should be heated.

11.4.2 Deceleration of the Detonation Propagation

As described above, the detonation propagation velocity depends on the gas flow velocity behind the combustion zone of the detonation wave. In order to decelerate the detonation propagation, therefore, it is necessary to decelerate the gas flow behind the detonation wave. There are several methods for it. We try two methods, one of them is to set some obstacles in the flow for hindering it and another one is to set some leaks in the gas flow path to take the flow energy off.

Obstacle Method for Suppression of the Detonation Wave[119]

Experiments

For the experiments to suppress the detonation wave an apparatus schematically illustrated in Fig. 11.68 is applied. A 1,000-mm long steel channel having a rectangular cross-section of 10 mm × 50 mm is prepared. An end of it is connected with a steel tube having an inner-diameter of 40 mm and a length of 1,200 mm, while the other is closed with a steel plate.

A 10-mm thick plexiglas plate coated with candle soot is inserted below the rectangular space to mark the cellular structure of the detonation wave. Eight steel blades are placed vertically on 250-mm long middle portion of the plexiglas plate BC. Each blade is 1-mm thick and 10-mm high. The blades are set so as to be parallel to the detonation propagation direction and their spacing is 6 mm,

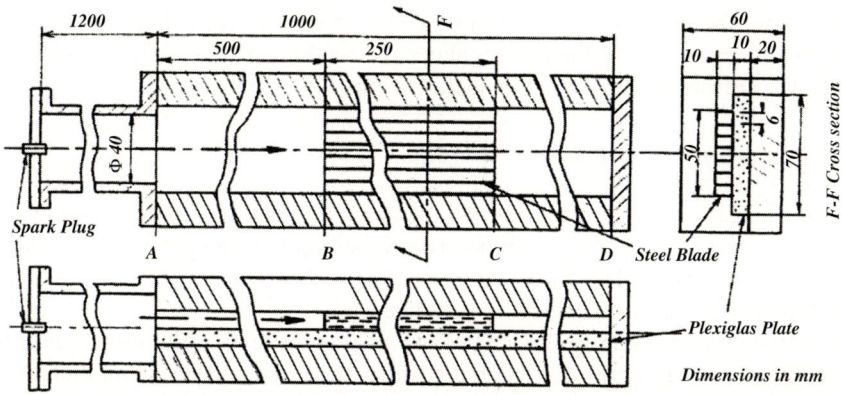

Fig. 11.68. Sketch of the channel for retarding the detonation propagation

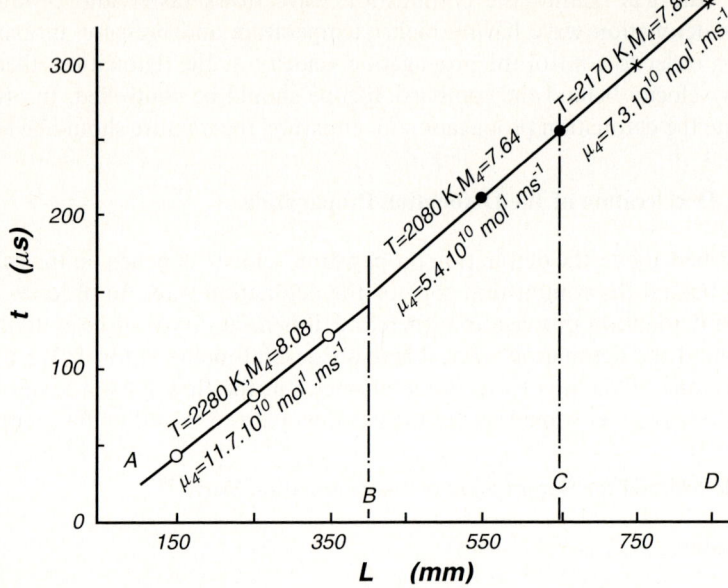

Fig. 11.69. Time–distance diagram of a detonation front propagating through the channel shown in Fig. 11.68

only a few times the cell width in the detonation waves. Thus, a multichannel section is established to hinder the flow behind the detonation front.

A stoichiometric propane–oxygen mixture filled in both the rectangular cross-section space and the steel tube under a pressure of 13.3 kPa at room temperature is ignited by an electric spark at the end of the steel tube so that a flame initiated from it is transformed into detonation waves. The detonation waves propagate further into the rectangular cross-section space along the direction of the arrow. The time–distance diagram of the detonation front at eight electrodes set on the center line of the rectangular cross-section space at different distance from the entrance of space A, as illustrated in Fig. 11.69.

Retardation of the Detonation Propagation

The propagation velocity is obtained from such time–distance diagrams, while the pressure P, density ρ and temperature T behind the shock wave at the detonation front are calculated according to (5.13)–(5.15). The experimental error of the temperature is estimated to be within ±3%.

In the section A–B, the detonation waves propagate with the usual constant velocity as in a conventional tube, but at the entrance edge B of the multichannel section the detonation waves are stagnated and then propagate further through the section B–C with a lower velocity, while they recover the propagation velocity to some degree in the section C–D. By the blades set in the channel the detonation propagation having a Mach number of 8.08 is decelerated to 7.64.

Probability of Apex Point Formation

Figure 11.70 shows a photograph of the soot traces marked by the detonation waves near the entrance to the multichannel region. The cellular structure of the detonation is observed in the section A–B and B–C. The size of the cells in the section B–C is much larger than that in the section A–B. The interval between two successive apex points on which two lines of each cell intersect never shows a constant value, but always fluctuates, because of the stochastic character of the detonation, as described in Sect. 11.2.2.

In Fig. 11.71 some histogram are shown of the intervals between two successive apex points in each section, A–B, B–C, and C–D, respectively. From such histograms we can calculate the probability μ_d of the apex point formation, according to (6.5) in Sect. 6.1 or 11.2.2. Some examples of the probability experimentally obtained are also given in Fig. 11.69. The error of the value is estimated to be within ±16%.

Figure 11.72 represents the relation between the logarithm of the probability μ_d of the apex point formation and the reciprocal mixture temperature $1/T$ behind the shock waves at the detonation front under an almost constant mixture density of 1.70 kg m^{-1}. All points fall on a straight line expressed by an Arrhenius' relation:

$$\mu_d = A_d \exp\left(-\frac{E_{ed}}{RT}\right), \tag{11.20}$$

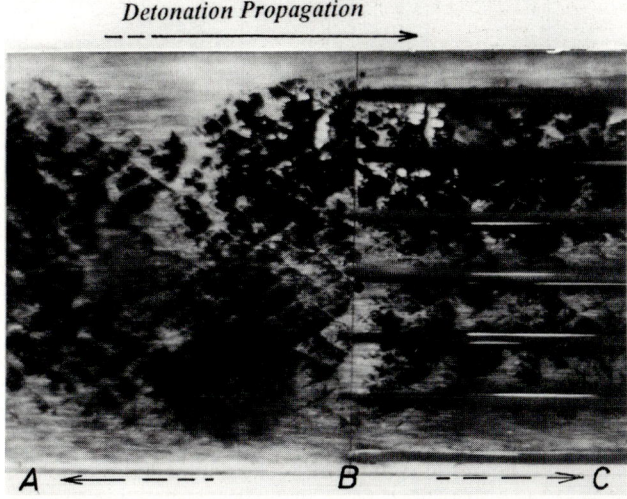

Fig. 11.70. Soot trace marked by a detonation wave at the entrance of the multichannel section B–C

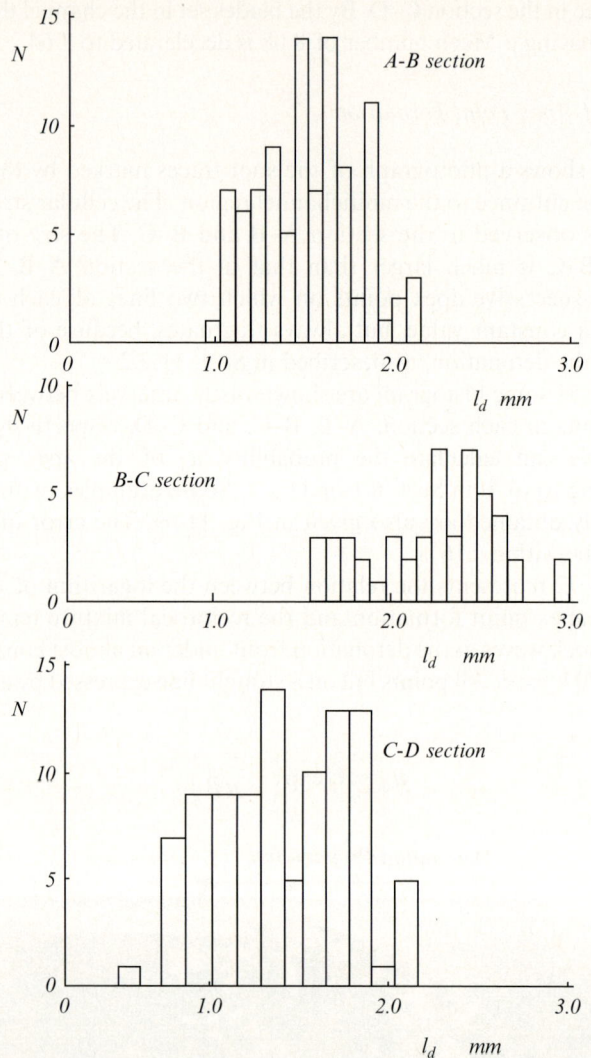

Fig. 11.71. Histograms of the interval l_d between two successive apex points in the detonation wave propagating in a ($C_3H_8 + 5O_2$) mixture in the section A–B, B–C, and C–D of the channel

as introduced in Sect. 11.2.2. The effective activation energy E_{ed} is estimated to be about 230 kJ mol^{-1}, which also agrees with that in Sect. 11.2.2.

The flow behind the detonation front is thus, retarded by the obstacles composed of many blades set in the detonation channel and at the same time the propagation of the detonation waves is decelerated from a velocity with a Mach number of 8.08 (2,490 m s^{-1}) to 7.64 (2,350 m s^{-1}).

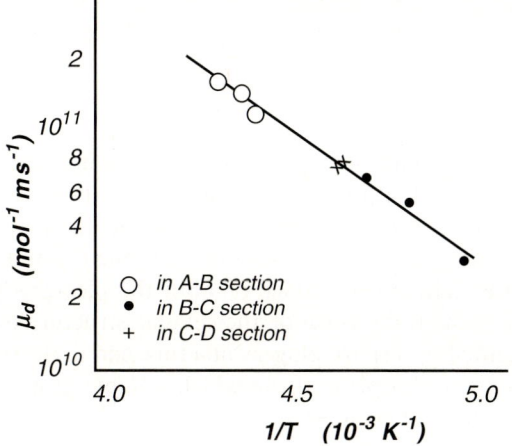

Fig. 11.72. ln μ_d with respect to the reciprocal mixture temperature $1/T$ behind the shock wave at the detonation front

Flow Leak Method for Retarding the Detonation Propagation

Apparatus of the Flow Leak Method

In order to decelerate the flow behind the detonation wave, an apparatus schematically illustrated in Fig. 11.73 is prepared. In a 1,000-mm long steel tube having an inner-diameter of 102 mm two plexiglas plates having a thickness of

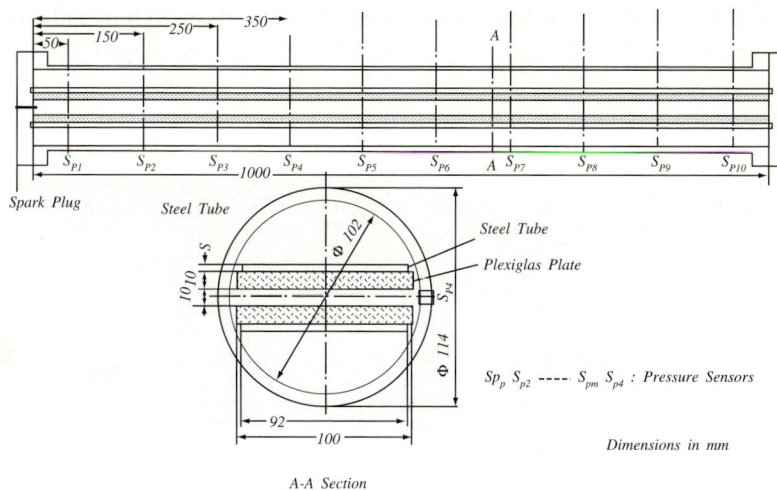

Fig. 11.73. Experimental apparatus for retarding the detonation propagation applying a flow leak method

10 mm are set in the center parallel, keeping a distance of 10 mm with each other, each supported by a steel plate, as shown in this picture. Between the both sides of the plexiglas plates and steel tube, a gap of about 1.0 mm is kept, so that the combustion gas between both the plexiglas plates leaks to the wider space of the tube.

A stoichiometric propane–oxygen mixture is filled in the space between both the plexiglas plates and that in the steel tube under a pressure of 13.3 kPa at a room temperature of 288 K. The propagation of the detonation waves is measured by piezoelectric pressure sensors set on the tube facing the space between the plexiglas plates. The inner surface of both the plexiglas plates are coated by candle soot to obtain the trace of the cellular structure of the detonation. The mixture is ignited by a spark plug set at a tube end in the center of the space between both the plexiglas plates, from which a flame propagates to the right, transforming to a detonation wave.

Soot Traces and the Propagation Velocity of the Detonation Waves

The experiment is repeated several times under the same conditions. An example of the soot traces marked by the detonation waves is shown in the photograph of Fig. 11.74, in which we observe cellular patterns marked by the detonation, though they vanish on both sides because of the leak of the combustion gas.

The detonation wave propagates not stably, but fluctuates much as shown in Fig. 11.75. Most of the observed values of average value, however, suggest that the detonation wave propagates almost with Chapman–Jouguet velocity just after the transition from the combustion waves, then with an almost constant one, but much lower velocity.

The reason for it should be attributed to the deceleration of the flow behind the detonation wave by the leak of the combustion gas through both sides of the space formed by the plexiglas plates. Even if the detonation wave propagates with a velocity lower than C–J one, the detonation wave is sustained, as the soot trace having the cellular patterns suggests.

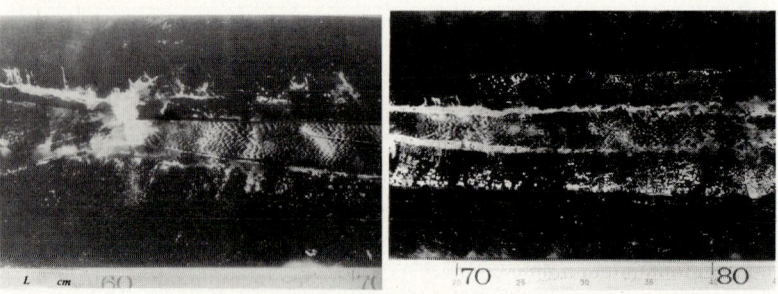

Fig. 11.74. Soot traces marked by the detonation propagating in a stoichiometric propane–oxygen mixture in a channel having an open gap of 1.0 mm on both the sides from two different experiments. L: distance from the spark plug

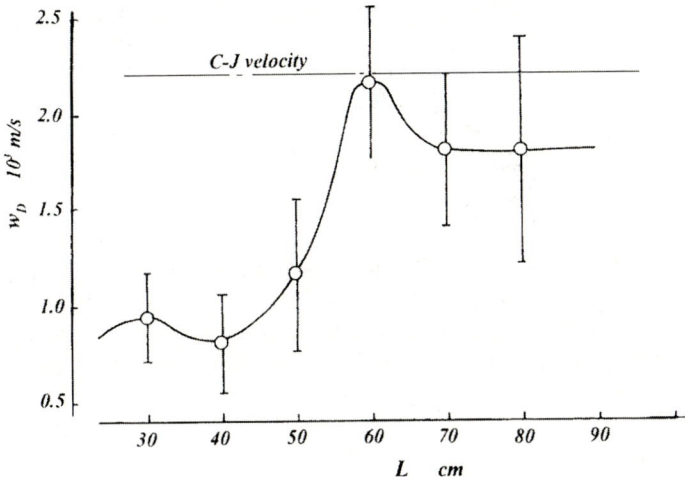

Fig. 11.75. Detonation propagation velocity w_D through the channel corresponding to the distance L from the spark plug

11.4.3 Concluding Remarks on the Propagation Velocity of the Detonation Wave

According to the classical detonation theory, the self-sustained detonation waves propagating in a certain combustible mixture propagate with a certain definite velocity under the so-called Chapman–Jouguet condition.

Against the theory introduced from the Chapman–Jouguet theorem, however, we observe that the propagation velocity of the self-sustained detonation waves increases with increase of the initial mixture temperature, as in Sect. 11.4.1. Besides, the propagation velocity lower than the C–J velocity also takes place at the transition from shock to detonation waves, as in Sect. 11.3, or in the detonation waves whose flow behind themselves are retarded. From such facts we can conclude that the Chapman–Jouguet condition cannot be applied to the detonation propagation and the C–J velocity does not exist.

In a tube having a closed end the flow behind the combustion wave propagating in the tube reflects from the closed end and goes to the same direction of the combustion wave. The flow velocity driving the combustion wave depends on the heat release velocity of the mixture behind the combustion wave governed by the combustion velocity. In the front of the combustion wave a shock wave is formed, as long as the flow velocity is high enough. A pseudodetonation is thus established and propagates faster by the accelerated combustion velocity in the mixture behind the shock wave increasing its temperature and density.

If the pseudodetonation is further accelerated so fast that a stronger shock wave is formed in the front of the combustion wave, some spontaneous ignitions can take place in the mixture behind the shock wave. Then a real detonation wave is established by the gas flow heated by the combustion caused by the spontaneous ignition. In the real detonation, therefore, some cellular structure is always observed, while in the pseudo-detonation no cellular pattern is recognized.

The propagation velocity of the detonation wave, therefore, depends on the flow velocity behind itself, which depends further on the heat release velocity governed by the combustion velocity.

11.5 Ionization of Gases Behind Detonation Waves

In detonation waves both the combustion and shock waves appear at the same time interacting with each other. Besides, the shock waves propagate with a velocity having a Mach number larger than 5. As the ionization of the gases behind combustion waves and that behind shock waves are stimulated by each other, a strong ionization in a nonequilibrium takes place, showing some character of irreversible phenomena.

Several methods for investigating the ionization have been developed in plasma physics. We apply here double probe method and the laser light scattering method described already in the investigation of the ionization in flames and shock waves in Chaps. 9 and 10. From different reasons, however, the correct values of the ionization cannot be obtained by the double probe method, that is, with the results measured by the double probe method, we can discuss the phenomena not quantitatively, but only qualitatively. On the other hand relatively correct values can be obtained by the laser light scattering method, but measured not continuously.

Nevertheless we can observe the behavior of the shock waves behind the detonation waves by measuring the ionization using a double probe method, since the ionization of gases behind combustion waves is strongly stimulated and increased by the interaction with shock waves, as explained already in Sect. 10.3.

11.5.1 Investigation Applying a Double Probe Method[120]

For the experiments a rectangular channel of plexiglas having a length of 2 m and a cross-section of inner space of 10 mm × 5 mm is prepared as illustrated above in Fig. 11.76. The outside of the channel is supported by four steel plates.

A propane–oxygen mixture ($C_3H_8 + 10.O_2$) is filled in this channel under atmospheric pressure at room temperature and ignited by a spark at the left closed end of the channel, so that the combustion and detonation waves propagate to the right open end. The propagation of the flame and detonation waves are observed by taking their luminescence into a photomultiplier through photoguides set at the position of $G_0, G_1, G_2, \ldots, G_{13}$. A double probe schematically shown below in Fig. 11.76 is set also at one of the same positions. A 100-kHz pulse potential of six steps between ±4 V is given to an electrode, while the current corresponding to each of five potentials is supplied to an oscilloscope. The experiments are carried out under the same condition, changing the measurement positions.

In the front of the flame propagating from the ignition point a strong explosion takes place at a position between G_2 and G_3, where the detonation wave is

11.5 Ionization of Gases Behind Detonation Waves

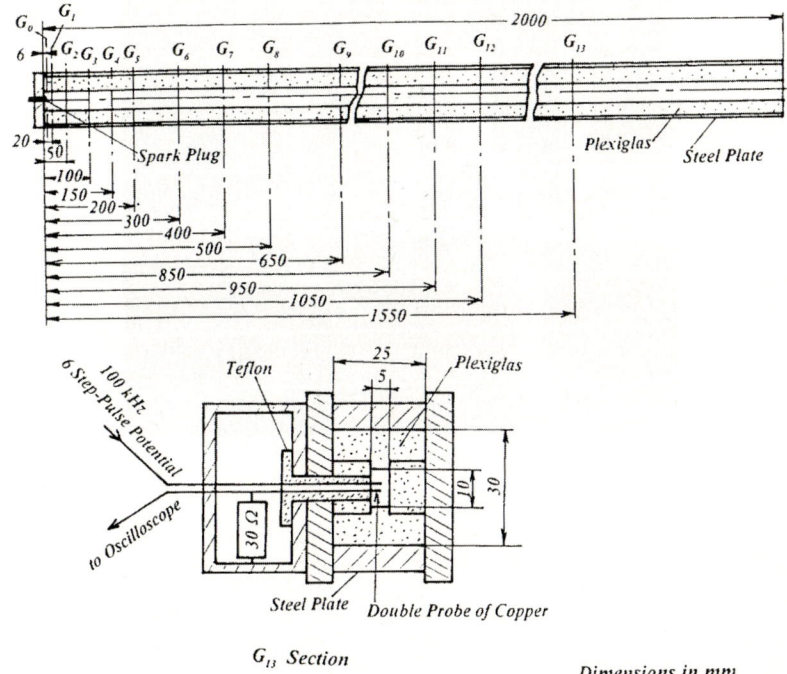

Fig. 11.76. Detonation channel for the ionization measurement using a double probe method

initiated and at the same time a shock wave, so-called retonation, propagates to the opposite direction of the detonation, too.

The oscillogram of Fig. 11.77 above shows the probe current of the double probe corresponding to each step of the probe potential measured at G_1. First we observe the shock wave produced by the propagating flame, then the ionization behind the combustion wave, and that behind the retonation.

From the probe current the probe characteristic at each instant after the passage of the shock front at the measurement point is obtained, from which we can further calculate the electron temperature T_e and ion density n_i, as described already in Sect. 9.2.

The variation of the electron temperature T_e and ion density n_i calculated from the probe characteristics obtained at G_1 are illustrated in Fig. 11.77 below. The error of T_e is estimated to be ±7%, but that of n_i depends on T_e and is estimated to be ±40% near $T_e = 5{,}000$ K and ±30% near $T_e = 10{,}000$ K. Behind the shock wave produced by the combustion wave the electron temperature T_e as well as the ion density n_i are relatively low, while both T_e and n_i behind the flame front increase rapidly.

In Fig. 11.78 above, an example of the probe current measured at the measuring point G_2 and in Fig. 11.78 below the electron temperature T_e and ion density n_i

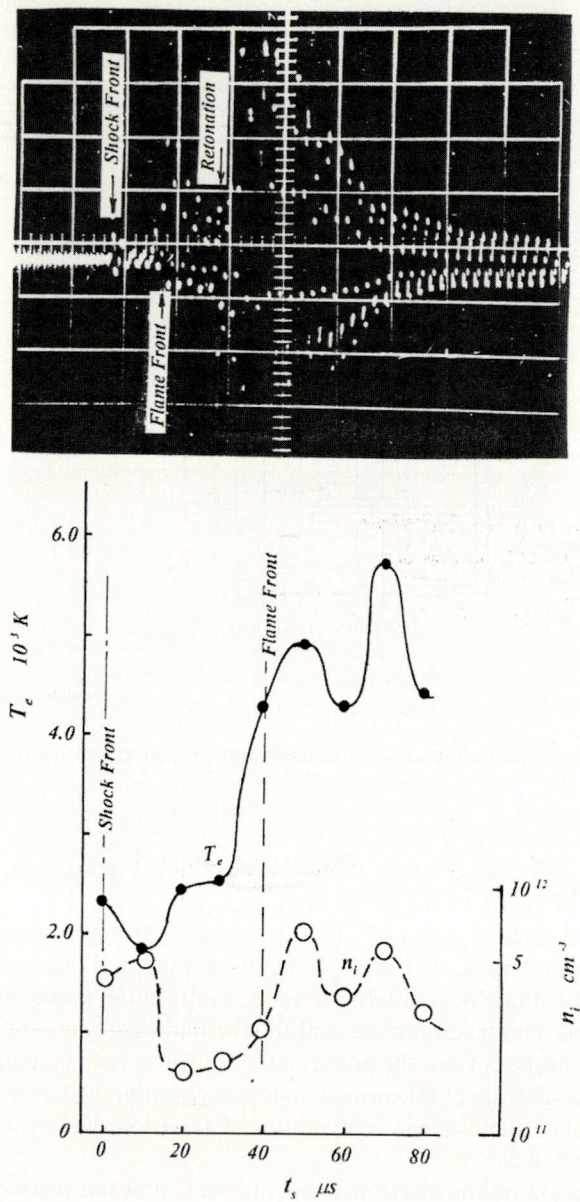

Fig. 11.77. *Above*: Probe current corresponding to the six-step pulse potential of the double probe observed at the measurement point G_1 behind the flame front propagating in a $(C_3H_8 + 10O_2)$ mixture before the transition to detonation wave 0.17 mA per div., 50 μs per div. *Below*: Electron temperature T_e and ion density n_i estimated from the results measured above in relation to the time t_s after the passage of the shock front

11.5 Ionization of Gases Behind Detonation Waves

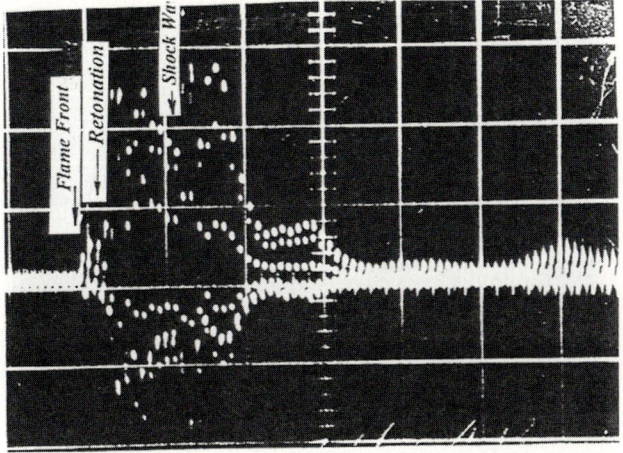

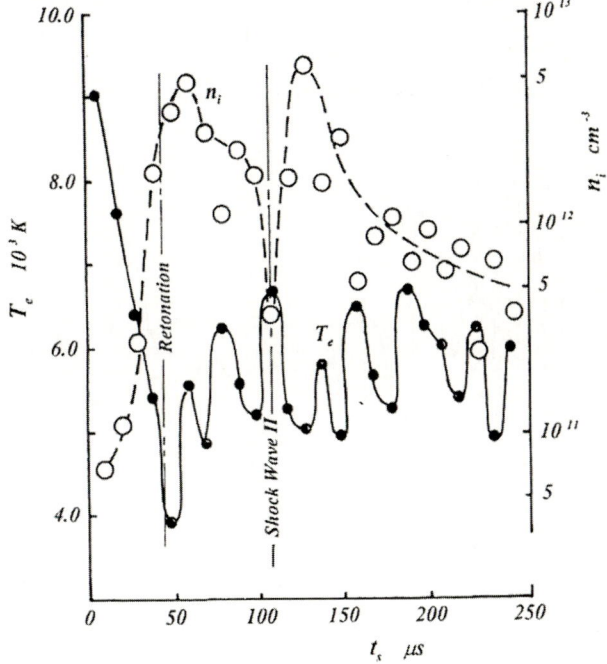

Fig. 11.78. *Above*: Probe current measured at G_2 at the transition to the detonation, current: 0.34 mA per div., time: 100 μs per div. *Below*: Electron temperature T_e and ion density n_i in relation to the time t_s after the passage of the flame front

calculated from this probe current at G_2. The first sudden increase of the ion density means the arrival of the retonation, while the second one that of a shock wave.

In Fig. 11.79 above the probe current measured at G_{11} is shown, while below the variation of the electron temperature T_e and ion density n_i calculated from

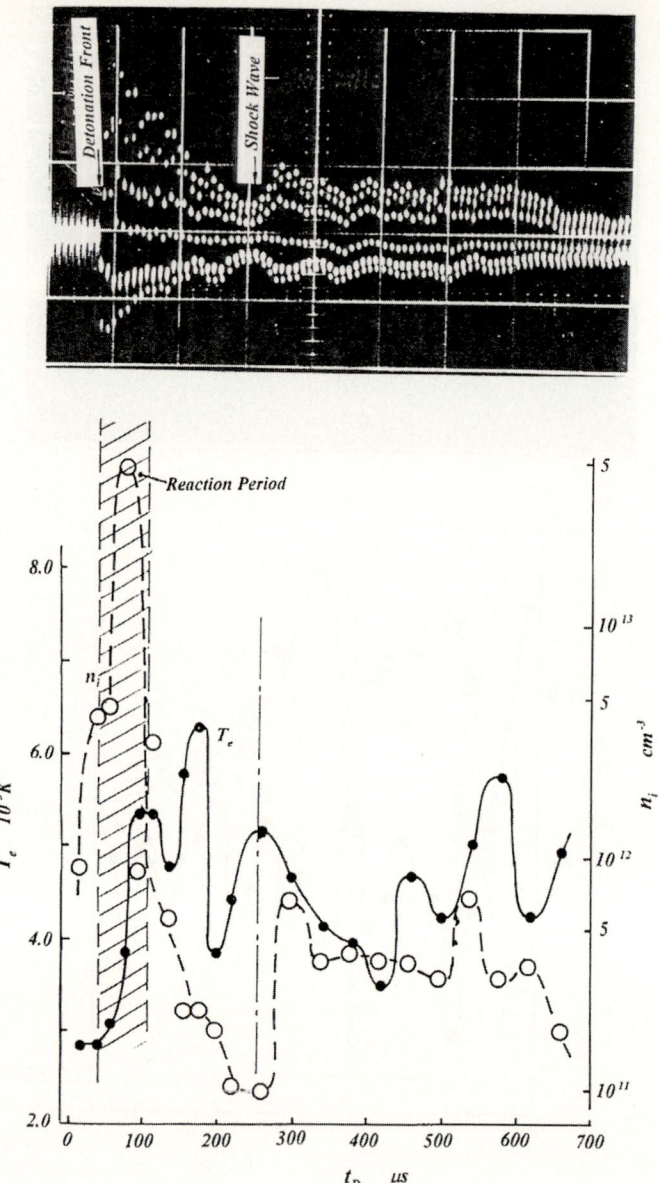

Fig. 11.79. *Above*: Probe current measured at G_{11} behind a detonation wave propagating in a $(C_3H_8 + 10O_2)$ mixture. Current: 0.17 mA per div., time: 100 µs per div. *Below*: electron temperature T_e, ion density n_i in relation to the time t_D after the passage of the detonation front

11.5 Ionization of Gases Behind Detonation Waves

the probe characteristic obtained from the probe current are presented. We recognize the reaction zone at the detonation front having an extremely high ion density and a shock wave behind the detonation wave, too.

From each oscillogram of the probe current measured at $G_3, G_4, \ldots, G_{13}$, the variations of T_e and n_i at each measurement position can be obtained, from which the arrival instant of the flame front, retonation, and shock fronts can be recognized. According to the arrival instants of the flame, retonation, detonation, and different shock fronts behind the detonation waves, as shown in Fig. 11.80

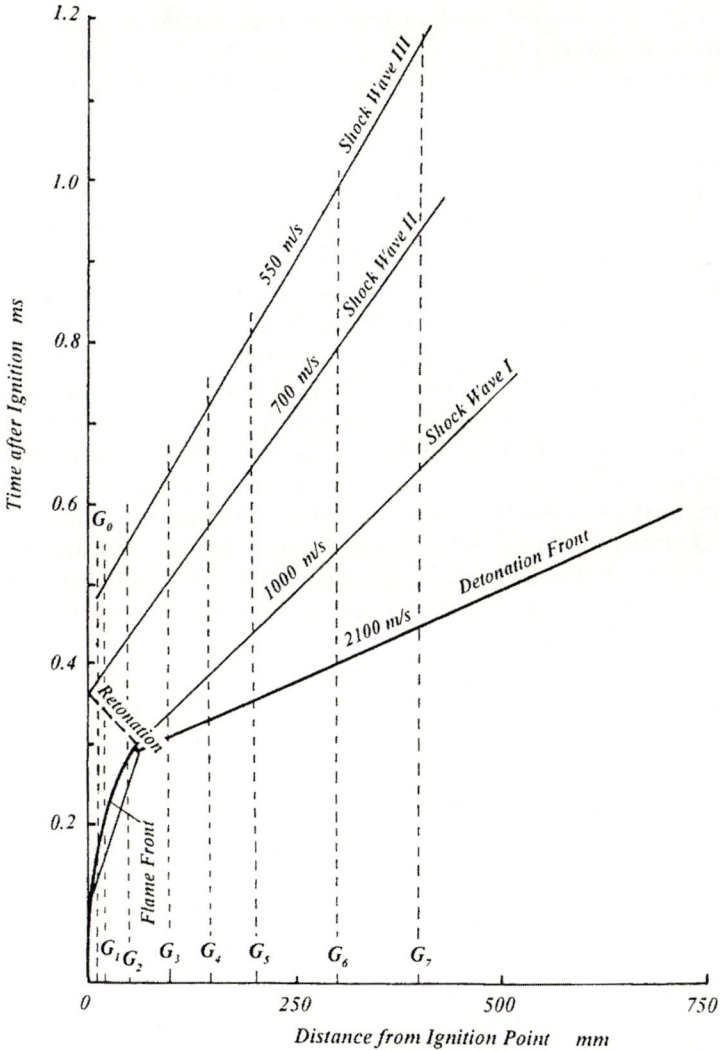

Fig. 11.80. Time–distance diagram of the flame front, detonation and different shock fronts behind the detonation wave propagating in a $(C_3H_8 + 10O_2)$ mixture in the channel shown in Fig. 11.76

This time–distance diagram suggests:

1. At the strong explosion point in the front of the propagating flame between G_2 and G_3, a detonation wave is initiated and propagates toward the open tube end with a velocity of 2,100 m s^{-1}
2. Behind the detonation wave at the explosion point a shock wave is also initiated and propagates with a velocity of 1,000 m s^{-1} following the detonation wave
3. From the explosion point a shock wave, called retonation, propagates in the opposite direction of the detonation, reflects from the closed tube end, and propagates as second shock or pressure wave toward the open end with a velocity of 700 m s^{-1}
4. Third shock or pressure wave propagating toward the open end is also observed

With the results measured using the double probe method, thus, several different shock or pressure waves behind or following the detonation wave are observed.

Behind the detonation waves propagating in a larger tube the same phenomena are observed. Namely, in the same measurement carried out in the detonation wave propagating in the same mixture (C_3H_8 + 10.O_2) in a steel tube having an inner-diameter of 25 mm and length of 2 m, a probe current of the same double probe as described above at the point 255 mm from the ignition point at the closed tube end is obtained as shown in Fig. 11.81 above. In Fig. 11.81 below the electron temperature T_e and ion density n_i obtained from the above measured probe current are illustrated with respect to the time t_f after the passage of the flame front at the measurement point. In the oscillogram of the probe current as well as in the diagram of T_e and n_i the retonation and shock waves are again observed.

In comparison with the ionization behind propagating flames measured by the double probe method (in Chaps. 9 and 10), the ion density behind detonation waves is much higher than that behind the flame corresponding to the reaction velocity, but the electron temperature behind the detonation front is lower than that behind the flame, because the mixture density behind the density front is much higher than that behind the flame, that is, the collision number with neutral particles of lower temperature per unit time, consequently the cooling effect of the free electron behind the detonation front is much larger.

11.5.2 Investigation by a Laser Light Scattering Method[121]

As mentioned already, the values of ionization measured by the double probe method behind detonation waves are not correct enough for the evaluation of the phenomena, the same laser light scattering method used for the measurement of the ionization behind a propagating flame explained in Chap. 9.3.3. is applied to the detonation as well.

A detonation wave vertically upward propagating in a stoichiometric propane–oxygen mixture in a channel having a square cross-section of 26 mm × 26 mm is measured by the laser light scattering method. The measurement is carried out many times at different positions in the front of and behind the detonation wave.

11.5 Ionization of Gases Behind Detonation Waves

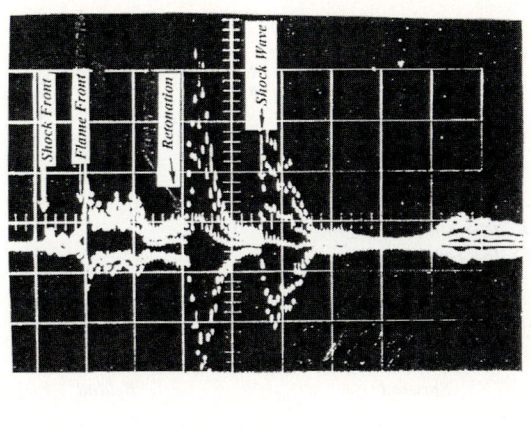

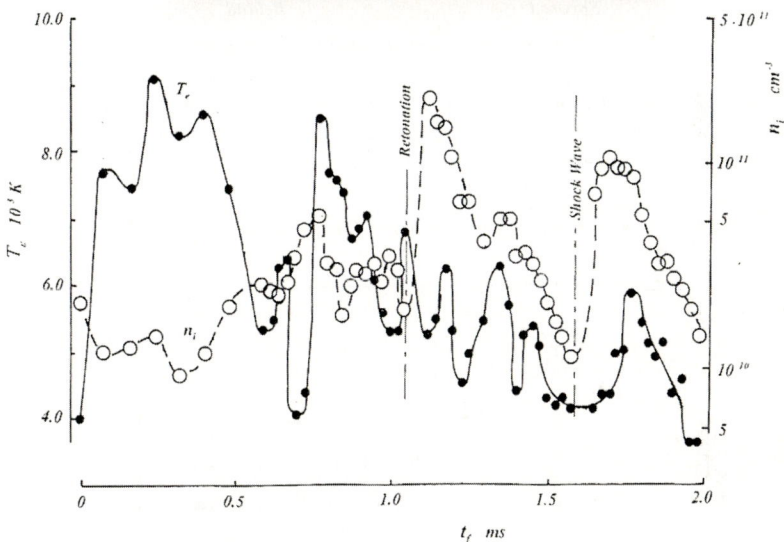

Fig. 11.81. *Above*: Probe current of the double probe at the transition to detonation waves propagating in a ($C_3H_8 + 10O_2$) mixture in a steel tube of 25 mm inner-diameter at 255 mm from the ignition point, current: 0.2 mA per div., time: 400 µs per div. *Below*: Electron temperature T_e, ion density n_i in relation to the time t_f after the passage of the flame front at the measurement point

The detonation is an irreversible phenomenon in which there is no homogeneous and equilibrium state. The measured results, therefore, fluctuate over a wide range.

The electron and ion temperature T_e and T_i as well as the electron density n_e estimated from the laser beams scattered by the charged particles in the gas near the detonation front are shown in Fig. 11.82. The measurements observing no scattered light are expressed by the points on the line of T_e and $T_i = 0$ showing only the measurement position, as the gas temperature is too low to be ionized.

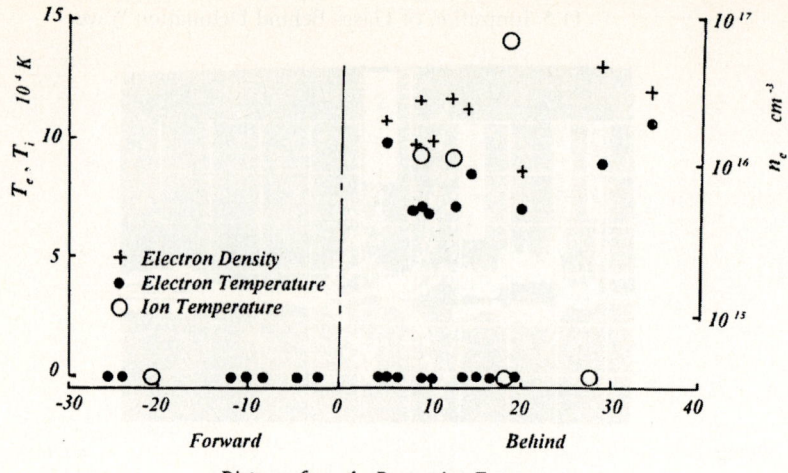

Fig. 11.82. Electron and ion temperature T_e, T_i and electron density n_e measured by the laser light scattering method ahead of and behind the detonation front propagating in a stoichiometric propane–oxygen mixture. Distance 0 means the detonation front

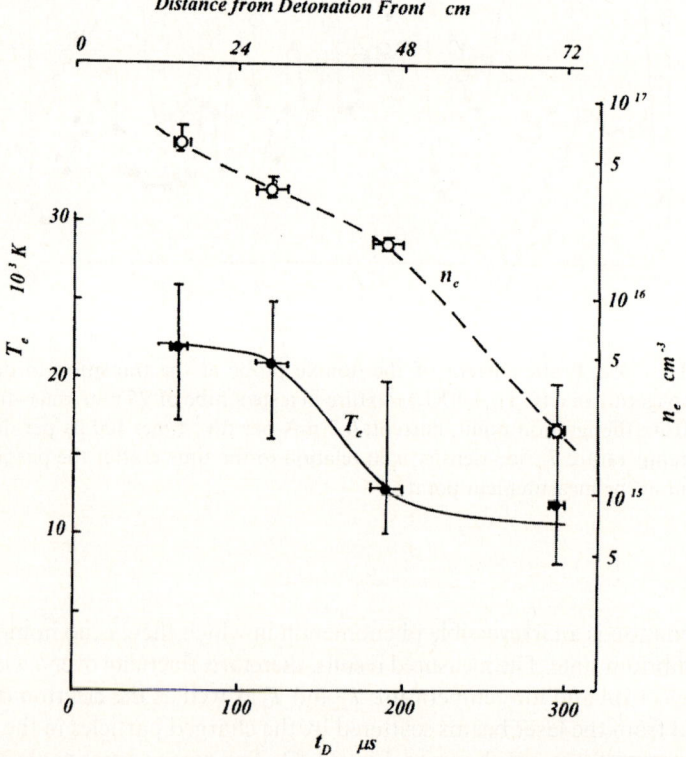

Fig. 11.83. Average electron temperature T_e and density n_e with respect to the time t_D after the passage of the detonation front at the measurement position and distance from the detonation front

11.5 Ionization of Gases Behind Detonation Waves

Behind the detonation front some particles having an electron temperature T_e of $6 \times 10^4 - 12 \times 10^4$ K, ion temperature T_i of $9 \times 10^4 - 15 \times 10^4$ K and the electron density n_e of 10^{16} per cm^3 to 10^{17} per cm^3 are observed, which are almost the same as those behind the combustion wave.

In the front of the detonation, however, we observe no ionized particles. This means, that the detonation wave propagates by self-ignition behind the shock at the detonation front, while the propagation of the combustion wave is led by the ions and free electrons of high temperature penetrating into the front of the flame.

Figure 11.83 illustrates the average electron temperature T_e and density n_e measured by the laser light scattering method with respect to the time t_D after the passage of the detonation front at the measuring point and distance from the front. The variation tendency of the electron density shows almost the same behavior as the ion density measured by the double probe method.

12

Industrial Applications of Detonation Waves

Because of the extremely high propagation velocity and pressure, the destroying power of detonation waves is very large and dangerous. Therefore, many investigations and efforts for the suppression of detonation waves have been carried out and the results have been reported. Up to date, however, the practical application of detonation waves for the industry have rarely been tried.

Detonation waves propagate in a combustible mixture accompanying shock waves having high pressure, and temperature, high combustion velocity and flow behind them. These three characters, namely, high speed combustion and flow can be applied to some practical purposes.

The shock waves can be produced using a shock tube, but its propagation as well as the pressure and temperature behind it are decreased during the propagation along a long or curved tube. On the contrary, the shock wave at the detonation front is not damped during its propagation, because it is supported by the combustion just behind the detonation wave. On the other hand, however, the temperature behind the detonation wave is much higher than that behind the shock wave produced in a shock tube. Therefore, the influence of gravity or buoyancy must be considered.

In this chapter the following three examples are presented:

1. Imploding detonation waves, in which the shock wave at the detonation front is concentrated into a point to obtain an extremely high temperature aiming to apply the method to initiate a nuclear fusion
2. A hypersonic combustion supported by the high-speed combustion behind the detonation waves with an intention to apply the method to hypersonic vehicles
3. A shock tube applying a free piston driven by the high-speed flow behind detonation waves to produce a very strong shock wave propagating with an extremely high Mach number with an intention to produce an isotropic gaseous plasma

12.1 Imploding Shock Waves

The imploding shock waves have been investigated by many scientists to obtain an extremely high temperature and pressure in the imploding focus with intention to produce some very hard materials like diamond or to initiate a nuclear fusion.

Perry and his coworkers tried to produce an imploding shock wave using a shock tube as schematically illustrated in Fig. 12.1.[121] The stronger the shock wave, the higher the temperature and pressure in the implosion focus. For it, a very large shock tube must be prepared, but using detonation waves accompanying a strong shock wave we can obtain much higher temperatures and pressures in the implosion focus.

In this chapter, first some theoretical treatments of imploding shock and detonation waves are discussed and then some experiments and their results are presented.

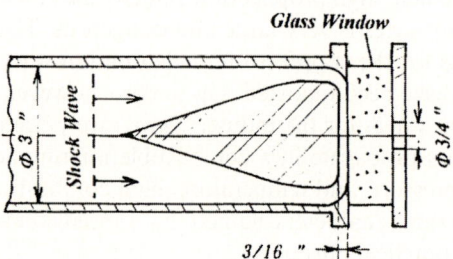

Fig. 12.1. Perry's shock tube for producing a cylindrical imploding shock wave[122]

12.1.1 Theoretical Calculation of Imploding Shock Waves[123]

Many works on theoretical calculation methods of imploding shock waves have already been reported. In these works there are two directions: one of them is the self-similarity method and the other the calculation of shock waves propagating along ducts of varying cross-section. Between the results calculated by the both methods there is practically only little discrepancy.

The work of Guderley in 1942 should be recognized as the first theoretical pioneer work of an imploding shock wave according to the self-similarity method, which is titled "Starke kugelige und zylindrische Verdichtungstoesse in der Naehe des Kugelmittelpunktes bzw. Zylinderachse" His essential idea is based on the self-similarity to solve the basic partial differential equations. His work is here briefly explained, referring the scheme in Fig. 12.2.

The theory starts from the basic formula of hydrodynamics and thermodynamics for the ideal gas. The sound velocity a is expressed by the following equation:

$$a^2 = \gamma \frac{P}{\rho}, \tag{11.5}$$

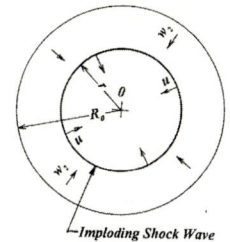

R_0 : initial radius
r: radial distance of the shock front from the implosion center 0
u : shock propagation velocity
w_2 : flow velocity behind the shock wave

Fig. 12.2. Scheme of a cylindrically or spherically imploding shock wave. The larger circle means the shock front at the beginning of the implosion

where P is the pressure, ρ the density, and γ the ratio of specific heats, and the entropy is given by:

$$s = \frac{R}{\gamma - 1} \ln \frac{P}{\rho^\gamma}, \tag{12.1}$$

where R is the gas constant.

The flow without any shock wave should satisfy the following equation,

$$\frac{\delta(\rho w)}{\delta r} + \frac{\delta \rho}{\delta t} + \frac{\alpha \rho w}{r} \quad \text{(continuity)}, \tag{12.2a}$$

$$\frac{1}{\rho}\frac{\delta P}{\delta r} + w\frac{\delta w}{\delta r} + \frac{\delta w}{\delta t} = 0 \quad \text{(Newton's principle)}, \tag{12.2b}$$

$$w\frac{\delta}{\delta r}\frac{P}{\rho^\gamma} + \frac{\delta}{\delta t}\frac{P}{\rho^\gamma} = 0 \quad \text{(constant entropy)}, \tag{12.2c}$$

where w is the particle velocity, t the time before the implosion of shock wave, r the distance from the implosion center, $\alpha = 1$ for the cylindrical implosion, and $\alpha = 2$ for the spherical one. On the other hand, the following equations can be applied for a shock wave:

$$P_2 = P_1 + \frac{2}{\gamma + 1} \rho_1(w_1 - u)^2 \left\{1 - \frac{a_1^2}{(w_1 - u)^2}\right\}, \tag{12.3a}$$

$$\rho_2 = \rho_1 \frac{\gamma + 1}{\gamma - 1 + 2\left(\frac{a_1}{w_1 - u}\right)^2}, \tag{12.3b}$$

$$w_2 = u + (w_1 - u)\frac{\gamma - 1 + 2\left(\frac{a_1}{w_1 - u}\right)^2}{\gamma + 1}, \tag{12.3c}$$

where u is the propagation velocity of the shock wave, while the subscript 1 means the state ahead of the shock wave and 2 that behind it.

For a strong wave, i.e., in the case of $(w_1-u) \gg a_1$ and $P_1 = \rho_1 a_1^2/\gamma \ll \rho_1(w_1-u)^2$, the following relations are obtained:

$$P_2 = \frac{2}{\gamma+1}\rho_1(w_1-u)^2, \tag{12.3a'}$$

$$\rho_2 = \rho_1 \frac{\gamma+1}{\gamma-1}, \tag{12.3b'}$$

$$w_2 = u + (w_1-u)\frac{\gamma-1}{\gamma+1}. \tag{12.3c'}$$

Assuming the self-similarity of the flow, the partial differential equations can be transformed to ordinary ones, namely, putting $r = a_n(-t)^n$ (parabola of nth order), where r the distance from the implosion center, the following equation for the propagation velocity of the shock wave can be obtained:

$$\frac{dr}{dt} = u = -na_n(-t)^{(n-1)} = -na_n^{1/n} r^{(1-1/n)}. \tag{12.4}$$

If t approaches 0, then u approaches ∞, therefore $n < 1$.

As the flow velocity w_1 before the arrival of the shock wave is 0,

$$P_2 = \left(\frac{2}{\gamma+1}\rho_1 n^2 a_n^{2/n}\right) r^{(2-2/n)}, \tag{12.3a''}$$

$$\rho_2 = \rho_1 \frac{\gamma+1}{\gamma-1}, \tag{12.3b''}$$

$$w_2 = \left(-\frac{2}{\gamma+1} n a_n^{1/n}\right) r^{(1-1/n)}. \tag{12.3c''}$$

Putting

$$\frac{r}{(\pm t)^n} = \xi,$$

where + sign means in the area $t > 0$, while – sign in $t < 0$, then we obtain the following relations:

$$P = P(\xi) r^{2-2/n} \tag{12.5a}$$

$$\rho = \rho(\xi) \tag{12.5b}$$

$$w = w(\xi)\, r^{1-1/n}. \tag{12.5c}$$

As the sound velocity a_0 at the initial state is expressed with the following equation:

$$a_0^2 = \frac{\gamma P_0}{\rho_0}, \tag{12.6}$$

$$a^2 = a_0^2 r^{2-2/n}. \tag{12.5d}$$

Applying these equations, the partial differential equations expressed by (12.2) can be transformed to ordinary ones

$$\frac{d\ln\rho}{d\ln\xi}\left(1\pm n\frac{\xi^{1/n}}{w}\right)+\frac{d\ln w}{d\ln\xi}\left(1-\frac{1}{n}+\alpha\right)=0, \quad (12.7a)$$

$$\frac{1}{\gamma}\frac{a^2}{w^2}\frac{d\ln P}{d\ln\xi}+\frac{d\ln w}{d\ln\xi}\left(1\pm n\frac{\xi^{1/n}}{w}\right)+\left(\frac{2}{\gamma}\frac{a^2}{w^2}+1\right)\left(1-\frac{1}{n}\right)=0, \quad (12.7b)$$

$$\left(\frac{d\ln P}{d\ln\xi}-\gamma\frac{d\ln P}{d\ln\xi}\right)\left(1\pm n\frac{\xi^{1/n}}{w}\right)+2\left(1-\frac{1}{n}\right)=0. \quad (12.7c)$$

Putting $\mu = a/(n\xi^{1/n})$ and $\nu = \pm w/(n\xi^{1/n})$, these differential equations are further transformed to a first-order one and solved. The results for $\gamma = 1.4$ suggest the following:

$n = 0.834$ at $\alpha = 1$, i.e., in a cylindrical implosion
$n = 0.717$ at $\alpha = 2$, i.e., in a spherical implosion
$P \propto r^{-0.396}$ for the cylindrical implosion
$P \propto r^{-0.792}$ for the spherical implosion

Next, we want to introduce the theory proposed by Chisnell[124], Chester, and Witham[125] for the shock wave propagating along a duct of a varying cross-section area.

The formula of a small change of Mach number δM due to a small area change δA is:

$$\frac{\delta A}{A}=\frac{-2M\,\delta M}{(M^2-1)\cdot K(M)}, \quad (12.8)$$

where A is the cross-section area of the duct and M the Mach number of the shock and

$$K(M)=2\left[\left\{1+\frac{2}{\gamma+1}\frac{1-\mu^2}{\mu}(2\mu+1+M^2)\right\}\right]^{-1} \quad (12.9)$$

and

$$\mu^2=\frac{(\gamma-1)M^2+2}{2\gamma M^2-(\gamma-1)}. \quad (12.10)$$

Integrating (12.8), we obtain the following relation:

$$Af(M) = \text{constant}, \quad (12.11)$$

where

$$f = z^{1/\gamma}(z-1)\left(z+\frac{\gamma-1}{\gamma+1}\right)^{-1/2}\left(\frac{1+R}{1-R}\right)^{\{\gamma/(2\gamma+1)\}^{1/2}}$$
$$\left[\frac{R-\sqrt{(\gamma-1)/2\gamma}}{R+\sqrt{(\gamma+1)/2\gamma}}\right]\exp\left\{\sqrt{\frac{2}{\gamma-1}}\tan^{-1}\left(\frac{2\sqrt{\gamma}\times R}{\gamma-1}\right)\right\}\cdots \quad (12.12)$$

$$R = \left\{1+\frac{\gamma-1}{(\gamma+1)z}\right\}^{-1/2} \quad (12.13)$$

$$z = \frac{P_2}{P_1} = \frac{2\gamma}{\gamma+1}M^2 - \frac{\gamma-1}{\gamma+1}. \quad (12.14)$$

For strong shock, $M \to \infty$, then $K(M)$ is almost constant and (12.8) should be

$$A^K(M^2-1) = \text{constant} \quad (12.15)$$

and

$$K \to K_\infty = 2\left[\left\{1+\sqrt{\frac{2}{\gamma(\gamma-1)}}\right\}\left\{1+\sqrt{\frac{2(\gamma-1)}{\gamma}}\right\}\right]^{-1}. \quad (12.16)$$

Equation (12.15) is expressed by:

$$M \propto A^{-K_\infty/2}. \quad (12.17)$$

For cylindrical implosion

$$M \propto A^{-K_\infty/2}. \quad (12.18)$$

And for spherical implosion

$$M \propto A^{-K_\infty}. \quad (12.19)$$

In the following Table 12.1 the values $m = 1-1/n$ obtained according to this method are compared with those by Guderley's method.

Table 12.1. $m = 1-1/n$ obtained by Guderley and Chisnell

γ	cylindrical implosion		spherical implosion	
	Chisnell	Guderley	Chisnell	Guderley
6/5	0.163112	0.161220	0.326223	0.320756
7/5	0.197070	0.197294	0.394141	0.394364
5/3	0.225425	0.226054	0.450850	0.452692

12.1 Imploding Shock Waves

Table 12.2. P/P_0 at implosion for different γ

γ	cylindrical implosion P/P_0 proportional to	spherical implosion P/P_0 proportional to
1.2	$r^{-0.362} \sim 0.366$	$r^{-0.631 \sim 0.652}$
1.4	$r^{-0.394 \sim 0.394}$	$r^{-0.788 \sim 0.789}$
1.67	$r^{-0.451 \sim 0.452}$	$r^{-.902 \sim 0.905}$

From these values we can deduce the relation between the pressure ratio P/P_0 and the radial distance r from the implosion center, as shown in Table 12.2, where P is the pressure behind the imploding shock wave and P_0 the initial one.

The logarithmic relation between the radial distance r of the shock front from the implosion center and the shock arrival time t at r before the implosion instant and that between the radial distance r and the pressure ratio P/P_0 are illustrated in Fig. 12.3a, b, where P is the pressure behind the imploding shock wave and P_0 the initial mixture pressure.

A theoretical treatment of the cylindrically and spherically imploding detonation waves has been made by Lee.[126, 127] Assuming a detonation wave propagating in a stoichiometric propane–oxygen mixture under the Chapman–Jouguet condition overdriven by the implosion, he tried to calculate the pressure at the C–J point, rising along the Hugoniot-curve by the implosion. At the beginning of the implosion the combustion heat plays an important role, but near the implosion center where a much stronger detonation wave propagates, the pressure rises almost in the same way as the imploding shock wave. The logarithmic

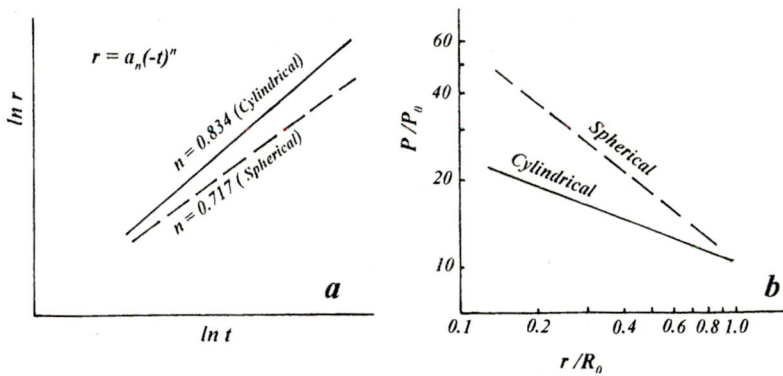

Fig. 12.3. (a) (*left*) Logarithmic relation between r and t, (b) (*right*) that between P/P_0 and r/R_0 in a cylindrical and spherical implosions according to Guderley's theory. r is the radial distance of the shock front from the implosion center O, t the shock arrival instant at r before implosion, P the pressure behind the shock, P_0 the initial gas pressure, R_0 the radius of the space where the implosion starts

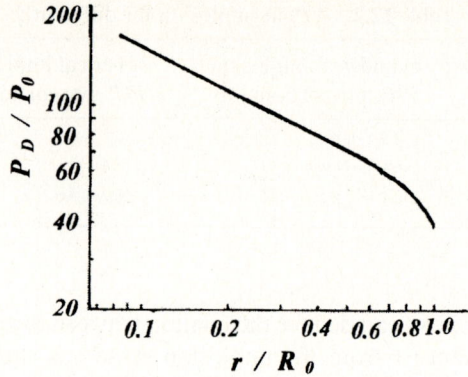

Fig. 12.4. Logarithmic relation between P_D/P_0 and r/R_0 of spherically imploding detonation wave propagating in a stoichiometric propane–oxygen mixture according to the Lee's theory, P_D: Pressure behind the detonation wave

relation between the pressure ratio P/P_0 and r/R_0 is shown in Fig. 12.4, where P_D is the pressure behind the detonation wave and r the radial distance of the detonation front from the implosion center.

12.1.2 Radially Divergent Detonation Wave[128]

According to the theories explained above in Sect. 12.1.1, the larger the initial radius of the cylindrically or spherically imploding detonation waves, the higher the temperature and pressure in the imploding focus. In order to produce an extremely high temperature and pressure in the implosion focus, therefore, we have to initiate the detonation at the same time on the surface of the large cylinder or sphere filled with an explosive mixture. Using a lot of spark plugs set on the surface of the cylinder or sphere, however, we cannot initiate the detonation at the same time, because the ignition as well as detonation initiation are stochastic, the ignition delay as well as the induction time and distance of the detonation initiation fluctuates over a wide range. It is, therefore, necessary to introduce a detonation wave everywhere on the surface of the cylinder or sphere at the same time. In order to enlarge a detonation wave propagating in a combustible mixture in a tube, a large cylindrical space is connected to the tube keeping a right angle to the tube, so that the detonation wave propagates into the cylindrical space center and turns the propagation direction by a right angle and radially diverges toward the circumference of the space.

In order to examine, if the detonation wave is really enlarged by such a method, some experiments are carried out applying an apparatus of steel shown in Fig. 12.5. Considering the buoyancy, the straight tube is kept exactly vertically, while the cylindrical space horizontally. A stoichiometric propane–oxygen mixture is filled in the tube and cylindrical space under an initial pressure between 13 and 22 kPa at room temperature.

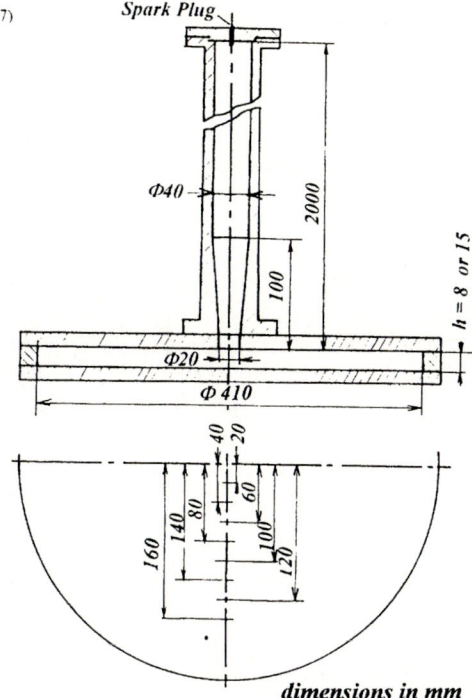

Fig. 12.5. Detonation chamber for producing radially diverging detonation waves

The mixture is ignited by a spark plug set at the top of the tube, from which a detonation wave transformed from the propagating flame propagates downward, then enters into the center of the cylindrical space, where the detonation should further propagate radially toward the circumference of the cylindrical space. The propagation of the detonation is observed by several ion probes or piezoelectric pressure transducers set at the points on the bottom plate of the cylindrical space having different radial distance r from the cylinder center, as shown in Fig. 12.5.

Measuring the passage instant t of the detonation front at each ion probe or pressure transducer, time–distance diagrams of the detonation propagation under different conditions are obtained. An example is shown in Fig. 12.6.

From such diagrams the variation of the detonation propagation velocity can be obtained as shown in Fig. 12.7. The pressure variation of the detonation front can be obtained by the piezoelectric pressure transducers which is also shown in Fig. 12.7. From the variation of the propagation velocity and that of the detonation front, we can recognize that a self-sustained detonation wave is produced about at the radial distance $r = 100$ mm and propagates toward the circumference of the cylinder vessel.

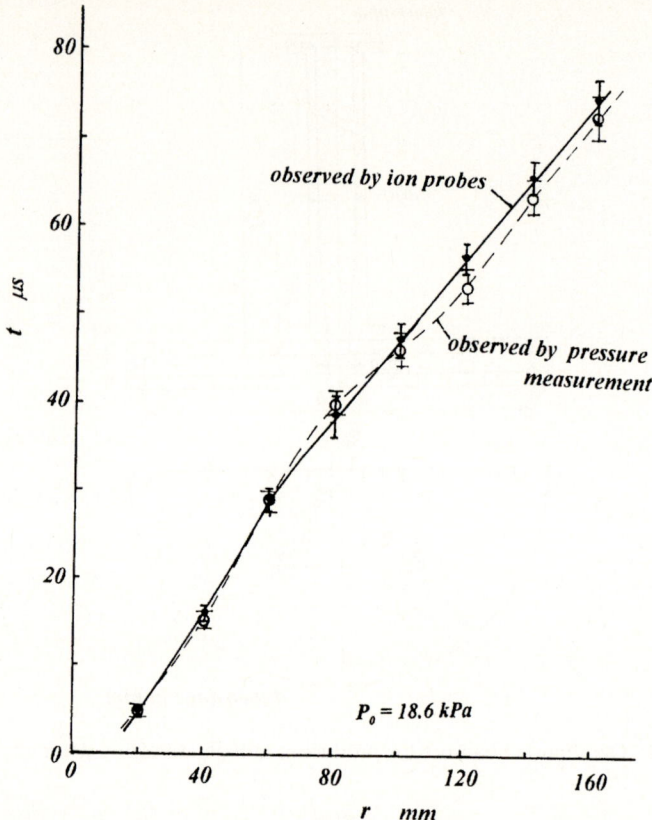

Fig. 12.6. Time–distance diagram of the divergent detonation front propagating radially in a stoichiometric propane–oxygen mixture in a flat cylindrical space. t is the passage instant of the front at the measurement point of a radial distance r

In Fig. 12.8 a photograph of the bottom plate (above) and that of the upper plate of the cylindrical space after several experiments of the detonation wave propagating in the cylindrical space having a height of 8 mm. On both the plates several circles marked by a bright line of different radial distance are observed. These bright circles and distances suggest that a shock wave produced by the detonation propagates radially from the cylinder center to its circumference, reflecting from both the plates reciprocally and marking the bright circles on the upper and bottom plates, as schematically shown in Fig. 12.9.

Setting a plexiglas plate coated with candle soot on the bottom plate, some traces of the shock or detonation wave are marked on the plate after each experiment. Cellular patterns having different cell sizes are also observed. On the circular band where the bright rings are marked on the bottom plate the cellular pattern is especially fine and the density N_a of the apex point, *the number of the apex point/cm²*, is much larger than those on the other area. In Fig. 12.7 the

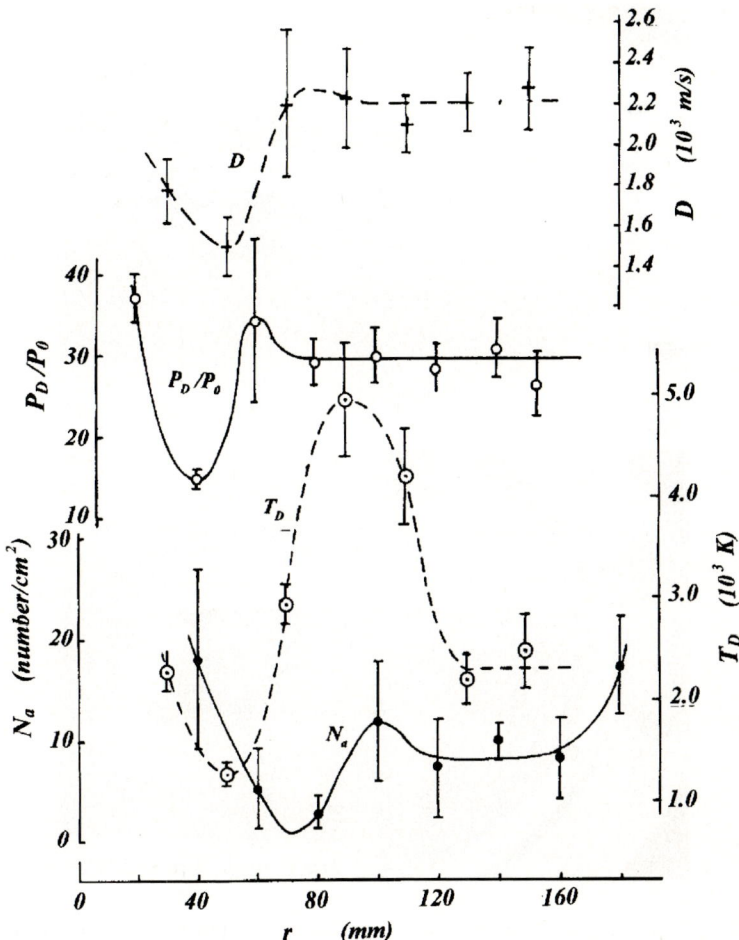

Fig. 12.7. Detonation propagation velocity D, pressure ratio P_D/P_0, mixture temperature T_D behind the shock wave at the detonation front estimated from the apex point formation probability and density N_a of the apex point in the cellular structure of the radially divergent detonation wave propagating in a stoichiometric propane–oxygen mixture in the cylindrical space in relation to the radial distance r. P_D: pressure behind the detonation wave, P_0: initial mixture pressure

density of the apex points N_a of the cellular pattern is illustrated with respect to the radial distance r.

From the distribution of the apex points in the cellular pattern the probability of the apex point formation can be calculated as described already in Sect. 11.2.2, from which the temperature T_D of the mixture behind the shock wave at the detonation front can be estimated according to (11.20). The temperature T_D is also illustrated in Fig. 12.7 with respect to the radial distance r.

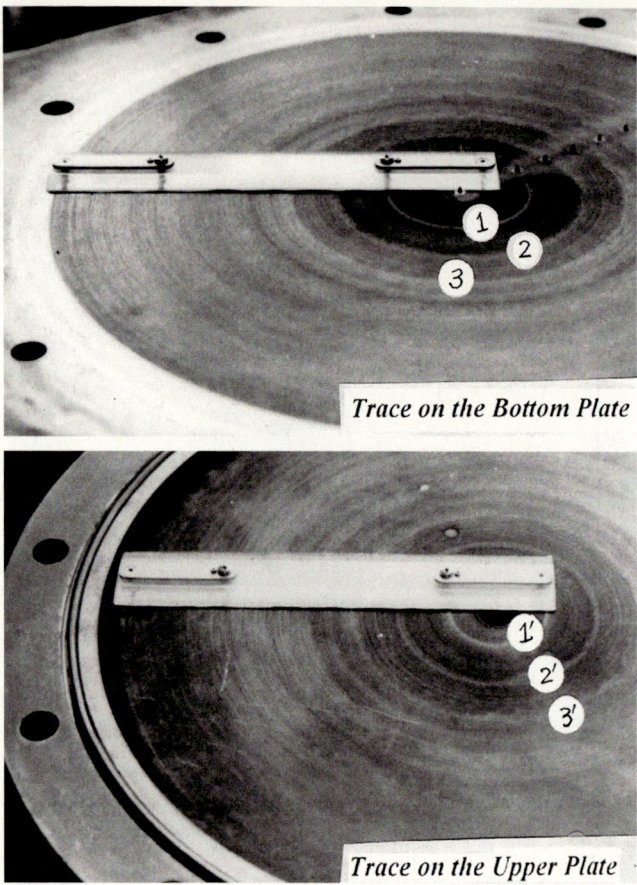

Fig. 12.8. Bright circular traces marked on the upper and bottom plates of the detonation chamber by reflected shock waves produced at the front of the detonation waves

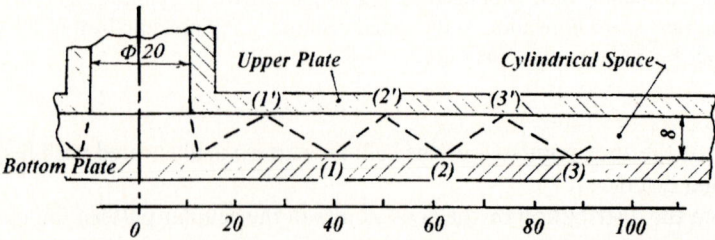

Fig. 12.9. A model of shock reflection by the detonation wave propagating from the detonation tube into the center of the cylindrical space. The numbers on the upper and bottom plates correspond to those in Fig. 12.8, dimensions in mm

The experimental results shown in Fig. 12.7 suggest that the detonation wave propagating from the tube into the center of the cylindrical space cannot directly propagate further radially in the cylindrical space turning its direction by a right angle, but the shock wave at the detonation front reflects on the center area of the bottom plate, then propagates upward to an oblique direction and downward after a reflection from the upper plate. The shock wave, thus propagates radially further to the circumference of the cylinder vessel between both the plates repeating the reflection. At each reflection the flow behind the shock wave is stagnated, increasing the temperature of the mixture, where many spontaneous ignition take place forming the cellular pattern. This cellular pattern produces the circular detonation wave propagating in the cylindrical space toward the circumference.

Applying this method a cylindrical detonation wave can be produced as large as one wants.

12.1.3 Cylindrically and Spherically Imploding Detonation Waves of Small Size[128]

The experimental apparatus, experiments and results of cylindrically and spherically imploding detonation waves of small size produced in cylindrical and hemispherical vessels having a diameter of about 400 mm are explained in this chapter, while those of spherically ones of larger size in hemispherical vessels having a diameter of about 800 mm are described in "Experimental Analysis" of Sect. 12.1.4.

A lot of theoretical and experimental investigations of cylindrically imploding shock or detonation waves have been reported by many scientists based on the theories of Guderley, Whitham, Chisnell, and Chester. Most of the experiments, however, carried out using a shock or detonation tube kept horizontally and a flat cylindrical space vertically in which the imploding detonation propagates along the vertical plane. Because of the buoyancy effect the imploding detonation waves cannot focus in the center well enough upon such a vertical plane. In this section as well as in the next, the experiments are carried out keeping the detonation tube vertically, while the bottom plates of the cylindrical and hemispherical spaces are kept horizontally to avoid the buoyancy effect.

Experimental Apparatus

The experimental apparatus for cylindrically imploding detonation waves is illustrated in Fig. 12.10, while that having a hemispherical space for spherical implosion is shown in Fig. 12.11. In order to initiate the imploding detonation the apparatus of radially divergent detonation described above in Sect. 12.1.2 is applied to both the detonation chambers.

The apparatus for the cylindrical implosion is composed of two cylindrical spaces having a diameter of 410 mm on both sides of a steel block, using two steel plates and one cylindrical vessel, as shown in Fig. 12.10. The upper

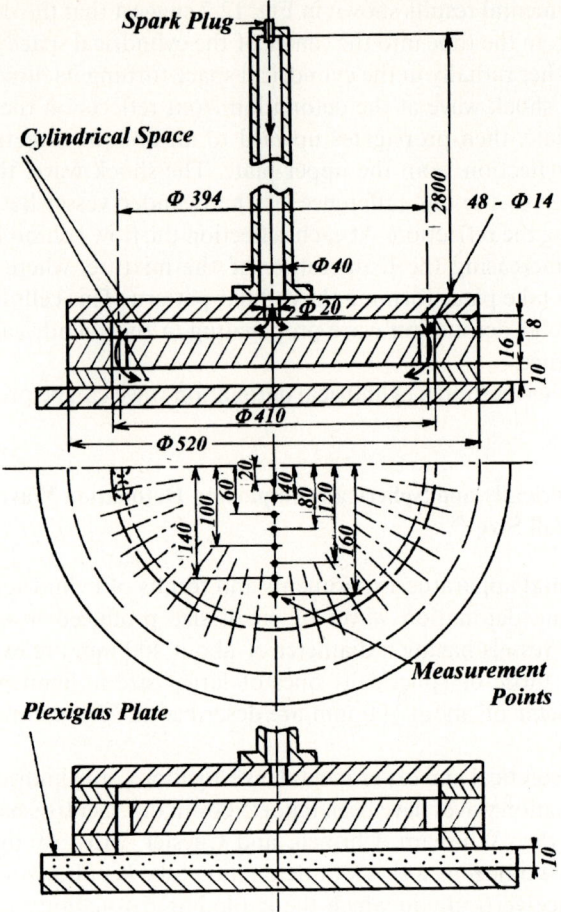

Fig. 12.10. Detonation chamber for the cylindrical implosion, dimensions in mm

cylindrical space is 8 mm high, while the lower one is 10 mm and both the spaces are interconnected with each other through 48 bore holes of 14 mm in diameter on the perimeter of the intermediate block.

The apparatus for the spherical implosion is also constructed like that of cylindrical implosion, but the cylindrical spaces have a diameter of 444 mm and the upper space is 10 mm high, while the lower one is 66 mm high. Both the spaces are also interconnected with each other through 45 bore holes of 14 mm in diameter on the perimeter of the intermediate block. The intermediate block has a conical shape on its lower side, as Fig. 12.11 shows, in order to converge the space from the circumference toward the center. In the lower space a steel ring is set, having a spherical surface and 90 bore holes each of 14 mm diameter equidistantly but alternately on two coaxial circles at the circumferences. In the center of the convergent space, a space of 2 mm thickness is reserved, so that the detonation waves can propagate further to the center.

12.1 Imploding Shock Waves

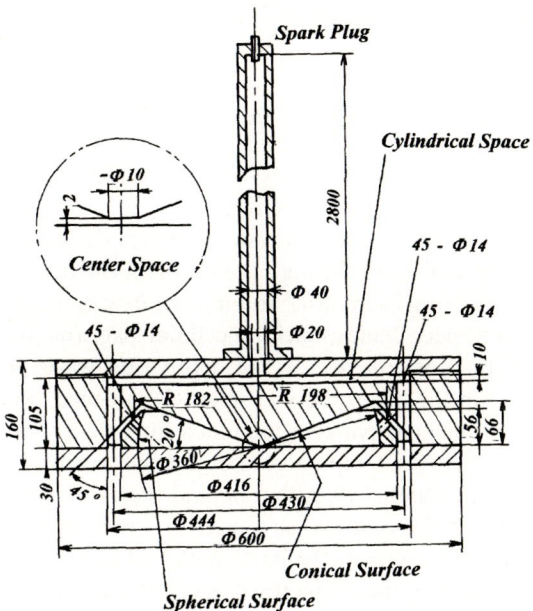

Fig. 12.11. Detonation chamber for the spherical implosion, dimensions in mm

A straight steel tube of 40 mm inner-diameter and 2,800 mm length is attached to the center of the cap of the upper cylindrical space of both the apparatus. A stoichiometric propane–oxygen mixture is introduced into the tube and both the spaces through an intake at the tube end after evacuating it. The initial pressure of the mixture is set at different values between 13 and 80 kPa at room temperature of about 20°C.

A detonation wave transformed from a flame initiated by the spark plug at the tube end propagates first through the tube into the center of the upper cylindrical space, then radially toward the circumference and stagnates at the cylinder wall. Then the shock waves produced by the detonation stagnation propagate through the holes on the perimeter of the intermediate block and also on the steel ring in the case of the spherical implosion into the lower space, where the ignition takes place everywhere at the circumference of the space by the shock waves almost at the same time. A circular detonation wave, thus, implodes toward the center, forming a cylindrically or spherically converging shock wave.

The passage time t of the imploding detonation front corresponding to the radial distance r from the implosion center is measured by the ion probes, while the pressure variations are recorded by piezoelectric pressure transducers having a time constant of 1.5 μs at the measurement points shown in the middle of Fig. 12.10, but the measurements are carried out on radii in different direction. At the imploding center the pressure is too high to measure.

In order to obtain some soot traces of the cellular structure of the imploding detonation waves, a 10 mm thick plexiglas plate coated with candle soot is

322   12 Industrial Applications of Detonation Waves

set on the bottom plate of the convergent space as shown in the lower part of Fig. 12.10.

Convergence of Imploding Detonation Waves

Figure 12.12A, B represent two examples of soot traces marked by the cylindrical (A) and spherical (B) implosions, respectively.

In Fig. 12.12A we observe that the shock waves driven by the detonation enter into the convergent space through the holes on the perimeter of the intermediate block, intersecting each other and initiate new detonation waves, which converge toward the space center marking cellular patterns. In these soot trace we recognize that the convergence of the detonation waves in the cylindrical space (A) as well as in the hemispherical space (B) proceeds well. In the spherical implosion (B) the cellular patterns are much finer than those in the cylindrical one, namely, the mixture temperature behind shock waves at the spherically imploding detonation is higher than that behind the cylindrical one (A).

We recognize the implosion focus as a trace of a point marked on the surface of a teflon plug set in the center of the bottom plate. The implosion center in each case does not coincide with the geometrical one, but shifts a few millimeters. The fluctuation of the implosion focus depends on the initial mixture pressure P_0, as Fig. 12.13 shows. The higher the initial mixture pressure, the smaller is the fluctuation range of the implosion focus. Under the initial mixture pressure higher than 80 kPa, the spherical imploding detonation wave focus is within a circle of 2 mm diameter.

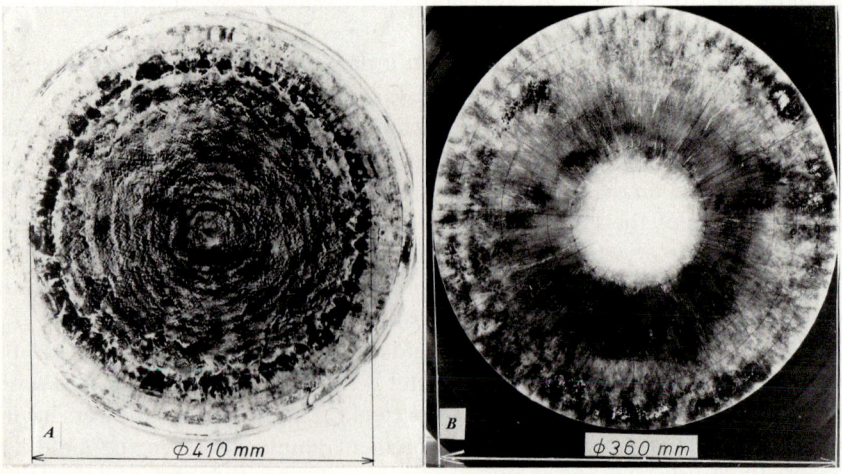

Fig. 12.12. (**A**) (*left*) Soot trace of a cylindrically propagating detonation wave and (**B**) that of spherically one observed in the detonation chambers in Figs. 12.10 and 12.11

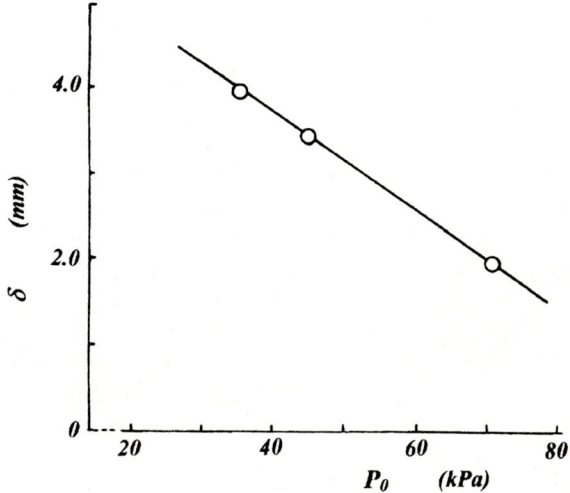

Fig. 12.13. Fluctuation range δ of the spherical implosion center in relation to the initial mixture pressure P_0

Propagation of the Imploding Detonation Waves

Observing the passage instant of the detonation front at each ion probe set on the outer surface of the bottom plate, the time–distance diagrams of the cylindrical and spherical imploding detonation fronts can be obtained as a relation between the radial distance r_c or r_s, from the implosion center and the passage instant t of the detonation front at r_c or r_s, as shown in the left side of Fig. 12.14, where r_c is the radial distance of the front of the cylindrical implosion and r_s that of the spherical one. The shift of the implosion center from the geometrical one is considered and t having a minus sign is the time before the implosion. The measurement error is estimated to be ±3%.

Transforming the time–distance diagrams of the imploding detonation waves in Fig. 12.14 into logarithmic relation, we obtain the diagram of $\ln r_c$–$\ln t$ and $\ln r_s$–$\ln t$ as illustrated in Fig. 12.15. In accordance with the proposal of Guderley we approximately obtain the following relations from the logarithmic relations in Fig. 12.15, considering the measurement error and fluctuation:

$$r_c \propto t^{0.83-0.89}, \tag{12.20}$$

$$r_s \propto t^{0.76-0.84}. \tag{12.21}$$

Pressure Behind Imploding Detonation Waves

Each pressure history in an oscillogram measured by a piezoelectric pressure transducer set on the bottom plate is always overlapped with mechanical noise. Eliminating such mechanical noise, some examples of the pressure variations behind imploding detonation waves are shown in Figs. 12.16 and 12.17. In each

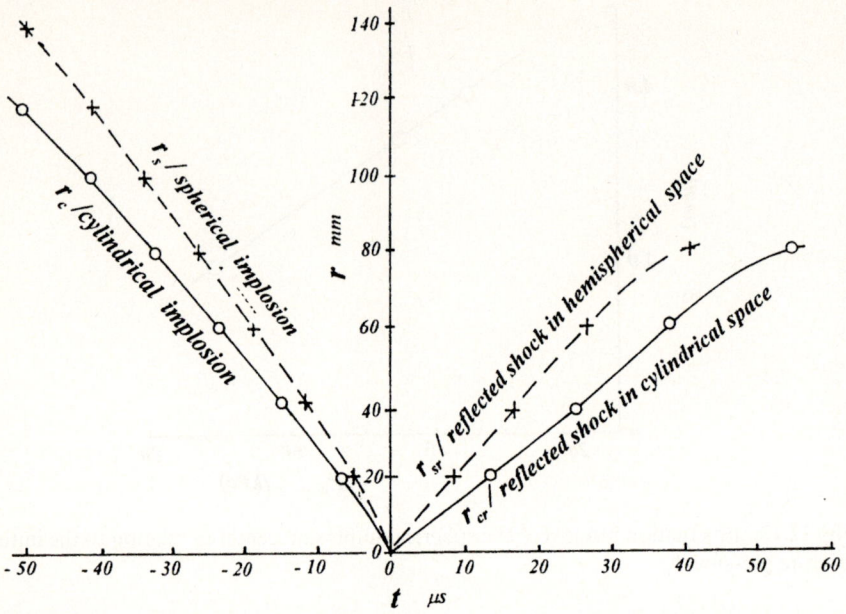

Fig. 12.14. The position r_c and r_{cr} of the fronts of the cylindrical detonation and its reflected shock, those r_s and r_{sr} of the spherical one and its reflected shock, respectively, in relation to the time $-t$ before and t after the implosion. r_c, r_{cr}, r_s, and r_{sr}: distances from the implosion center

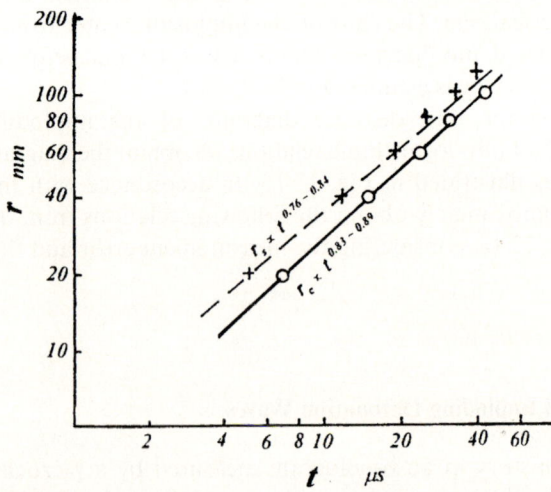

Fig. 12.15. Logarithmic time–distance diagrams of the cylindrically and spherically imploding detonation waves

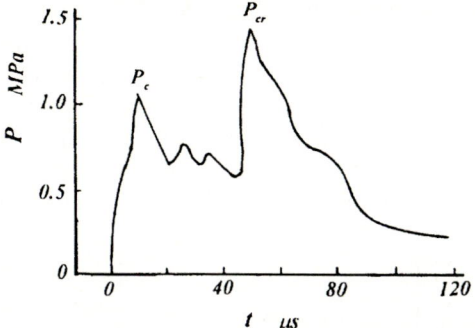

Fig. 12.16. Pressure history behind the cylindrically imploding detonation wave measured at $r = 40$ mm, initial mixture pressure: 19 kPa

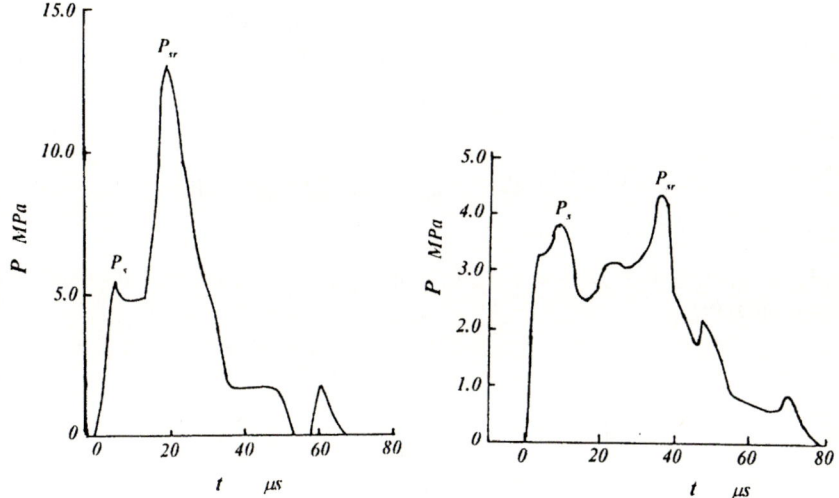

Fig. 12.17. Pressure variations behind the spherically imploding detonation waves measured at $r = 20$ mm (*left*) and 40 mm (*right*). Initial mixture pressure: 40 kPa

pressure diagram we always observe two peaks P_c and P_{cr}, or P_s and P_{sr} following one after another.

Assuming that the first peak P_c or P_s is the pressure behind the imploding detonation, we can obtain a time–distance diagram of the second peak pressure P_{cr} or P_{sr} of each imploding detonation front shown in the right side of Fig. 12.14, from which we recognize that the second peak pressure suggests the pressure behind shock waves produced by the reflection of the imploding detonation waves from the implosion center with each other.

Figure 12.18 shows logarithmic diagrams of the pressure ratios P_c/P_0 and P_{cr}/P_0 with respect to the radial distance r from the implosion center and Fig. 12.19 those of P_s/P_0 and P_{sr}/P_0, where P_c is the peak pressure behind the cylindrically imploding detonation wave, P_{cr} that behind the reflected shock wave, P_s the peak pressure behind the spherically imploding detonation wave, P_{sr} that behind the reflected shock wave, and P_0 the initial mixture pressure. P_{th} is the pressure behind the imploding detonation obtained according to Lee's theory.

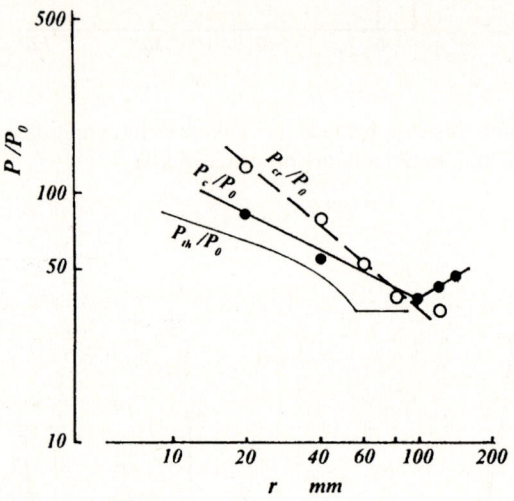

Fig. 12.18. ln P_c/P_0, ln P_{cr}/P_0 and ln P_{th}/P_0 with respect to ln r in cylindrically imploding detonation waves

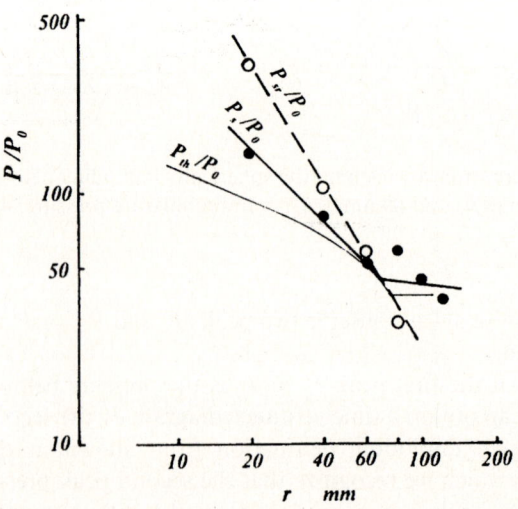

Fig. 12.19. ln P_s/P_0, ln P_{sr}/P_0 and ln P_{th}/P_0 with respect to ln r in spherically imploding detonation waves

12.1 Imploding Shock Waves

From the results the following relations are obtained:

$$\frac{P_c}{P_0} \propto r^{-0.5 \pm 0.1} \text{ for the cylindrical implosion,} \qquad (12.22)$$

$$\frac{P_s}{P_0} \propto r^{-1.0 \pm 0.15} \text{ for the spherical implosion.} \qquad (12.23)$$

According to Guderley's theory the exponent of t in $r = a_n(-t)^n$ is

$n = 0.83$–0.89 for the cylindrical implosion and
$n = 0.76$–0.84 for the spherical implosion.

Therefore

$$\frac{P_c}{P_0} \propto r^{-0.3 \sim 0.4} \text{ for the cylindrical implosion,}$$

$$\frac{P_s}{P_0} \propto r^{-0.4 \sim 0.6} \text{ for the spherical implosion.}$$

The experimental results suggest, however, the exponents of r in (12.5) are much larger than those obtained according to the theory of Guderley as well as those of Whitham and others, as shown in (12.22) and (12.23).

In Fig. 12.20 $\ln(P_c/P_0)$, $\ln(P_s/P_0)$, $\ln(P_{cr}/P_0)$, and $\ln(P_{sr}/P_0)$ with respect to r/R_0 are illustrated, where R_0 is the radius of the cylindrically or spherically imploding detonation waves at their start.

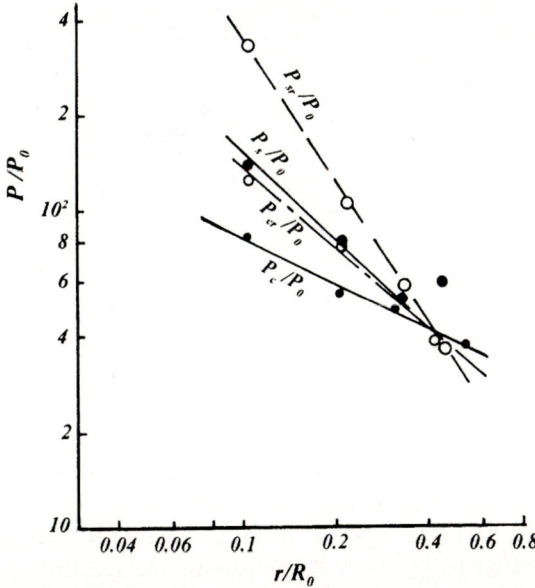

Fig. 12.20. Logarithmic relations of P_c/P_0, P_{cr}/P_0, P_s/P_0, or P_{sr}/P_0 to r/R_0

The results suggest the following relations:

$$\frac{P_c}{P_0} \propto \left(\frac{r}{R_0}\right)^{-0.5 \pm 0.1}, \tag{12.24}$$

$$\frac{P_{cr}}{P_0} \propto \left(\frac{r}{R_0}\right)^{-0.8 \pm 0.1}, \tag{12.25}$$

$$\frac{P_s}{P_0} \propto \left(\frac{r}{R_0}\right)^{-1.0 \pm 0.15}, \tag{12.26}$$

$$\frac{P_{sr}}{P_0} \propto \left(\frac{r}{R_0}\right)^{-1.5 \pm 0.15}. \tag{12.27}$$

Temperature Behind Imploding Detonation Waves[129]

There are several different temperatures to be measured behind imploding detonation waves, namely, gas temperature, ion and electron temperature, and spectroscopic temperature measured by observing the intensity of several spectral lines of gas added to the mixture.

As the phenomena are very unstable, the direct measurement of the gas temperature is very difficult. Therefore we tried to measure the spectroscopic temperature through an optical system, the electron temperature by a double probe method and gas temperature behind the shock wave at the detonation front estimated from the probability of the apex point formation observed in the cellular pattern of the detonation waves.

Spectroscopic Temperature

The same method in Sect. 4.1 developed for measuring the spectroscopic temperature of argon behind reflected shock waves is applied to the spherically imploding detonation waves propagating in a stoichiometric propane–oxygen mixture in the same detonation chamber explained in "Experimental Analysis" in Sect. 12.1.3. The propane–oxygen mixture contains 5% argon and has an initial pressure of 32 kPa at 15°C.

As schematically shown in Fig. 12.21 the intensity of the spectral lines of argon in the mixture having the wavelength 5,912.1 Å, 5,373.5 Å, 6,043.2 Å and 6,965.4 Å are measured at radial distance r = 1.5 mm, 20 mm, and 40 mm through a hole of 2 mm diameter behind sapphire window by means of a monochromator, photomultiplier, and oscilloscope.

In order to eliminate the stray light, several spectral lines having wavelengths near those of argon, emitted from the mixture with and without argon gas behind the detonation waves are measured, respectively. Comparing the two results shown in Fig. 12.22, the real intensity of the spectral lines of argon are able to be obtained.

12.1 Imploding Shock Waves   329

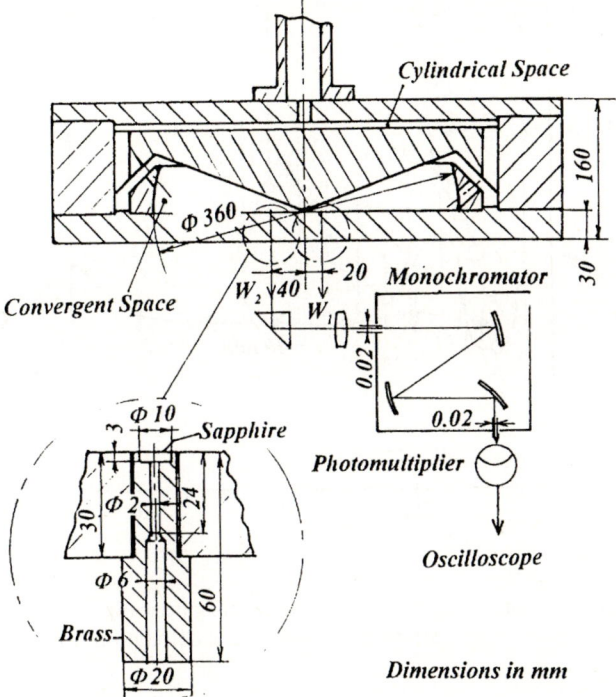

Fig. 12.21. Sketch of the experimental apparatus for measurement of the spectroscopic temperature behind spherically imploding detonation waves

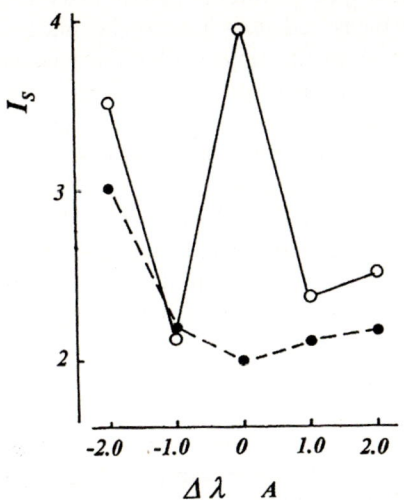

Fig. 12.22. Intensity of spectral line I_s in an arbitrary unit. $\Delta\lambda$: deviation of wavelength from argon. Spectral line of 6,043.2 Å *Solid line*: from mixture with 5% argon, *broken line*: from that without argon

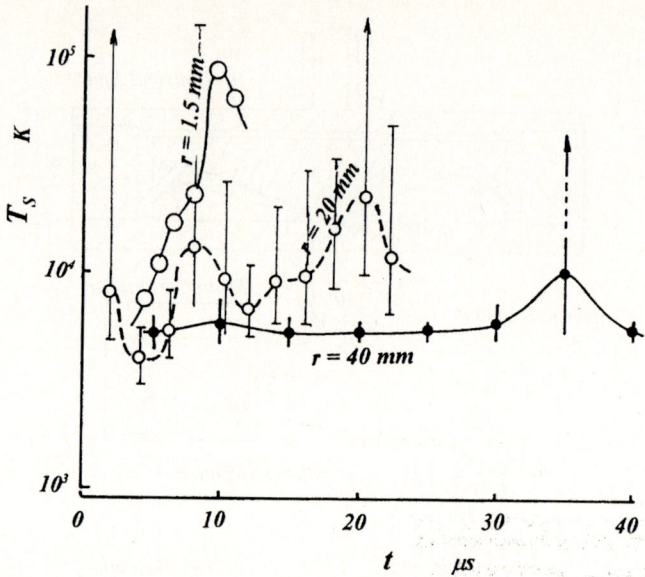

Fig. 12.23. Spectroscopic temperature T_s behind the spherically imploding detonation wave with respect to the time t after the passage of the detonation front at the measurement point

Applying (9.47) in Sect. 9.4.1 the spectroscopic temperature of an imploding detonation wave is calculated at each measurement position from the measured intensities of the several spectral lines of argon mixed in the mixture. The spectroscopic temperature measured at $r = 1.5$ mm, 20 mm and 40 mm are illustrated in Fig. 12.23 in relation to the time t after the passage of detonation front at each measurement position together with a confidence interval of 40% at $r = 20$ mm and 60% at $r = 40$ mm.

In each temperature variation two peaks are observed, the first at $t = 8$–10 μs for $r = 20$ mm and 40 mm, and the second at $t = 20$ μs for $r = 20$ mm, while at $t = 35$ μs for $r = 40$ mm. As the central space of $r < 4$ mm is cylindrical, the temperature measured at $r = 1.5$ mm should show a different tendency from those at $r = 20$ mm and 40 mm. It is not evident if the value in the central space is that behind the imploding detonation or reflected shock wave.

Electron Temperature

Applying the double probe method explained already in Sect. 9.2.3, the electron temperature behind detonation waves can be measured. The electron temperature measured at $r = 16$ mm and 20 mm behind the spherically imploding detonation waves propagating in a stoichiometric propane–oxygen mixture in this apparatus described above is illustrated with respect to the time t after the passage of the detonation front at the measuring point in Fig. 12.24. The electron temperature T_e measured by the double probe method usually shows a

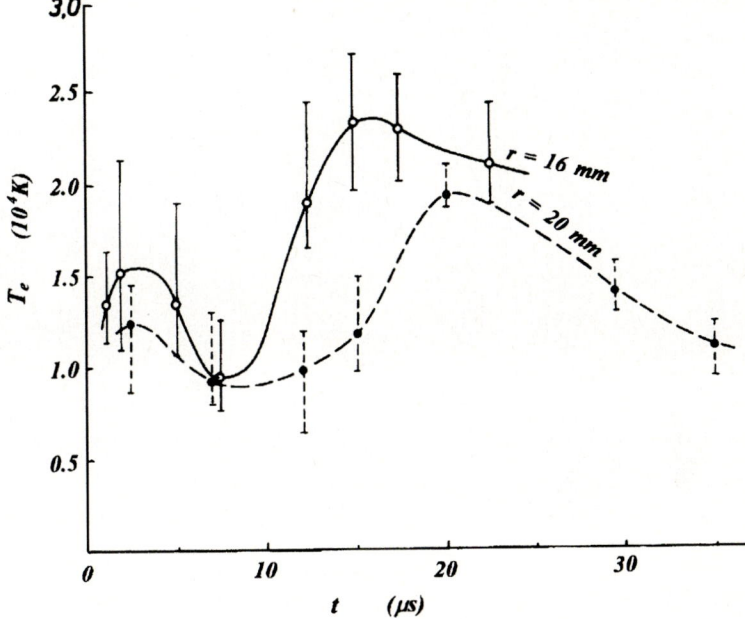

Fig. 12.24. Electron temperature T_e behind the spherically imploding detonation wave with respect to the time t after passage of the detonation front at the measurement point

value much higher than the gas temperature, but it closes to the gas temperature with increase of gas density. As the gas density behind the imploding detonation is very high, the electron temperature measured by the double probe method may be close to the gas temperature T_g.

Mixture Temperature Behind the Shock Wave at the Detonation Front

As presented in Fig. 12.12A, B, soot traces of the cylindrically or spherically imploding detonation waves are obtained on plexiglas plates coated with candle soot. Figure 12.25 shows an example of soot trace photographs printed by a cylindrical implosion, in which we observe that the pattern of cellular structure becomes finer with propagation of the detonation waves toward the implosion center.

From the histogram of distance between two successive apex points in the cellular pattern on both sides of a coaxial circle having a radius from the implosion center, the probability μ_d of the apex formation at the radius r can be calculated by the method explained in section "Cellular Structure of the Detonation Waves" of Chap. 11. According to (11.20) presented in section "Cellular Structure" of Chap. 11, the mixture temperature T_μ behind the shock waves at the detonation front is further calculated from the apex point formation probability μ_d as shown in the Table 12.3.

$$\mu_d = A \exp\left(-\frac{E_{ed}}{RT}\right). \tag{11.20}$$

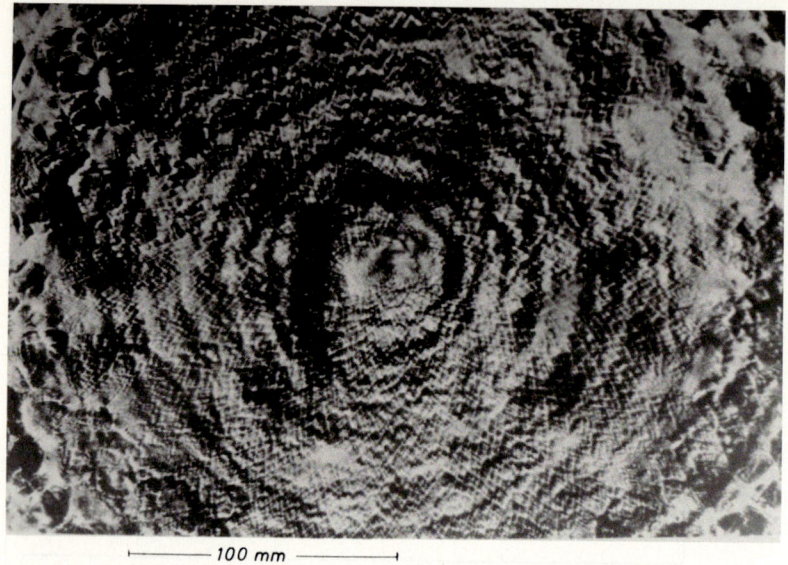

Fig. 12.25. Soot trace printed by a cylindrically imploding detonation wave

Table 12.3. Mixture temperature behind implosion front estimated from the probability of apex point formation

radial distance r (mm)	cylindrical implosion		spherical implosion	
	μ_c (10^{10} ms^{-1} mol^{-1})	$T_{\mu c}$ (K)	μ_s (10^{13} ms^{-1} mol^{-1})	$T_{\mu r}$ (K)
20	9.7 ± 0.4	2,160 ± 15	6.8 ± 0.3	4,140 ± 30
40	2.2 ± 0.2	1,970 ± 15	3.3 ± 0.2	3,770 ± 30
60	1.1 ± 0.1	1,880 ± 15	1.9 ± 0.1	3,530 ± 30

In this table r means the radial distance from the implosion center, μ_c the apex point formation probability and $T_{\mu c}$ the temperature estimated from μ_c behind the shock wave at the cylindrically imploding detonation front, while μ_s and $T_{\mu r}$ are those behind the spherically imploding ones. For the calculation of the temperature according to (11.20) the frequency factor A_d is assumed to be 10^{17} and the effective activation energy E_{ed} to be 250 kJ mol^{-1} as the diagram in Fig. 11.28 suggests.

The temperature $T_{\mu c}$ or $T_{\mu s}$ estimated from the apex point formation probability is the mixture temperature behind the shock wave at the detonation front ahead of the combustion zone, should be much lower than the maximum temperatures behind the imploding detonation waves and is rather comparable with the gas dynamic temperature T_{gr} or T_{gs} calculated from the shock propagation velocity.

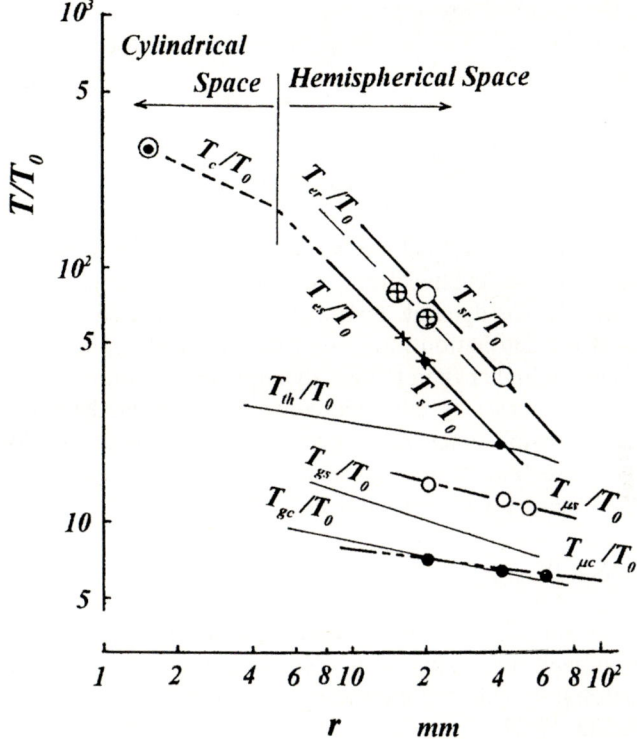

Fig. 12.26. Ratios of the measured and theoretical peak temperatures T to the initial one T_0 with respect to the radial distance r in logarithmic relation, subscript c means cylindrical implosion, s spherical implosion, r shock reflected in implosion center and th theoretical one

The ratio of the all measured peak temperatures to the initial mixture temperature T_0 with respect to the radial distance r are presented in a logarithmic diagram in Fig. 12.26.

Concluding Remarks on the Temperatures Behind Imploding Detonation Waves

The gas is often in a nonequilibrium state during a certain period after the arrival of shock waves and consequently the spectroscopic temperature behind the shock waves does not agree with the gas temperature, while the electron temperature measured by a double probe method usually shows a higher value than the gas temperature in a low pressure gas, but can approach the gas temperature in such high-pressure gas as behind the imploding detonation waves, as explained in Sect. 9.2.4.

As the spectroscopic temperatures measured behind spherically imploding detonation waves, however, well agree with the first and second peak electron temperatures, respectively, the measured spectroscopic temperatures as well as the electron temperatures are approximately equal to the gas temperature. From the diagram in Fig. 12.26 the following relations are obtained:

$$\frac{T_s}{T_0} \propto r^{-1.0 \pm 0.2} \text{ and} \tag{12.28}$$

$$\frac{T_{sr}}{T_0} \propto r^{-1.1 \pm 0.2}. \tag{12.29}$$

The mixture temperature $T_{\mu c}$ behind the shock wave at the cylindrically imploding detonation front calculated from the apex point formation probability μ_d agrees well with the gas dynamic temperature T_{gc} calculated from the propagation velocity of the detonation wave, while that $T_{\mu r}$ at the spherically imploding detonation is much higher than the gas dynamic temperature T_{gs}. This suggests that the entropy increases by the rapid convergence of the detonation waves in the spherical implosion much more than in the cylindrical one. Without any consideration of such an entropy increase, the simple self-similarity cannot be applied to the spherically imploding detonation waves.

Pressure–Volume Diagram

Considering the variations in the detonation propagation velocity, pressure and temperature at each measuring point, the process of the gaseous state in the imploding detonation can be followed by a diagram of pressure vs. volume as illustrated in Fig. 12.27.

In this diagram, the coordinate P/P_0 means the ratio of the pressure behind the spherically imploding detonation waves or that behind their reflected shock waves to the initial mixture pressure of 40 kPa, and that of the specific volume v/v_0. H_S is the Hugoniot-curve of the shock wave propagating in a stoichiometric propane–oxygen mixture, while H_D is the Hugoniot-curve corresponding to the self-sustained detonation wave and H_{D0} that after complete combustion without any dissociation. The spherical implosion starts from the circle of $r = 76$ mm, while the self-sustained detonation propagates from the circumference of the convergent space beyond this circle.

The so-called Rayleigh line at $r = 76$ mm, therefore, passes the C–J point contacting the H_D-curve, while those at $r = 40$ mm and 20 mm intersect the H_D-curve because of their overdriven state. In the pressure–volume diagram, the gas state at the shock front of the detonation should be expressed as the point where the Rayleigh line intersects the H_S-curve, that is, S_4 at $r = 40$ mm and S_2 at $r = 20$ mm.

The measured pressure and temperature in relation to the time after the arrival of the detonation front at the measuring point suggest the following process of the gas state in the diagram:

at $r = 40$ mm,

$$S_4 \to P_{S-40} \to 4a \to 4b \to P_{r-40} \to 4c \text{ corresponding to } t \text{ } (\mu s)$$
$$= 0 - 10 - 15 - 30 - 35 - 40$$

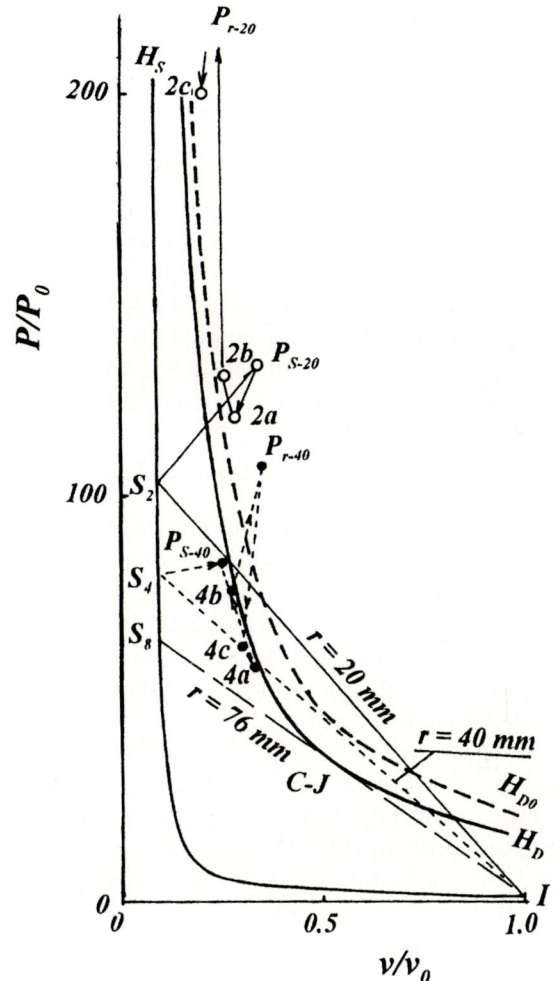

Fig. 12.27. Pressure–volume diagram of spherically imploding detonation waves. H_S, H_D, and H_{D0}: Hugoniot-curves for the shock and detonation waves with and without dissociation, respectively. I-S_2, I-S_4, and I-S_8: Rayleigh lines

and
at $r = 20$ mm

$$S_2 \to P_{S-20} \to 2a \to 2b \to P_{r-20} \to 2c \text{ corresponding to } t(\mu s)$$
$$= 0 - 8 - 10 - 15 - 20 - 22$$

At $r = 40$ mm, the gas state after the combustion changes along the H_D-curve and that behind the reflected shock waves is represented by a point on the right-hand side of the H_D-curve in the pressure–volume diagram, while the gas at

$r = 20$ mm is always in a state on the right-hand side of the H_D-curve after combustion as shown in the diagram of Fig. 12.27. Especially, the gas at the first peak and that behind the reflected shock waves are in states on the right-hand side of the H_{D0}-curve. These results consequently suggest that the heat released in the spherically imploding detonation at $r = 20$ mm must be higher than that released by the ordinary combustion.

Considering both the processes of pressure and temperature in the spherically imploding detonation waves, we conclude that an anomalous high pressure does not occur behind the shock waves at the detonation front but in the combustion zone of the detonation, deviating from the Rayleigh line and releasing more heat than the combustion heat. We should now remember that such an anomalous large heat release is also observed at the interaction between shock or pressure waves and combustion waves, as described in Sect. 10.3.3.

12.1.4 Spherically Imploding Detonation Waves of Large Size

In the experiments of the imploding detonation waves in the apparatus of small size explained in Sect. 12.1.3, the relation among the pressure, temperature rise, and the radial distance from the implosion center are expressed with (12.24)–(12.29). According to the relations we can obtain a higher temperature in the implosion center of a larger apparatus. It is, therefore, very important to confirm, if the imploding detonation waves have the same relations as above described also in an enlarged apparatus.

First the experiments were carried out using an enlarged detonation chamber having almost the same form as that described in Sect. 12.1.3, but an almost double size. As the experimental results suggest, a secondary detonation wave produced at the upper corner of the convergent space follows the first main detonation wave and the combustion energy cannot be well concentrated in the implosion center, an improved detonation chamber was developed and some experiments using the improved one were carried out.

G-Type Detonation Chamber for the Spherically Imploding Detonation Waves

As the experiments using the detonation chamber schematically shown in Fig. 12.28 were carried out in the Max-Planck Institut fuer Strömungsforschung in Göttingen, Germany, we call the apparatus G-type[130].

Just like the detonation chamber of small size in Fig. 12.11 in Sect. 12.1.3, the detonation chamber G-type is composed of a detonation tube having an innerdiameter of 25 mm and a length of 2 m, a flat cylindrical space having an inner-diameter of 950 mm and a height of 20 mm, 96 connecting holes of 18 mm diameter, a ring having a spherical inner surface and 96 holes of 18 mm diameter equidistantly on each circle but alternately spaced on five coaxial circles at the circumference as shown in Fig. 12.28b, and a conically convergent space having an inner-diameter of 800 mm. In the center of the convergent space, the top and bottom parts are 1.5 mm apart, so that the detonation waves can propagate into the center.

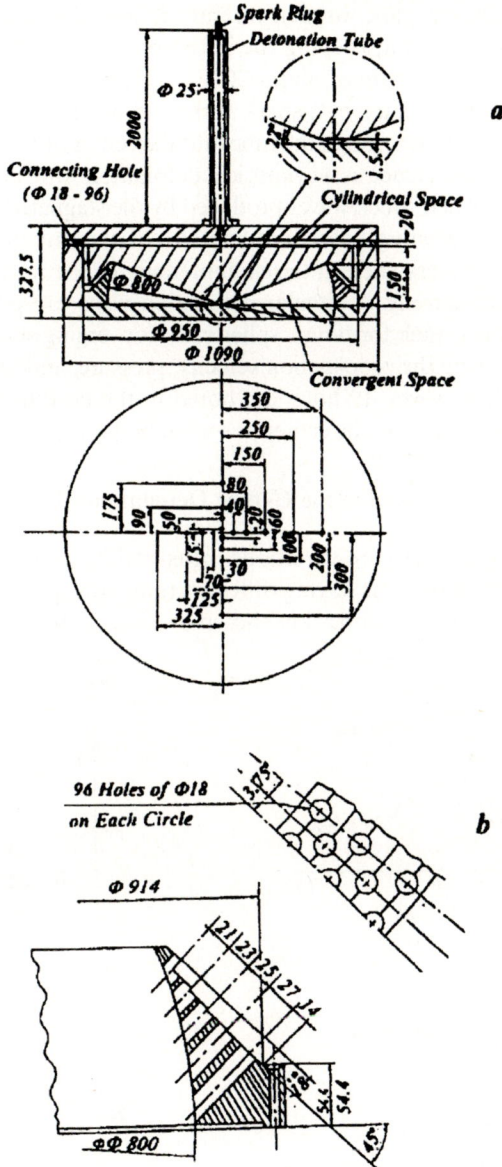

Fig. 12.28. (**a. Above**) Detonation chamber G-type for spherically imploding detonation waves, dimensions in mm. (**b. Below**) Location of connecting holes in the ring having a spherical surface, dimensions in mm

338   12 Industrial Applications of Detonation Waves

A stoichiometric propane–oxygen mixture is introduced into the chamber after evacuating it to 1.0 Pa, keeping the initial pressure of the mixture around 40.5 kPa at room temperature of about 20°C.

A detonation wave initiated by a spark at the end of the detonation tube develops and propagates at first through the tube into the center of the cylindrical space, then toward the circumference as a cylindrical detonation wave and is stagnated on the cylindrical wall. Then shock waves produced by the stagnated detonation propagate through the connecting holes and the holes of the ring into the convergent space where ignitions everywhere at the circumference of the convergent space take place almost simultaneously. A circular detonation wave caused by the ignitions implodes toward the center, forming a spherically converging shock wave.

In order to measure the propagation velocity, pressure, and temperature of the imploding detonation waves, 19 holes are bored in the bottom steel plate having different radial distances from the center as shown in the lower part of Fig. 12.28a.

Pressure Variations Measured in the G-type Detonation Chamber

Three examples of the pressure variations measured by a piezoelectric pressure transducer at several positions having different radial distance r from the implosion center in the G-type chamber are shown in Fig. 12.29a–c.

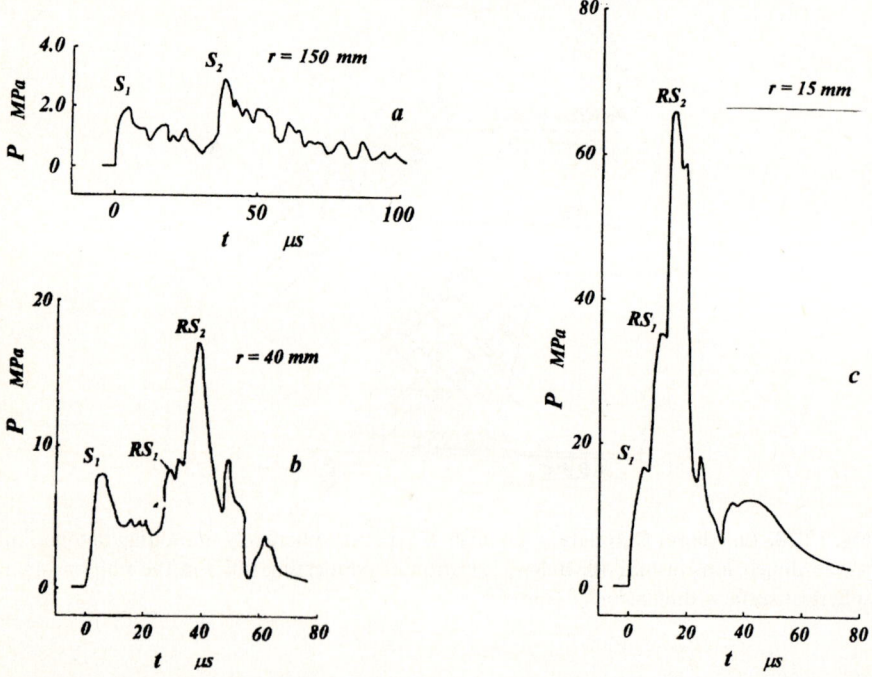

Fig. 12.29. (a) Pressure variation at $r = 150$ mm, (b) pressure variation at $r = 40$ mm, (c) pressure variation at $r = 15$ mm

12.1 Imploding Shock Waves

In these diagrams we observe not only the peak pressure at S_1 which appears at the detonation front, but also other peak pressures at S_2, RS_1 and RS_2 which are caused by different shock waves following the detonation wave. Combining the pressure diagrams with the time–distance diagram obtained by observing the passage instant of the detonation front at each measurement point using ion probes, a time–distance diagram of each shock wave is obtained as illustrated in Fig. 12.30.

In the time–distance diagram the imploding detonation front S_1, a shock wave S_2 propagating behind the detonation wave and a reflected shock wave propagating from the focus are observed. The propagation velocity of the shock wave S_2 relative to the bottom steel plate is 17,000 m s^{-1} which is the propagation velocity of the wave in the normal direction divided by the inverse of the sine of the angle between the normal propagation direction and the bottom plate. At $r = 150$ mm this relative velocity is altered to 6,200 m s^{-1}. This transformed portion of the shock wave is termed S_3, which follows the detonation waves and merges with it at about $r = 40$ mm as shown in Fig. 12.30.

From the ratio of pressure ahead of and behind the shock wave S_2, the normal propagation velocity is determined to be approximately 2,500 m s^{-1}.

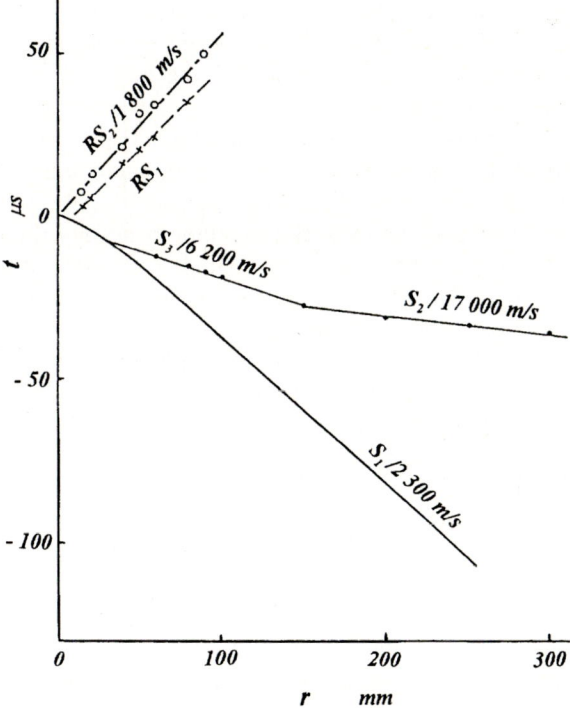

Fig. 12.30. Time–distance diagrams of the spherically imploding fronts and shock waves following the detonation or shock waves reflected at the center together with propagation velocity

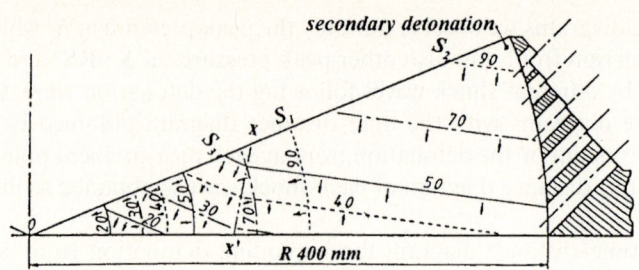

Fig. 12.31. Schematic aspect of the propagation of the imploding detonation front S_1 (*circular curve*) and that of the second and third waves S_2, S_3 (*straight lines*). Number on each line means the time before the implosion instant at the center

Vector combination of this velocity with 17,000 m s^{-1}, as determined from the pressure measured on the bottom plate, yields the oblique shock diagram of Fig. 12.31. In a similar fashion, the oblique wave diagram for S_3 is determined and is also presented in Fig. 12.31.

This suggests that the imploding shock wave are generated in two steps, in the first step the detonation is initiated on the spherical surface in the lower part of the convergent space and produces the first shock wave S_1, while an explosion takes place in the upper corner of the convergent space in the second step, delayed a few microsecond after initiation of the first main detonation wave, producing the second shock wave S_2.

RS$_2$ means the propagation of the shock wave reflected at the focus center, while RS$_1$ is reflected at the edge where the conical surface changes to a flat one in the center.

Figure 12.32 illustrates the logarithmic relations among the ratios of each peak pressure P_{g1}, P'_{g1}, P_{g2}, P_{gr1}, and P_{gr2} behind S_1, S_2, RS$_1$, and RS$_2$, respectively, to the initial pressure P_0 and that of the radial distance r to the radius R_0 of the convergent space. The first pressure ratio P'_{g1}/P_0 at the detonation front increases with the propagation of the imploding detonation almost in proportion to $(r/R_0)^{-0.6\pm0.1}$, but jumps from 80 to 160 at $r = 40$ mm, where the second pressure joins with the detonation front. On the other hand, the second pressure ratio P_{g2}/P_0 increases in proportion to $(r/R_0)^{-1.1\pm0.1}$.

After joining with S_3 the peak pressure at the front of the imploding detonation wave approximately takes the mean value of the pressure, and then P_{g1}/P_0 increases almost proportionally to $(r/R_0)^{-0.9\pm0.1}$. The exponent agrees with that in the smaller vessel of 360 mm inner-diameter, as described in Sect. 12.1.3, in which the exponent of the ratio r/R for the pressure increase is 1.0 ± 0.15. The pressure ratio of the reflected shock P_{gr2}/P_0 as well as P_{gr1}/P_0 decreases proportionally to $(r/R)^{-1.4\pm0.2}$, which agrees well with that in the smaller vessel.

Spectroscopic Temperature Measurement in *G*-type Detonation Chamber

Spectroscopic temperatures behind the spherically imploding detonation and reflected shock waves in the *G*-type detonation chamber are measured using the

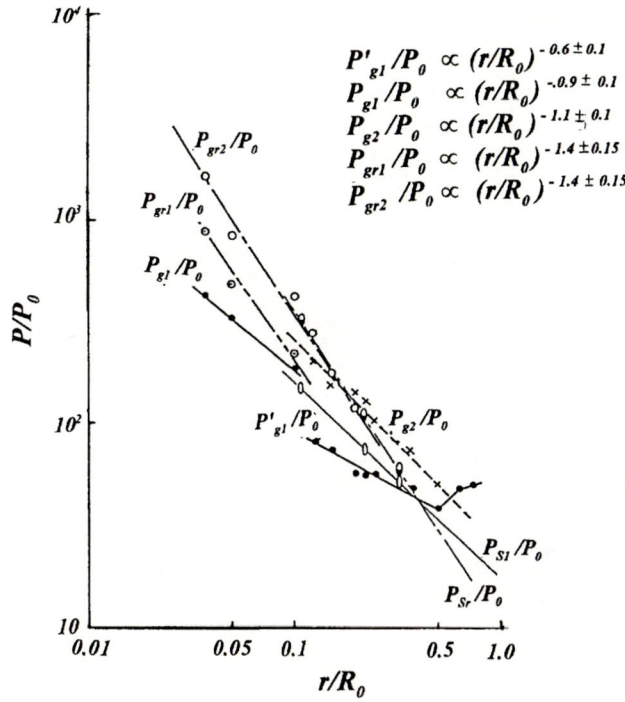

Fig. 12.32. Logarithmic relation between the pressure ratio P/P_0 in the spherical implosion and the ratio r/R_0. P_{g1}: peak pressure at the imploding detonation front, P_{g2}: that behind the second shock S_2, P_{r1}: that behind the first reflected shock, P_{gr2}: that behind the shock reflected at the implosion focus, P_0: initial mixture pressure, r: radial distance from the implosion center, R_0: radius of the convergent space. *Thin lines* are results obtained in the smaller convergent space of 360 mm inner-diameter. P_{s1}: the peak pressure at the detonation front, P_{sr}: that behind shock wave reflected at the center

same method shown in Fig. 12.21 which is applied in the detonation chamber of small size, as described in Sect. 12.1.3. Argon gas is mixed into the stoichiometric propane–oxygen mixture by a rate of 5% and the spectral lines of the argon are observed during the propagation of the imploding detonation and reflected shock waves.

In the experiments carried out under the same condition as those for the pressure measurement, the intensity of the spectral lines of argon having the wave length of 5,373.5 Å, 5,912.1 Å, 6,043.2 Å, and 6,052.7 Å are measured through a monochromator, photomultiplier, and an oscilloscope. The measurement of each spectral line at measuring points $r = 15$ mm, 20 mm and 40 mm is repeated 3–5 times and the mean value of these are used in the calculation. Considering the characteristic of the photomultiplier, the error in the measurement is relatively large (up to a factor of 2).

As illustrated in Fig. 12.33, each curve of the measured spectroscopic temperature T with respect to the time t after the passage of the detonation front at

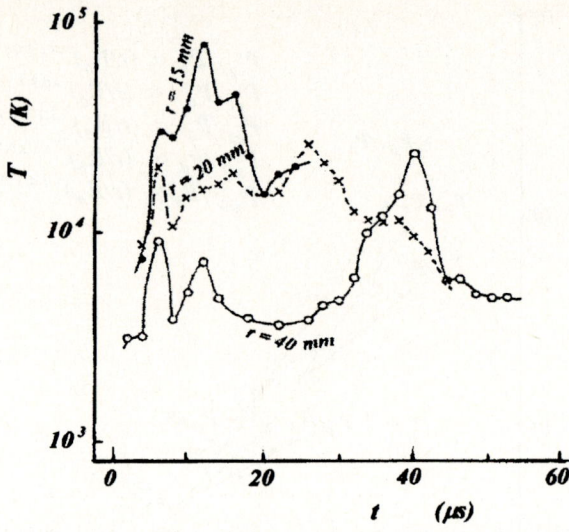

Fig. 12.33. The mean spectroscopic temperatures T measured at r = 15, 20, 40 mm with respect to the time t after the passage of the detonation front at the measurement point

the measurement point shows two peak temperatures, which should correspond to the first peak pressure at the imploding detonation front and the second peak pressure behind the reflected shock wave, respectively.

In Fig. 12.34, logarithmic relations between the ratio of the average spectroscopic temperature T to the initial mixture temperature T_0 and the ratio r/R_0 are shown. T_{gl} is the average temperature measured at the imploding detonation front and T_{gr} that behind the reflected shock waves, while T_{Sl} and T_{Sr} those measured in the small detonation chamber having an inner-diameter of 360 mm described in Sect. 12.1.3. The temperature ratio T_{gl}/T_0 varies proportionally to $(r/R_0)^{-1.2 \pm 0.2}$, while T_{gr}/T_0 proportionally to $(r/R_0)^{-1.3 \pm 0.2}$.

Summarizing the results, the following relations are obtained:

$$\frac{P_{gl}}{P_0} \propto \left(\frac{r}{R_0}\right)^{-0.9 \pm 0.1}, \tag{12.30}$$

$$\frac{P_{gr}}{P_0} \propto \left(\frac{r}{R_0}\right)^{-1.4 \pm 0.15}, \tag{12.31}$$

$$\frac{T_{gl}}{T_0} \propto \left(\frac{r}{R_0}\right)^{-1.2 \pm 0.2}, \tag{12.32}$$

$$\frac{T_{gr}}{T_0} \propto \left(\frac{r}{R_0}\right)^{-1.3 \pm 0.2}. \tag{12.33}$$

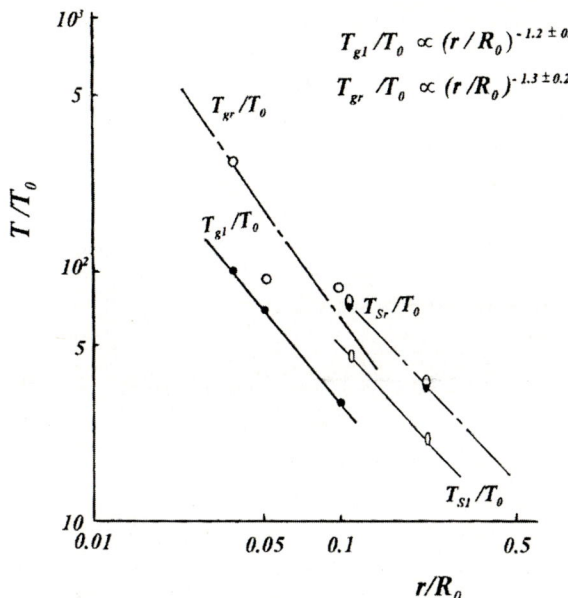

Fig. 12.34. Logarithmic relation between the ratio of the average spectroscopic temperature T_{gl} measured at the imploding detonation front, that T_{gr} behind the reflected shock to the initial mixture temperature T_0 and r/R_0. Thin lines are results obtained in the smaller convergent space of 360 mm inner-diameter. T_{Sl}: temperature at the detonation front, and T_{Sr}: that behind the reflected shock

12.1.5 Spherically Imploding Detonation Waves Initiated by Two-Step Divergent Detonation[131]

Spherically imploding detonation waves propagating in a stoichiometric propane–oxygen mixture in the hemispherical convergent space of the enlarged apparatus explained in Sect. 12.1.3 cannot concentrate the combustion energy into the focus, because of the secondary detonation following the first main detonation wave.

In order to focus the detonation waves more effectively, the enlarged apparatus for imploding detonations is improved and a detonation chamber having a divergent–convergent space is prepared, in which an imploding detonation wave is initiated by a two-step divergent detonation and focused well into the center.

Detonation Chamber having a Divergent–Convergent Space

The improved apparatus made of chrome–molybdenum steel for spherically imploding detonation waves is schematically illustrated in Fig. 12.35.

Just like the G-type detonation chamber developed in Göttingen, it is composed of a detonation tube, a flat cylindrical space, 96 connecting bore holes

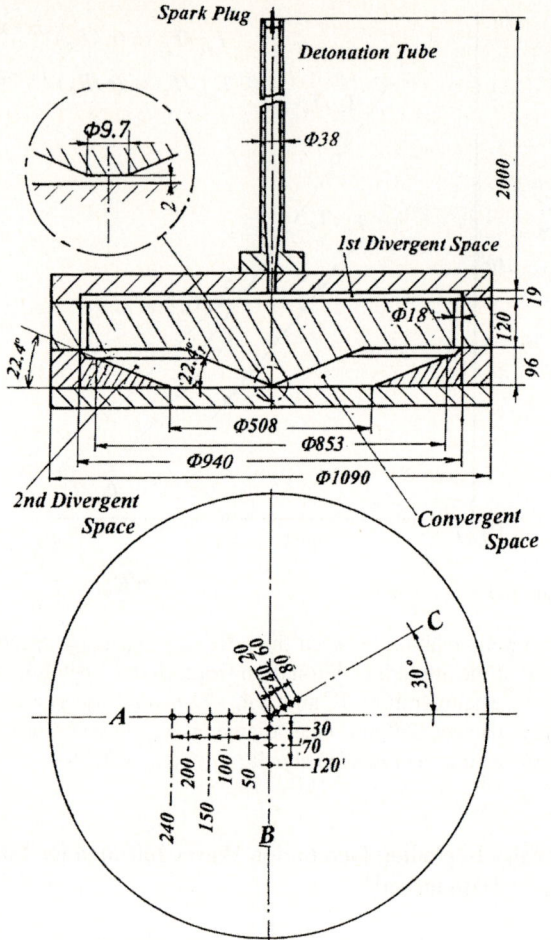

Fig. 12.35. Scheme of the detonation chamber having a divergent–convergent space for spherically imploding detonation waves, dimensions in mm

and a divergent–convergent space under the flat cylindrical space. Except the divergent–convergent space, all parts of the apparatus have the same form and size as in the G-type. The divergent–convergent space has an external diameter of 940 mm and a radial section of rhomboidal form having a height of 94 mm and a side length of 250 mm.

A stoichiometric propane–oxygen mixture is filled in the entire space of the apparatus under a pressure of 53.3 kPa at room temperature of about 20°C. A detonation wave initiated by a spark plug at the end of the detonation tube propagates through the tube into the center of the flat cylindrical space, is reflected there and propagates further toward the circumference, forming a cylindrical detonation wave, which is stagnated on the cylindrical wall at the circumference. Shock waves produced by the stagnated detonation propagate through the 96 connecting holes into the external corner of the divergent–convergent space having a rhomboidal radial section, where ignitions take place almost simultaneously. A circular detonation wave caused by the ignition, thus, propagates toward the space center, first diverging till the middle of the rhomboid diagonal and then spherically converging to the center.

In order to measure the propagation velocity, pressure, and temperature of the imploding detonation waves, 13 holes are bored in three different directions A, B, and C in the steel bottom plate of the divergent–convergent space, as shown in the lower part of Fig. 12.35. As the detonation waves begin to converge at a radial distance of 250 mm, we can observe each state of the spherically imploding detonation waves at different radial distances from the center, by setting measurement devices into these holes.

Propagation and Convergence of the Spherically Imploding Detonation Waves

Using ion probes the passage instant t of the detonation front is measured at each measuring point having a radial distance r from the implosion center. The measured results are represented in Fig. 12.36 as a logarithmic relation between r and t of the detonation front expressed by a time before the focusing instant at the center, according to Guderley's formula $r = at^n$, where a is a proportional constant. This diagram suggests r is proportional to $t^{0.84 \pm 0.03}$.

The exponent n of t estimated to be 0.84 ± 0.03 is almost equal to that observed in the G-type chamber. Considering that all results measured in three different directions fall on the same straight line, the imploding detonation wave must well converge into the focus.

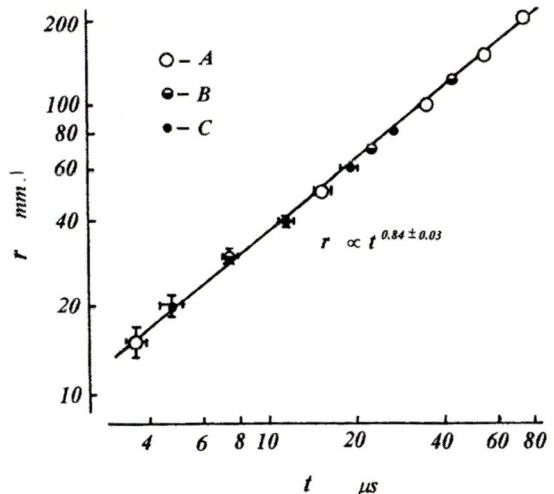

Fig. 12.36. Time–distance diagram of the imploding detonation wave. Passage time t of the detonation front at a point of radial distance r. t: time before focusing instant

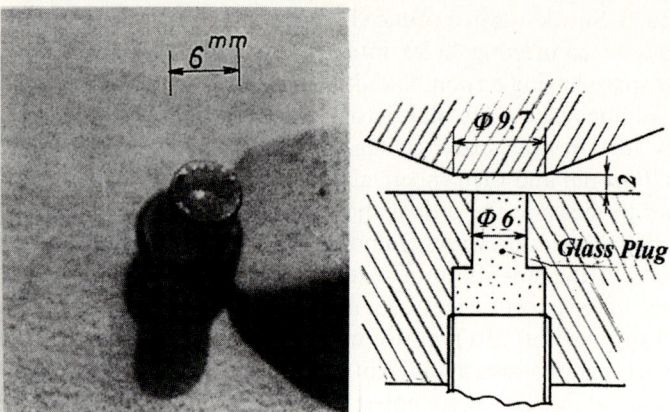

Fig. 12.37. Trace of implosion focus marked on the glass plug set at the implosion center, scheme of the glass plug setting is shown *right* side

In Fig. 12.37 a photograph of a glass plug is shown having a diameter of 6 mm set at the implosion center. On the surface of the plug we observe a white circular trace having a diameter of about 0.5 mm, which is marked by the focusing of the imploding detonation waves, as the gas within the implosion focus has such a high temperature that the dust sticking to the glass surface is vaporized and cleaned up. The circle of the trace shows, therefore, the size of the implosion focus.

From the relation between the radial distance r and the passage instant t of the detonation front at r, the propagation velocity D of the imploding detonation waves and their Mach number M_D can be obtained, which are illustrated in Fig. 12.38 with respect to the radial distance r. We observe in this diagram that the propagation of the detonation is accelerated from the circle of $r = 80$ mm.

Pressure Behind the Imploding Detonation Waves

The pressure behind the imploding detonation waves is measured by a piezo-electric pressure transducer at each measurement point. In Fig. 12.39 two examples of the pressure variation are shown measured at $r = 20$ mm and 40 mm, respectively. In each of these diagrams two pressure peaks P_1 and P_r are observed. The measurement error is estimated to be about ±10%.

Combining these diagrams obtained at point of different radial distances r with the time–distance diagram of the detonation front shown in Fig. 12.36, we obtain a time–distance diagram of both the pressure peaks P_1 and P_r, as shown in Fig. 12.40, in which all points of P_1 fall on the curve of the detonation front, while those of P_r fall on the curve of the shock wave reflected from the implosion center. In this diagram we obtain no secondary detonation wave following the main imploding detonation which always appears in the *G*-type chamber. In the improved detonation chamber the detonation energy is concentrated to the implosion focus more effectively than in the *G*-type chamber.

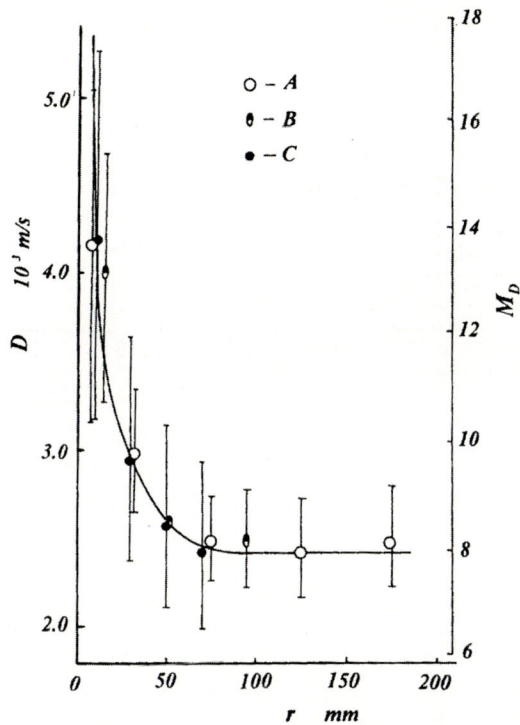

Fig. 12.38. Propagation velocity D and Mach number M_D as function of the radial distance r

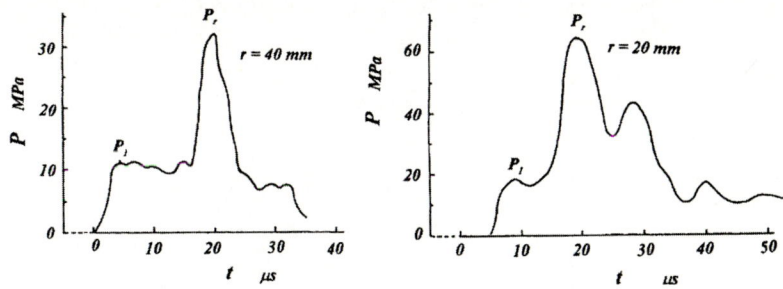

Fig. 12.39. Pressure history of the imploding detonation waves. *Left* at $r = 40$ mm, *right* at $r = 20$ mm

Figure 12.41 illustrates logarithmic relations of P_1/P_0 and P_r/P_0 with respect to the radial distance r, where P_0 is the initial mixture pressure. The diagram suggests that P_1/P_0 is proportional to $r^{-1.4\pm 0.1}$ and P_r/P_0 to $r^{-1.4\pm 0.15}$. Comparing these with the results obtained in the G-type chamber, the absolute value of the exponent of r for P_1/P_0 in the improved one is larger, while that for P_r/P_0 is

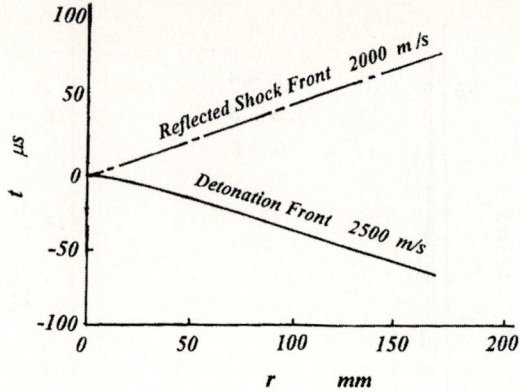

Fig. 12.40. Time–distance diagram of the first peak pressure P_1 of the imploding detonation and second peak pressure P_r behind the shock wave reflected from the implosion center, t is the passage time of each peak pressure at r. $t = 0$ means focusing instant in the center

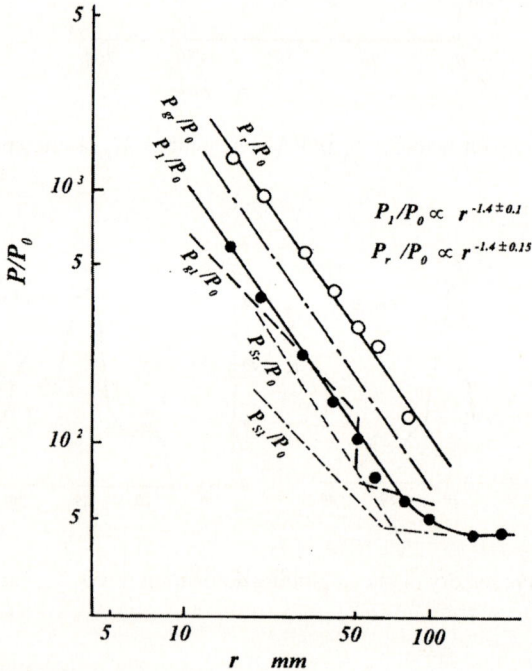

Fig. 12.41. Logarithmic relation of pressure ratios P_1/P_0, P_r/P_0, P_{gl}/P_0, and P_{gr}/P_0 to the radial distance r from the implosion center. P_1 and P_r: peak pressure at the imploding detonation front and that behind the reflected shock in the improved chamber, respectively, P_{gl} and P_{gr} those in G-type, P_0: initial pressure

12.1 Imploding Shock Waves

almost the same, but the convergence starts on a circle having a larger radius and the maximum pressure of the imploding detonation as well as that of the reflected shock waves at the implosion center is much higher.

Temperature in the Implosion Focus

According to the spectroscopic temperature observed in the *G*-type chamber, the temperature in the imploding focus is estimated to be higher than 10^5 K. The gas in the implosion center must be fully ionized. In order to investigate the temperature in the implosion focus, therefore, it is necessary to measure the ion and electron temperature here.

In order to investigate the ionized particles in the implosion focus, the same laser light scattering method developed by Kunze as that described in Sect. 9.3.3 is applied. In Fig. 12.42 the measurement method is schematically illustrated.

A laser beam having 10 mm in diameter, 40 MW in power, and 20 ns in duration of half intensity from a giant pulse ruby laser set outside the chamber is projected into the convergent space and focused in the implosion center under an incident angle of 4° to the horizon. The light scattered by the ionized particles in the implosion center falls vertically through a glass window of *BK-7* and a polarizer to a 50% beam splitter.

The scattered light passing through the beam splitter is observed by a photomultiplier together with the emission of the imploding detonation to confirm the delay time of the laser beam projection from the focusing instant of the

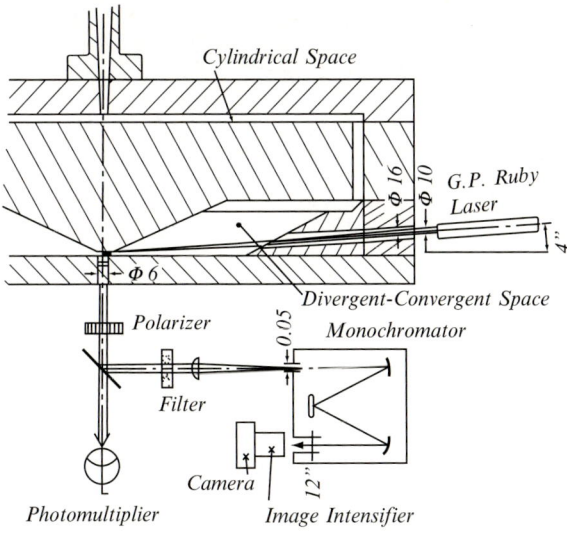

Fig. 12.42. Measurement system of the temperature in the implosion focus applying a laser light scattering method, dimensions in mm

implosion detonation waves, while that reflected from the beam splitter into a horizontal direction is introduced through a filter into a 0.05 mm wide slit of a monochromator having a grating line number of 1,200 per mm and a collimator radius of 250 mm. The scattered light dispersed by the monochromator is further introduced into an image-intensifier which can amplify the incident light about 20,000 times and recorded directly on a photographic film having a sensitivity of *ISO 400*.

Each photographic film on which the scattered light dispersed by the monochromator is recorded must be analyzed by a microphotometer to obtain a spectrum of the scattered light. The measurement is repeated many times, as the delay time of the laser light projection triggered by the light emission of the detonation front propagating through the detonation tube fluctuates over several microseconds. We can thus obtain many spectra of the scattered light at different times after the focusing instant of the imploding detonation waves.

The temperature of gas in the implosion focus is so high that only ion component is recorded on the photographic film, while the electron component is dispersed much wider than the film-width. In Fig. 12.43 an example is shown of the intensity I_λ of the scattered light normalized by the maximum intensity with respect to the logarithm of the wavelength $\Delta\lambda$ shifted from that of the incident laser beam. The curve of I_λ is composed of two different bell-shaped curves. Considering the components of the mixture, one bell-shaped curve having a wider range must be light scattered by the lighter ions, i.e., H-ions, while that having a narrower range by C- or O-ions, as already explained in Sect. 9.3.3.

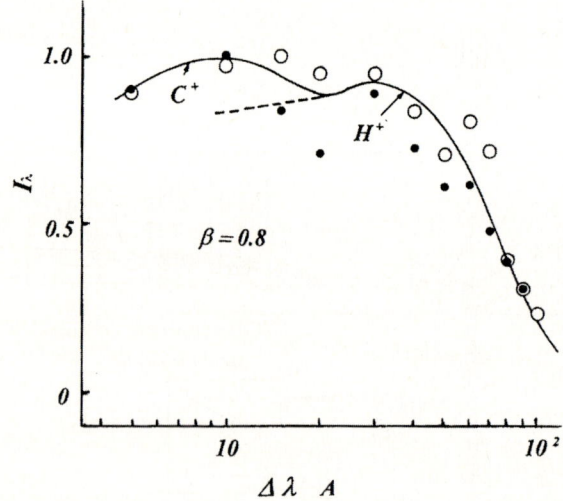

Fig. 12.43. Normalized intensity I_λ of the laser beam scattered in the implosion focus with respect to wavelength $\Delta\lambda$ deviated from that of incident laser beam 0.05 μs after the implosion. *White circle*: $+\Delta\lambda$ side, *black circle*: $-\Delta\lambda$ side, β: plasma parameter of H^+ and C^+

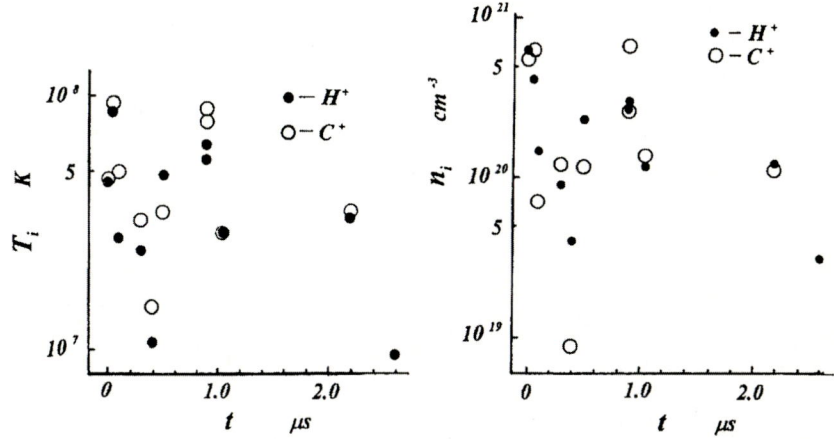

Fig. 12.44. Ion temperature T_i (*left*) and ion density n_i (*right*) of H- and C- or O-ions in the focus of spherically imploding detonation waves with respect to the time t after the implosion focusing

With the same method described in Sect. 9.3.3, the temperature of the H-ions and C- or O-ions can be estimated according to the set of graphs proposed by Kegel which gives the temperature of T_H = 100,000 K under a scattering angle $\theta = 90°$ together with the electron density n_0.

In Fig. 12.44 the thus obtained temperature T_i of H-ions and C- or O-ions, and the density n_i of those ions in the focus of the spherically imploding detonation waves propagating in a stoichiometric propane–oxygen mixture are illustrated with respect to the time t after the focusing instant. During the first period of 1 µs after the implosion focusing the ion temperature T_i has a value from 10^7 to 10^8 K and the density n_i has a value between 10^{19} per cm^3 and 10^{21} per cm^3 during the same period. As already mentioned, the gas in the implosion center is fully ionized, thus the ion temperature means the gas temperature.

In the improved detonation chamber in which the spherically imploding detonation wave is driven by two-step divergent detonation, almost the whole combustion energy of the mixture is concentrated into the main detonation wave, whose peak pressure at the detonation front rises with its propagation much more rapidly than that in the G-type chamber. Comparing the variation of the peak pressure ratio P/P_0 with respect to the radial distance r in both the G-type chamber and the improved one and considering the temperature behind the imploding detonation as well as that behind reflected shock in the G-type chamber, the spectroscopic temperature T_1 behind the spherically imploding detonation waves as well as that T_r behind the reflected shock in the improved chamber can be estimated.

Figure 12.45 represents logarithmic relations of all temperature ratios T_{s1}/T_0, T_{g1}/T_0, and T_1/T_0 at the imploding detonation front in the detonation

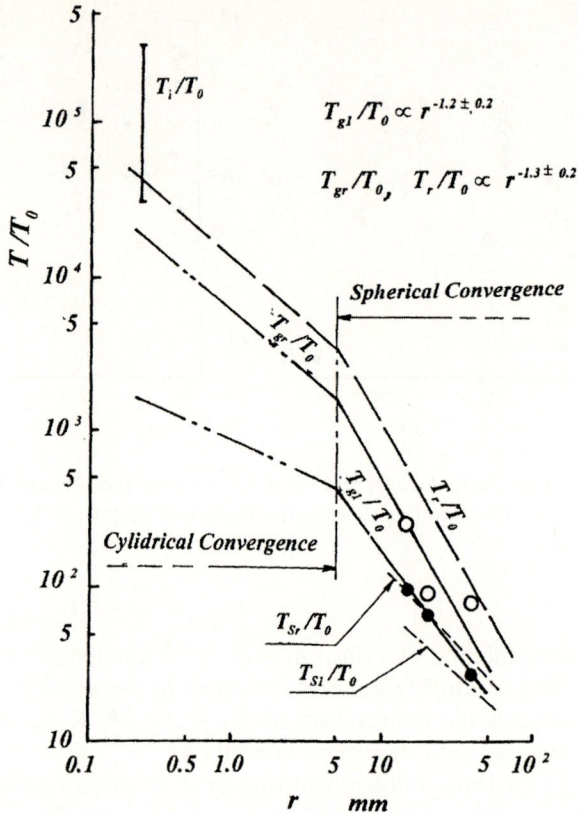

Fig. 12.45. Logarithmic relations of temperature ratios T/T_0 to the radial distance r in the convergence space T_{g1} and T_{gr}: spectroscopic temperature at the imploding detonation front and that behind reflected shock waves in G-type chamber, respectively. T_i: ion temperature in the implosion focus, T_1 and T_r: estimated spectroscopic temperature at the imploding detonation front and that behind the reflected shock waves in the improved detonation chamber, respectively, T_{S1} and T_{Sr}: those in small size chamber

chamber of small size, G-type and improved one, respectively, and T_{Sr}/T_0, T_{gr}/T_0, and T_r/T_0 behind the reflected shock wave, respectively in the small size, G-type and improved one and T_i/T_0 in the implosion focus in the improved chamber to the radial distance r. According to this diagram we can recognize such a high ion temperature of 10^8 K as a reasonable one.

As the photograph in Fig. 12.46 shows, a hemispherical hole of 9 mm in diameter and 1.1 mm in depth is observed on the surface of the top of conical block facing the implosion focus after the experiment has been repeated 80 times. The temperature in the implosion focus is really so high that the top of the chrome–molybdenum steel must be melted.

The gas in the implosion focus having a temperature from 10^7 to 10^8 K and density from 10^{19} per cm^3 to 10^{21} per cm^3 for a duration of 1.0 μs has a state

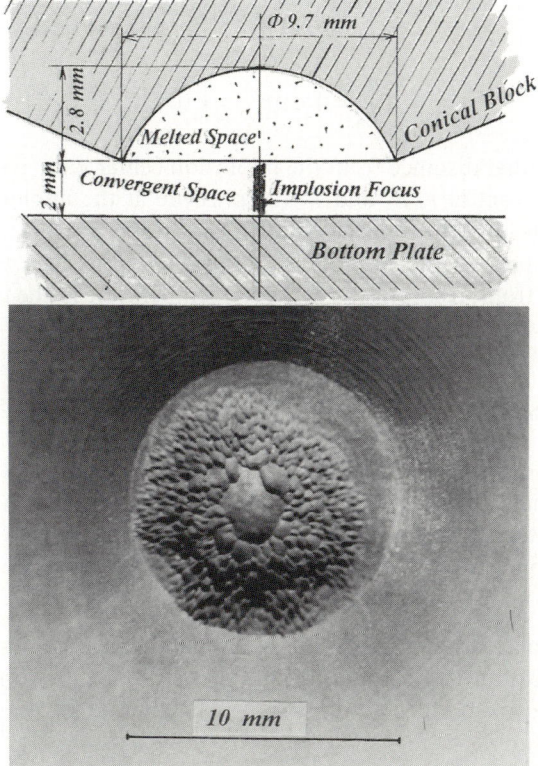

Fig. 12.46. Melted top part of the conical block facing the implosion focus

within the limit of the so-called Lawson's condition for a nuclear fusion in a T–D reaction.[132] This suggests that a nuclear fusion can be initiated using the same apparatus having a three times larger size.

Summary of the Experimental Results

Summarizing the results of the experiments carried out applying a detonation chamber having a divergent–convergent space of a rhomboidal radial section form in which the spherically imploding detonation wave initiated by a two-step divergent detonation propagates in a stoichiometric propane–oxygen mixture, the following relations are obtained:

$$r \propto t^{0.84 \pm 0.03}, \tag{12.34}$$

$$\frac{P_1}{P_0} \propto r^{-1.4 \pm 0.1}, \tag{12.35}$$

$$\frac{P_{r_r}}{P_0} \propto r^{-1.4 \pm 0.15}, \tag{12.36}$$

$$\frac{T_r}{T_0} \propto r^{-1.3 \pm 0.2}, \tag{12.37}$$

$$\frac{T_i}{T_0} = 4 \times 10^4 - 4 \times 10^5, \tag{12.38}$$

where r is the radial distance from the implosion center, t the passage instant of the detonation front at r, P_1 and P_r the pressure at the imploding detonation front and that behind the reflected shock front, respectively, T_1, T_r the spectroscopic temperature at the imploding detonation front and that behind the reflected shock front, respectively, T_i the ion temperature in the implosion focus, and T_0 the initial mixture temperature.

Discussion on the Convergence of Imploding Detonation Waves[133, 134]

Some scientists have reported about the convergence of imploding detonation waves and pointed out that the imploding detonation waves cannot well focus into a point, as their theoretical calculations suggest that a perturbation assumed at the detonation front develops with the propagation of the detonation waves. In the calculation, however, they have made two mistakes:

1. As the cellular structure behind the detonation wave shows, many ignition points appear at the detonation front, namely, there are so many perturbations that a few perturbations cannot play any role for the propagation of the detonation
2. The effect of buoyancy is neglected for the detonation propagation and structure, although it plays an important role especially for the detonation imploding in a space kept not horizontally

The imploding detonation waves can propagate and focus more stably than simple shock waves, as far as they implode in a space kept horizontally, since the shock waves at the detonation front are always continuously driven by the combustion waves developing from the spontaneous ignition taking place behind them one after another.

12.1.6 Nuclear Fusion Applying Spherically Imploding Detonation Waves[135]

As described above, the experimental results of the spherically imploding detonation waves propagating in a stoichiometric propane–oxygen mixture in the detonation chamber suggest the following relations:

$$\frac{P_r}{P_0} \propto \left(\frac{r}{R_0}\right)^{-1.4 \pm 0.13}$$

and

$$\frac{T_r}{T_0} \propto \left(\frac{r}{R_0}\right)^{-1.3 \pm 0.2}. \tag{12.37}$$

12.1 Imploding Shock Waves

Here P_r and T_r are the peak pressure and temperature behind the reflected shock wave, respectively, where the highest pressure and temperature appear in this imploding detonation system.

Each detonation chamber for spherically imploding detonation waves, however, has a cylindrical area having a diameter of about 10 mm in the center. In this area the detonation waves implode not spherically but cylindrically and the pressure increase as well as the temperature increase with propagation of the detonation is less than those in the spherical implosion. The experimental results of cylindrically imploding detonation waves propagating in the same propane–oxygen mixture suggest the following relation:

$$\frac{P_{cr}}{P_0} \propto \left(\frac{r}{R_0}\right)^{-0.8 \pm 0.1}, \quad (12.25)$$

where P_{cr} is the peak pressure behind the reflected shock waves of the imploding detonation in the cylindrically convergent space. The peak temperature T_{cr} behind the reflected shock waves of the cylindrical implosion should have the same tendency and the following relation is obtained

$$\frac{T_{cr}}{T_0} \propto \left(\frac{r}{R_0}\right)^{-0.9 \pm 0.1}. \quad (12.25')$$

In the focus of the spherically imploding detonation waves the ion have a temperature between 10^7 and 10^8 K and a density between 10^{19} per cm^3 and 10^{21} per cm^3. The energy state in the implosion center is thus, higher than the so-called Lawson's criterion[132] where a nuclear fusion reaction can start in a mixture of deuterium and tritium.

Size of Detonation Chamber of a Steady Fusion Reaction in a $D + T$ Mixture

In a deuterium–tritium mixture, the following fusion reaction takes place[136]:

$$D + T = {}^4He + n + 17.6 \text{ MeV},$$

in which only 3.52 MeV of the reaction heat can be used to heat the mixture, while the other 14.08 MeV is transported by neutrons. On the other hand, the heat W_f per second released by the fusion reaction depends on the following equation:

$$W_f = \frac{1}{4} n <\sigma v> Q. \quad (12.39)$$

And the heat loss W_r per second caused mainly by Bremsstrahlung is expressed by[136]

$$W_r = \frac{4}{3} \frac{\pi^3 Z^2 e^6}{m_e c^3 h} n_e v_\varepsilon = (1.71 \times 10^{-27}) Z^2 n_e T_e^{1/2}, \quad (12.40)$$

where n_i is the density of particles T (tritium) or D (deuterium), σ the collision cross-section of the particles, v_e the free velocity of the particles, Q the reaction heat 3.52 MeV, Z the charge of the ions, e the elementary electron charge, m_e the

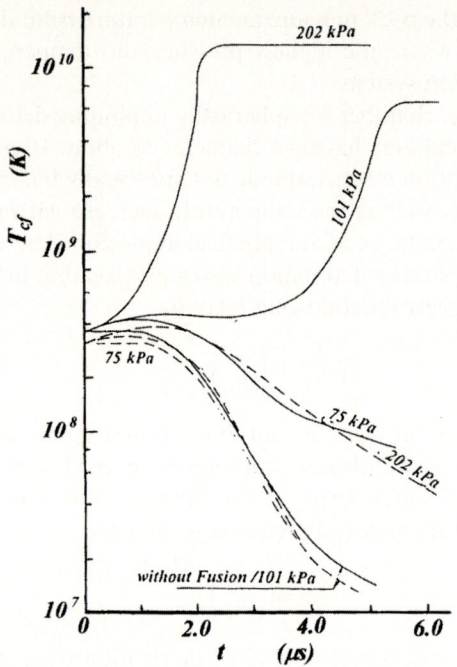

Fig. 12.47. Theoretical temperature T_{cf} of D + T mixture at fusion reaction with respect to the time t after the reaction start. Number on each curve means the initial mixture pressure

electron mass, c the light velocity, h Planck's constant, n_e the electron density, and T_e the electron temperature.

The mixture in a capsule put into the implosion focus is thus heated by the heat $\Delta Q = W_f - W_r$ after the initiation of the fusion reaction. In the diagram of Fig. 12.47 the temperature T_{cf} is illustrated in the T + D mixture under different initial pressures and two different implosion temperatures of 3.13×10^8 K and 3.6×10^8 K estimated according to the (12.39) and (12.40) with respect to the time t after the initiation of the reaction. As this diagram shows, if the temperature in the implosion focus is higher than 3.6×10^8 K and the initial mixture pressure higher than atmospheric, a steady nuclear fusion in a mixture of T + D can take place. The detonation chamber having a cylindrical divergent space explained in Sects. 12.1.2–12.1.5 can be enlarged as large as one wants. According to (12.37), the detonation chamber should be enlarged to more than three times to obtain such a high ion temperature and have a convergent space of radius larger than 760 mm. As the ion density is so high (10^{20} per cm^3–10^{21} per cm^3), it may not necessary to confine the plasma just like in the laser fusion.

A Proposal of Nuclear Fusion Reactor Applying Imploding Detonation Waves

According to the results obtained in Sect. 12.1.5, a reactor for a power plant of 1,000 MW under a thermal efficient of 20% as schematically illustrated in Fig. 12.48 is proposed, as an example. The fusion reactor is composed of a two-step divergent implosion chamber, having a convergent space of 850 mm in radius, a reaction guide tube of stainless steel having a length of about 2 m and an inner-diameter of 5 mm below the implosion center, several fuel nozzles of mixtures D + 9He and T + 9He in a reaction chamber set at the guide tube end, first wall, blanket composed of neutron multiplier, tritium breeder and reflector of graphite and others necessary for a fusion reactor around the reaction zone. The detonation chamber is filled with a stoichiometric propane–oxygen mixture, while a flat cylindrical capsule of about 6 mm in diameter, filled with a T + D mixture under atmospheric pressure is put into the implosion center.

The fusion reaction ignited in the capsule propagates accompanied by detonation waves having a Mach number of about 1.25×10^4 in the T + D mixture and 3.96×10^3 in the T + D + 18He mixture through the guide tube filled with a mixture of T + D + 18He to the reaction chamber at the tube end, where a steady fusion reaction is held by supplying the mixture D + 9He and T + 9He,

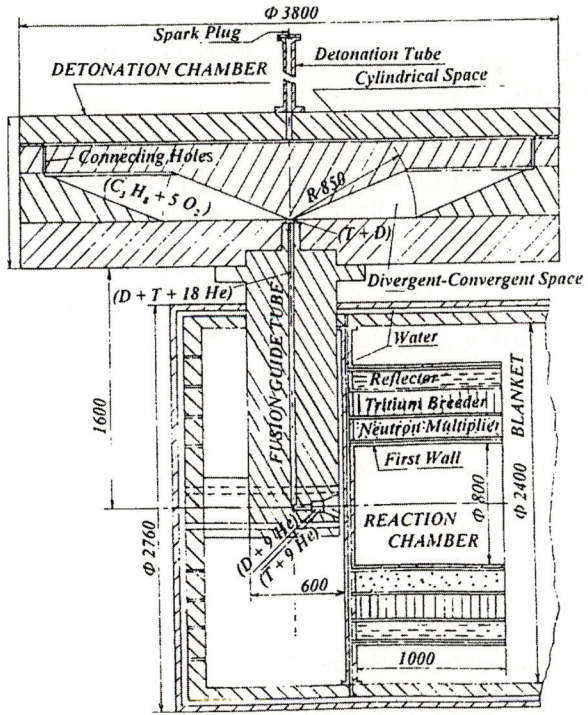

Fig. 12.48. Sketch of a nuclear fusion reactor proposed for a 1,000 MW plant

both together at a rate of 0.12 mol s^{-1}. Such mixtures are chosen to keep the temperature in the reaction zone at about 4.0×10^8 K at which the steady fusion reaction can continue.

In order to gain a mechanical power, a system of steam turbine driven by water vapor heated by the fusion reaction is proposed, as water can well decelerate the fast neutrons and absorb their energy occupying 80% of the whole reaction heat. The water used to cool the turbine case, the reaction chamber and blanket under a pressure of 10 MPa is injected into the reactor vessel and evaporated around the reaction chamber. The vapor is heated there to a temperature of about 900 K under a pressure of 5 MPa and drives the turbine, expanding to atmospheric pressure. The water to be supplied into the steam turbine is estimated to be about 1,400 kg s^{-1}.

Radioactivity by the Nuclear Fusion[137]

Some radioactive isotopes of carbon ^{15}C, nitrogen ^{16}N, ^{17}N, ^{18}N, and oxygen ^{19}O are produced from water by fast neutrons. They emit γ-ray, but all their half-lives are shorter than 30 s. They pose, therefore, no serious problem. By shielding with a stainless steel wall having a thickness of more than 15 cm, the intensity of the radioactivity is reduced to less than one-tenth. By effective shielding we can avoid the danger of such radioacitvities, α-particles, and neutrons.

12.1.7 Nuclear Fusion Rocket Engines Applying Imploding Detonation Waves[138]

As the nuclear fusion reactor ignited by spherically imploding detonation waves explained in the Sect. 12.1.6 is much smaller than those of Tokamak type or using strong power lasers, it well fits rocket engines.

The fusion reaction ignited in the capsule set into the implosion center propagates accompanied by detonation waves in a mixture of D + T + 18He through a guide tube to the reaction chamber at the tube end, where a steady fusion reaction can be sustained by supplying mixtures of D + 9He and T + 9He, keeping the temperature in the reaction zone higher than 4.0×10^8 K.

In order to obtain mechanical power from the fusion reactor, the reaction zone must be enclosed within walls, blankets, and neutron reflectors of graphite in which the high energy neutrons are decelerated and transformed into thermal energy, while gases of liquids are injected into the reaction chamber around the reaction zone to transform the fusion reaction energy to mechanical one. Moreover, the vessel of the reactor must be cooled by liquids or gases in order to keep its temperature lower than the melting point.

A Proposal of a Fusion Rocket Engine[138]

By applying such a detonation chamber and a fusion system, an example of fusion reactors for a rocket engine having a thrust of 1,000 ton (9.8×10^6 N) is schematically illustrated in Fig. 12.49.

12.1 Imploding Shock Waves

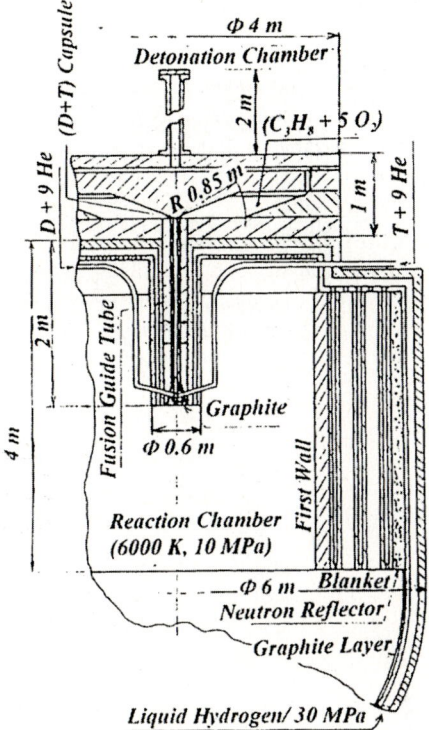

Fig. 12.49. Sketch of a proposed fusion reactor for rocket engines

From the center of the convergent space of the detonation chamber described above the fusion reaction is introduced through a guide tube of stainless steel coated with a thin graphite layer to a reaction chamber. A mixture of T + 9He and that of D + 9He are injected into the reaction chamber, so that the reaction can steadily continue.

Around the reaction zone a first wall and several blankets of ferritic steel and a neutron reflector of graphite are built in the reaction chamber. As the coolant and rocket driver gas, liquid hydrogen is used, as hydrogen can absorb the neutron energy quite well and has the highest flow velocity at the nozzle exit. The liquid hydrogen flowing along the outer surface of the reaction chamber as well as the power nozzle and through the blankets as coolant is injected into the reaction chamber, further heated by the reaction and drives the rocket, decelerating the fast neutrons and absorbing their energy.

A rocket engine using such a fusion reactor initiated by spherically imploding detonation waves is schematically illustrated in Fig. 12.50. The liquid hydrogen having a pressure of 30 MPa and a temperature lower than 30 K is injected into the reaction chamber of the rocket engine, where the hydrogen is

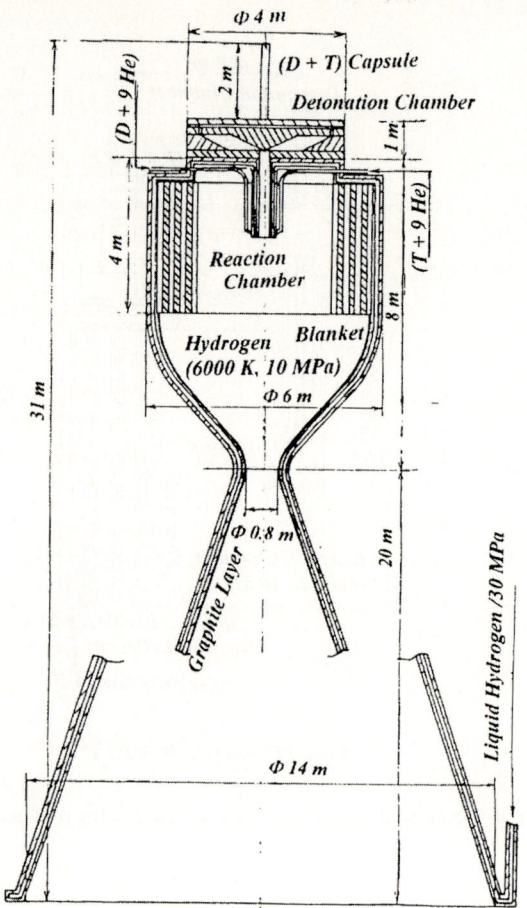

Fig. 12.50. Sketch of a proposed nuclear fusion rocket engine having a thrust of 1,000 ton (9.8×10^6 N) applying spherically imploding detonation waves

evaporated and heated to about 6,000 K under a pressure of 10 MPa. The heated hydrogen gas flows out through a Laval nozzle having a throat of 0.8 m in diameter and an exit of 14 m in diameter, expanding to the pressure of the atmospheric space with a velocity of 14,000 m s^{-1}.

In order to keep the engine state described above having a thrust of 1,000 ton in vacuum, the fuel mixture D + 9He and T + 9He should both be supplied together to the reaction zone at a rate of 1.0 mol s^{-1}, while the hydrogen stays at a rate of 700 kg s^{-1}. At sea level, the fuel mixture must be supplied at 1.5 mol s^{-1} and hydrogen at 1,080 kg s^{-1} in order to keep the same thrust of 1,000 ton (9.8×10^6 N).

Cooling System

The inner-side of the reaction chamber as well as that of the power nozzle of stainless steel is coated with a graphite layer having a thickness of 5 mm to protect them from the reaction heat. In order to keep the temperature of the graphite surface lower than 3,600 K, the reaction chamber and nozzle must be cooled by the liquid hydrogen flowing at the rear of the reaction chamber and nozzle from the nozzle exit to the top plate of the reaction chamber. The liquid hydrogen flowing at a rate of 700 kg s^{-1} is heated from 15 K to about 28 K during the cooling period.

Launch of the Rocket

At the launch of the rocket engine on the ground, the engine driving hydrogen gas of about 1,500 K flows out from the power nozzle into the atmospheric air, in which the high temperature hydrogen burns immediately mixing with air. In order to avoid such an explosion danger, an incombustible gas, for example, water vapor should be used for a short period at the start.

If H_2O is applied as the propellant, it must be supplied at a rate of 2,300 kg s^{-1} to keep the thrust of 1,000 ton. When the rocket reaches an altitude higher than 20,000 m, 1 min after the start and we have no danger of explosion, then hydrogen can be used as propellant instead of water vapor. During the drive by water vapor, the fuel mixture (D + 9He) + (T + 9He) should be supplied by a rate of 0.7 mol s^{-1}.

Besides, at the start of the rocket engine on the ground the divergent–convergent space of the detonation chamber must be kept horizontally to avoid the effect of buoyancy for the implosion.

12.2 Hypersonic Combustion for RAM Jet Engine[139]

In the so-called SCRAM (Supersonic Combustion RAM) jet engine for hypersonic airplanes, the air introduced into the engine with a flow velocity having a Mach number higher than 5.0 is heated by adiabatic compression to a temperature at which the injected fuel ignites and burns, decelerating the flow velocity to a Mach number between 2.0 and 3.0.

It may be possible to produce a standing detonation wave propagating against the introduced mixture flow in the combustion chamber by balancing the detonation with flow velocities which are moving in opposite direction of each other. An engine of this type combines the advantage of decreasing the compression loss with a rapid, uniform combustion. To study such a standing detonation in the laboratory, however, we need first to have a hypersonic flow faster than the detonation propagation, and then decelerate it to balance with the detonation propagation.

In this section an experimental method is explained how to produce a hypersonic flow with a Mach number higher than 5.0 by using a shock tube connected to a divergent nozzle and how to balance the flow with a detonation front propagating in the opposite direction to the flow in a convergent nozzle behind the divergent nozzle.

12.2.1 Standing Detonation Waves[139-141]

There are some proposals for the scram jets to apply the shock induced combustion waves, which are also called standing detonation waves. In such cases, however, the shock waves have been set-up in the intake flow by some mechanical methods using a wedge or cone, but the combustion behind the detonation waves is never produced by the flow itself. Therefore, such shock induced combustion system are quite different from the real detonation waves. Besides, the shock compression has a larger loss than the adiabatic compression usually applied to most scram jets because of a large entropy increase.

The intention in this section is to arrange self-sustained detonation waves propagating in the opposite direction to the intake flow. Namely, the absolute value of the detonation velocity D_x propagating in the opposite direction to the intake flow must be equal to the intake flow velocity w_f, that is, the detonation propagation velocity in laboratory fixed co-ordinates $D_{xe} = D_x - w_f = 0$. In order to form and hold a stable self-sustained detonation in the combustion chamber of a ram jet, the detonation must collide with the counter flow of the mixture with the same velocity as the detonation propagation. This is automatically selected in the flow which decelerates from a velocity higher than D_x to a lower one. The detonation propagating velocity fluctuates within a certain range ΔD_x according to the mixture state, as schematically illustrated in Fig. 12.51.

In order to carry out the experiments in the laboratory, therefore, we need

1. An apparatus to produce a hypersonic flow having a velocity higher than the detonation propagation
2. A convergent nozzle to decelerate the hypersonic flow to a lower velocity than the detonation propagation velocity
3. A detonation tube connected to the convergent nozzle

The standing detonation front appears somewhere in the convergent nozzle where $D_x - w_f = 0$, as also shown in Fig. 12.51. If we can have no hypersonic flow faster than the detonation propagation, we cannot form any standing detonation front. Nevertheless, we can say, that a standing detonation can be built in a hypersonic flow faster than the detonation propagation in a ram jet, if we could find the relation $D_{xe} = D_x - w_f$, where D_{xe} is the detonation propagation velocity in the laboratory fixed co-ordinates and w_f the flow velocity in the nozzle.

12.2.2 Advantage of the Ram Jet Using Standing Detonation Waves[140, 142]

In order to compare the performance of a scram jet with that of the ram jet having standing detonation waves, some roughly calculated results are shown as an example as follows:

12.2 Hypersonic Combustion for RAM Jet Engine

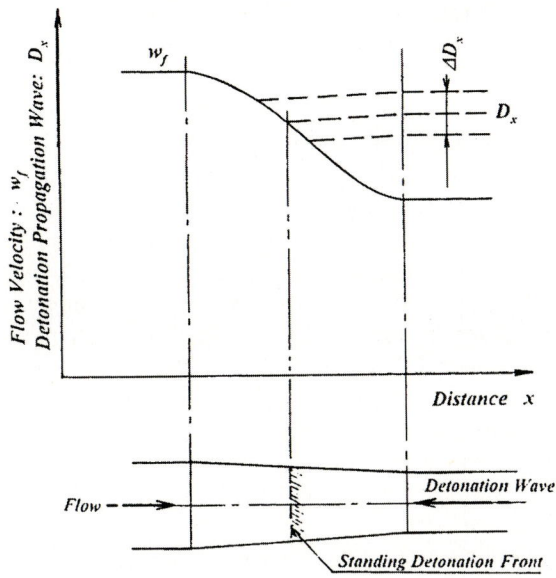

Fig. 12.51. Formation of a standing detonation front in a hypersonic counter flow. w_f: counter flow velocity, D_x: propagation velocity of the detonation, x: distance, ΔD_x: fluctuation range of the detonation propagation

Under the assumption that air of initial temperature of 217 K and pressure of 5.0 kPa corresponding to those at an altitude of 20 km is sucked into the engine flying at a velocity of 2,320 m s^{-1}, i.e., a Mach number of 8.0 and hydrogen is injected into the air to produce a stoichiometric mixture, the introduced air is heated by adiabatic compression to a temperature at which the injected fuel ignites and burns accompanied by a little dissociation as possible. The ignition temperature is estimated to be 900 K in the hydrogen–air mixture. The compression and combustion pressure in the normal scram jet reaches a value of about 1.21 MPa, while the introduced air is decelerated to a Mach number 2.1 the flow velocity of the exhaust gas at the exit of the power nozzle of the engine is estimated to be about 3,000 m s^{-1}. The thermodynamic process in the case of the supersonic combustion is shown by the broken line 1–2–3–4 in Fig. 12.52.

In the presence of a standing detonation wave, the flow should be compressed to a pressure of 80 kPa to balance with the detonation propagation. Pressure at the so-called Chapman–Jouguet point should be 1.17 MPa, while the flow velocity at the power nozzle exit is estimated to be about 3,300 m s^{-1}. As already explained in Chap. 11, there is no Chapman–Jouguet condition. The C–J point in this case means the end of the combustion behind the detonation wave, from which an adiabatic expansion should proceed.

Comparing the compression in the scram jet from 5 kPa to 1.21 MPa with that in the presence of a standing detonation to 80 kPa, it is found that the compression loss in the latter case is reduced.

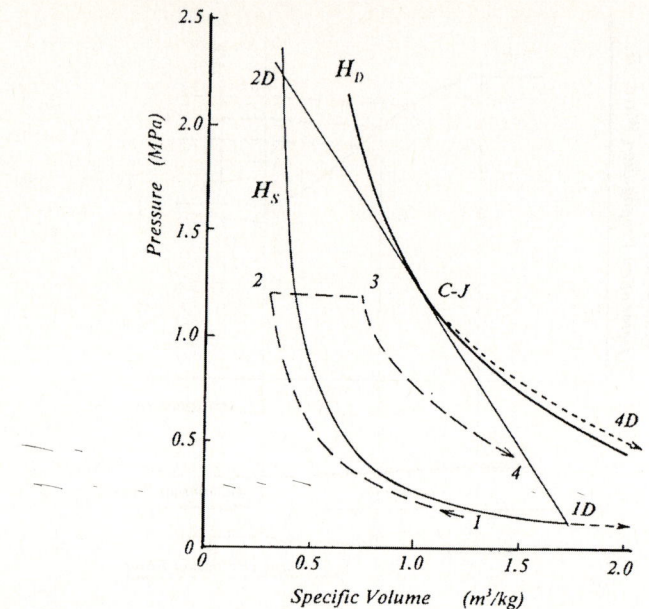

Fig. 12.52. Thermodynamic process in a scram jet shown by the *broken line* 1–2–3–4 and that in an engine driven by standing detonation shown by the *solid lines* 1D-2D-C-J-4D, H_S, H_D: Hugoniot curves of shock and detonation wave, respectively, C–J: Chapman–Jouguet point

12.2.3 Experimental Apparatus

The experimental apparatus for the standing detonation waves is illustrated in Fig. 12.53. As described already, the experimental apparatus must have three component, first a component for producing a hypersonic flow is composed of a shock tube connected to a divergent nozzle, second that for decelerating the flow to balance with the detonation propagation consists of a convergent nozzle, and third that for producing detonation waves of two tubes separated with a diaphragm.

In the first component the high pressure driver gas HS, low pressure tube LS, and divergent nozzle DN are separated by diaphragms F_1 and F_2, whilst the high pressure detonation tube DT_1 and low pressure detonation tube DT_2 are separated by the diaphragm F_3. Tube HS is filled with gaseous He at a pressure between 0.5 and 1.5 kPa, tube LS with a stoichiometric hydrogen–oxygen mixture at a pressure between 5.0 and 30 kPa and temperature of 20°C. A divergent nozzle DN accelerates the flow coming from LS.

In the second component the convergent nozzle CN is also filled with a stoichiometric hydrogen–oxygen mixture under the same pressure and temperature as that in DN and decelerates the flow to stagnate the detonation propagation.

12.2 Hypersonic Combustion for RAM Jet Engine

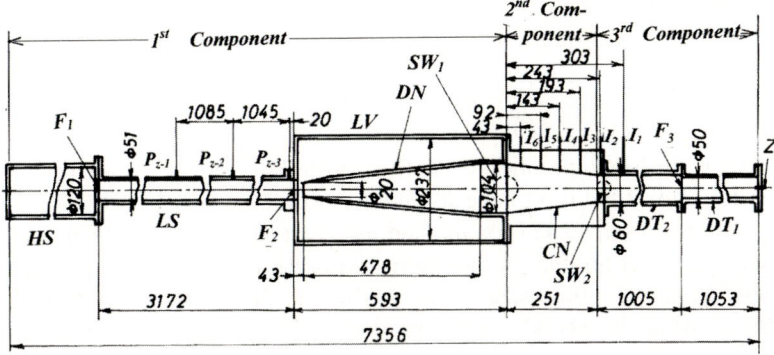

Fig. 12.53. Experimental apparatus for standing detonation. 1st component: for producing hypersonic flow, 2nd component: for decelerating flow to balance with detonation, 3rd: for producing detonation wave, HS: shock driver gas tube, LS: low pressure tube, DN: divergent nozzle, CN: convergent nozzle, DT_1: detonation initiation tube, DT_2: detonation propagation tube, SW_1 and SW_2: BK-7 glass windows, F_1, F_2, and F_3: diaphragms of polyester film, P_{z-1}, P_{z-2}, and P_{z-3}: piezoelectric pressure transducers, $I_1, I_2, \ldots, I_3$: ion probes, Z: spark plug

In the third component the tube DT_2 is filled with the same mixture under a pressure between 1.0 and 4.0 kPa, while DT_1 in which the detonation waves are produced with the same mixture under a pressure between 10 and 55 kPa and room temperature.

Rupture of the diaphragm F_1 results in a shock tube propagating along the low pressure tube LS with a Mach number of 2.8–3.8. The flow with a velocity of 750–1,000 m s^{-1} enters the divergent nozzle DN and accelerates. The temperature behind the shock wave is between 750 and 1,000 K. Below 800 K, the ignition delay is longer than 0.3 ms and no ignition takes place in the mixture during each experiment.

The stoichiometric hydrogen–oxygen mixture in the tube DT_1 is ignited by a spark plug set at the tube end and triggered by a shock arrival signal at the piezoelectric pressure transducer P_{z-1} and transformed to a detonation wave propagating into the tube DT_2.

12.2.4 Experiments

Production of Hypersonic Flow[143]

Shock velocities are measured by transit times between the pressure transducers P_{z-1}, P_{z-2}, and P_{z-3}, and those of the detonation by transit times between the ion probes $I_1, I_2, \ldots, I_6$.

Flow velocities at the exit of the divergent and convergent nozzles are measured by the schlieren method at a thin steel needle set in the flow at a pair of BK-7 glass windows to give Mach angles. The flow velocities at the divergent and convergent nozzles are estimated according to the following equation introduced from the equations for oblique shocks[144]

$$\frac{1}{M_0^2} = \sin^2 \beta - \frac{\gamma+1}{2} \frac{\sin\beta \sin\theta}{\cos(\beta-\theta)}, \tag{12.41}$$

where M_0 is the Mach number of the flow, β the Mach angle, 2θ the tip angle of the needle, and γ the ratio of the specific heats, as shown in Fig. 12.54.

In such a two-step expansion method used here, however, two shock waves are produced one after another in the divergent nozzle, DN. As shown in Fig. 12.55, the first shock S' is produced in the gas in the divergent nozzle propagating downward, while the second one S' is produced in the flowing from the shock tube LS into the divergent nozzle propagating upstream, as long as the initial pressure of the gas in the nozzle is almost the same as that in the low pressure tube LS.

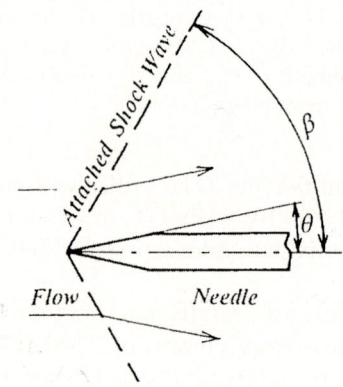

Fig. 12.54. Attached shock wave formed at the tip of a needle set in a hypersonic flow

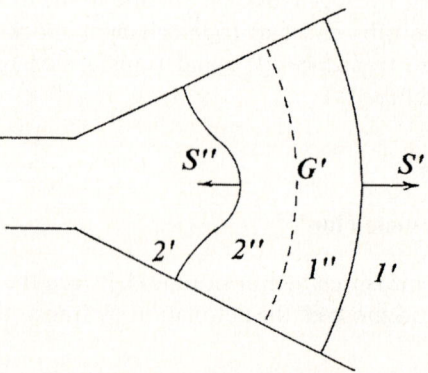

Fig. 12.55. Scheme of the first shock S' and second shock S'' produced in a divergent nozzle

12.2 Hypersonic Combustion for RAM Jet Engine

If the pressure P_n in the divergent–convergent nozzle is lower than that P_1 in the low pressure tube LS, as shown below in Fig. 12.54, the second shock S'' also propagates downward and behind it we can have a flow of high velocity. The flow velocity behind the second shock increases with increase of the pressure difference P_1-P_n, as also shown in Fig. 12.56.

In Fig. 12.57 three examples of schlieren photographs taken at the exit of the convergent nozzle CN are shown. Here the pressure of the driver gas He is 1.10 MPa, while that of the air in the low pressure tube LS is 10.0 kPa and that in the divergent nozzle DN is 1.0 kPa. In the first photograph a relatively slow flow

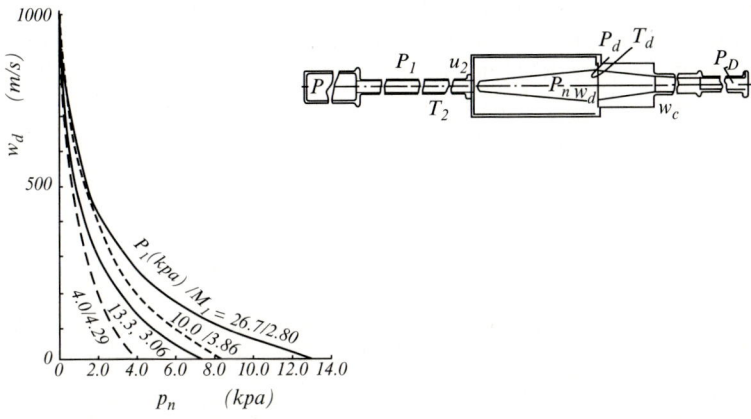

Fig. 12.56. Flow velocity w at the exit of the divergent nozzle DN in relation to P_n with a parameter of the ratio P_1/M_1. P_1: the initial pressure in LS, P_n: the initial pressure in DN, M_1: Mach number of the incident shock

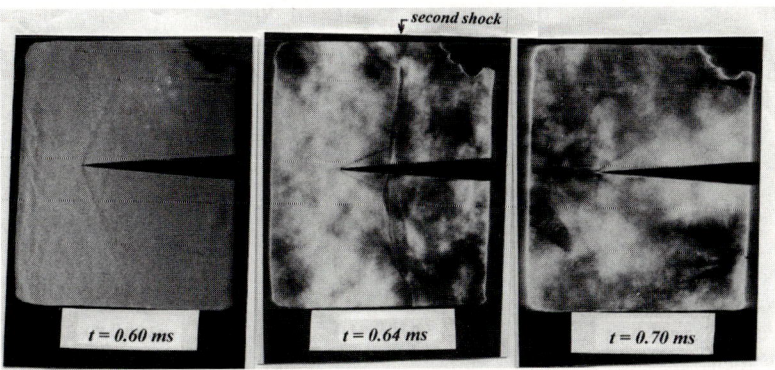

Fig. 12.57. Schlieren photographs in air flow at the exit of the convergent nozzle CN, P_4 (He) = 1.10 MPa, P_1 (air) = 10 kPa, P_n (air) = 1.0 kPa, t: time after the passage of the shock front at P_{z-1}

behind the first shock is observed, although the shock almost vanishes. The second shows a flow with a higher velocity behind the strong second shock.

Three different types of experiments are carried out, namely, using as test gas air (I), $2H_2 + N_2$ (II) and $2H_2 + O_2$ (III) in the low pressure tube, and the divergent and convergent nozzle. To obtain similar results in a stoichiometric hydrogen–oxygen mixture, avoiding a spontaneous ignition in the mixture, the measurement of shock waves propagating in a hydrogen–nitrogen mixture $2H_2 + N_2$ in this shock tube are carried out under the initial condition in which the pressure of the driver gas He is 2.0 MPa, the pressure P_1 of hydrogen–nitrogen in LS 20 kPa and P_n in DN 4.0 kPa under room temperature, as it has almost the same γ and sonic velocity (564 m s^{-1} at 293 K) near that (532 m s^{-1}) of $2H_2 + O_2$. Table 12.4 presents the experimental results in each case.

In Fig. 12.58 the flow velocities at the exits of the divergent and convergent nozzle are shown. The flow velocity increases suddenly to a Mach number higher than 5.0 during about 100 μs after the second shock passes the measurement point. But, it decelerates more slowly than predicted at the end of the convergent nozzle. The duration of the flow velocity >600 m s^{-1} is held for 200–300 μs.

Table 12.4. Experimental conditions and results

experiment	I measurement of flow velocity	II measurement of flow velocity	III measurement of detonation wave
pressure in HS P_4	1.1 MPa per He	2.0 MPa per He	600 kPa per He
pressure in LS P_1	10.0 kPa per air	20.0 kPa per $(2H_2 + N_2)$	13.3 kPa per $(2H_2 + O_2)$
pressure in DN–CN P_n	0.13 kPa per air	4.0 kPa/ $(2H_2 + N_2)$	2.7 kPa/ $(2H_2 + O_2)$
Mach number of incident shock M_1	3.86 ± 0.01	2.64 ± 0.01	2.49 ± 0.01
temperature behind incident shock T_2	1,090 ± 25 K	683 ± 20 K	618 ± 20 K
flow velocity behind incident shock u_2	1,060 ± 20 m s^{-1}	1,096 ± 20 m s^{-1}	1,030 ± 20 m s^{-1}
flow velocity at DN-exit w_c	1,670 ± 40 m s^{-1}	2,240 ± 50 m s^{-1}	
pressure at DN-exit P_d	0.85 kPa		
temperature at DN-exit T_d	245 ± 10 K		
Mach number of flow at DN-exit M_d	5.3 ± 0.1	4.0 ± 0.1	
flow velocity at CN at w_c	1,071 ± 30 m s^{-1}	760 ± 20 m s^{-1}	
pressure in DT_1			53.3 kPa per $(2H_2 + O_2)$

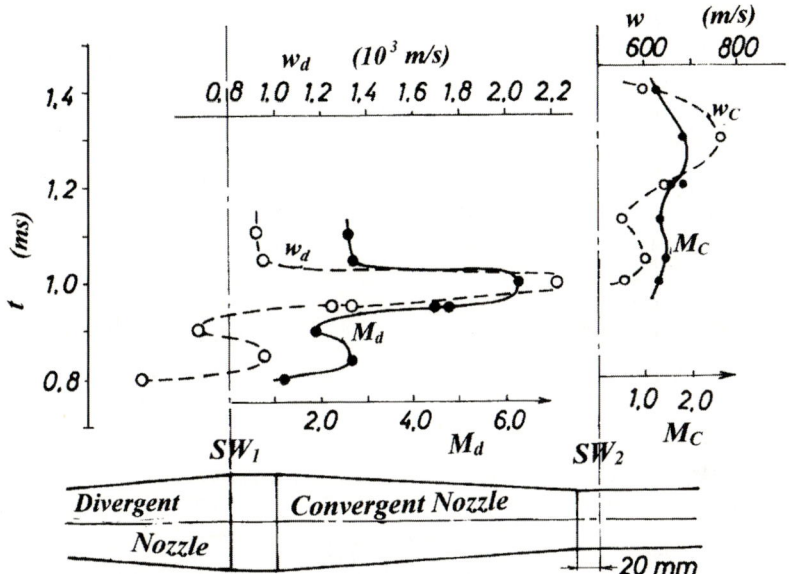

Fig. 12.58. Flow velocity w_d and Mach number M_d measured at the exit of divergent nozzle DN, flow velocity w_c and Mach number M_c at the exit of convergent nozzle CN in $(2H_2 + N_2)$

Figure 12.59 illustrates how the flow velocity varies between the divergent and convergent nozzle. The flow velocities measured at the exit of the divergent nozzle agree well with the theoretical ones, while the experimentally observed velocities in the convergent nozzle decrease much more than the theoretical ones. The reason for the large pressure drop in the convergent nozzle is not yet clear. Some reflections of the shock waves on the nozzle wall may disturb the flow, or a rapid development of the boundary layer may be possible.

Detonation Waves

The time of each experiment from the rupture of the diaphragm F_1 to the arrival of the shock front at the exit of the convergent nozzle is less than 0.3 ms, the whole delay of the spark ignition and establishment of the detonation must be shorter than 0.2 ms. As it is very difficult to establish a self-sustained detonation wave in such a short period under such a low mixture pressure of 4.0 kPa in the tube DT_2, the tube DT_1 is previously filled with a stoichiometric hydrogen–oxygen mixture under a relatively high pressure of 66.0 kPa and ignited by a spark plug set at the detonation tube end to initiate the detonation. The detonation wave propagates into the tube DT_2, breaking the diaphragm F_3. In this case, however, the detonation wave built in DT_2 propagates faster than the self-sustained detonation, because the detonation in DT_2 is driven by the higher pressure in DT_1.

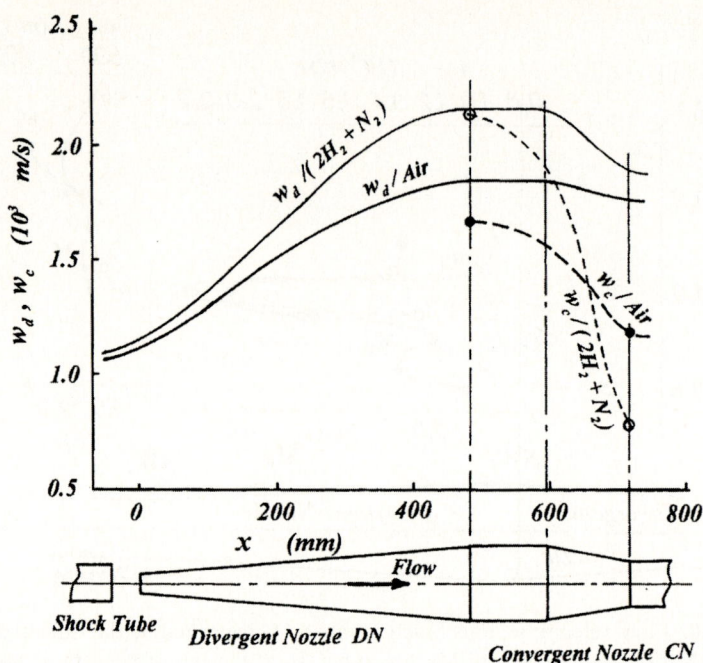

Fig. 12.59. Flow velocity w_d in the divergent nozzle DN and w_c in the convergent nozzle CN. *Solid lines*: theoretical velocity, *broken lines*: experimentally observed, test gas: $2H_2 + N_2$

Practically the experiment must be repeated many times, till the hypersonic flow from the divergent nozzle DN collides with the detonation front propagating against the flow somewhere in the convergent nozzle CN, as the fluctuation of the delay of ignition and transition to detonation is large.

Retardation of Detonation Propagation by Counter Hypersonic Flow

Figure 12.60 presents three experimentally obtained detonation propagation velocities over the length of the convergent nozzle CN. Although no standing detonation front is observed in this experiment, a retardation of 1,000–2,000 m s^{-1} is realized. The velocity of 1,000–2,000 m s^{-1} corresponds to the flow velocity, namely the detonation propagates with a velocity of $(D_w - w_c)$ in the laboratory fixed co-ordinates.

The flow velocity w_c estimated from the detonation propagation velocities D_w and D_{we} are illustrated in the lower part of Fig. 12.60. These correspond to the detonation velocity in each experiment, that is, w-1 to D-1, w-2 to D-2, and w-3 to D-3. These velocities may agree with those estimated from the experimentally obtained flow velocity distribution in the time–space co-ordinates illustrated in Fig. 12.58 or with the measured flow velocities in Fig. 12.59.

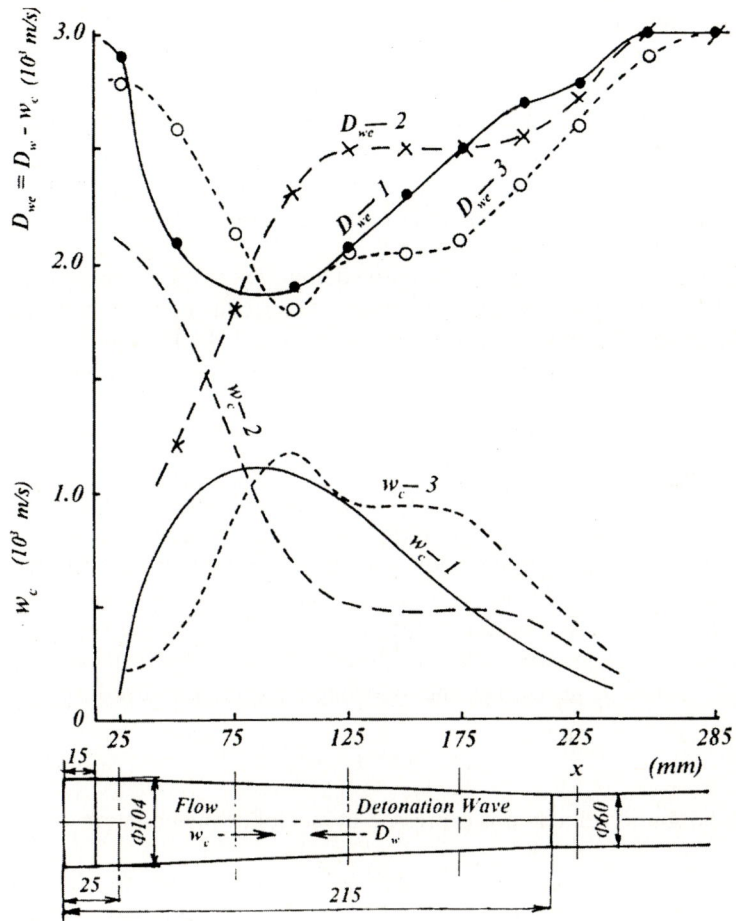

Fig. 12.60. Detonation propagation velocity D_{wc} and flow velocity w_c with respect to the distance x from the inlet of the convergent nozzle CN. $D_{wc} = D_w - w_c$: propagation velocity of the detonation front in the convergent nozzle CN in laboratory fixed co-ordinates. w_c: flow velocity estimated from the relation $D_{wc} = D_w - w_c$

These results suggest that the detonation propagation can be stagnated by the flow and a standing detonation can be established, if we could have a hypersonic flow faster than 3,000 m s^{-1}. Applying such a standing detonation waves, a hypersonic combustion ram jet engine having a higher efficiency might be realized.

12.3 Shock Tubes Driven by Detonation Waves

Considering (5.17), it is necessary to apply a shock driver gas having a high sound velocity and pressure in order to obtain strong shock waves in a shock tube, i.e., shock waves propagating with a high Mach number. For this purpose, a method using a free piston has been developed by Stalker,[145, 146] in which the driver gas is compressed by a free piston driven by a high-pressure gas in a tube upon which the diaphragm between the driver gas tube and low pressure tube is broken, producing a shock wave in the low pressure gas. Such a shock tube is called Stalker tube.

12.3.1 Rarefaction Waves Behind Detonation Waves (Taylor Expansion)

In a detonation wave the flow behind the detonation front in the same direction of the detonation propagation is necessary to sustain the detonation propagation. In order to apply the energy released by the detonation to mechanical work, thus, only the flow behind the rarefaction wave produced behind the detonation wave and flowing in the opposite direction to the detonation propagation can be applied. Now we investigate what flow behind the rarefaction waves is observed.

Assuming a detonation wave propagating along a straight tube having a constant inner diameter and satisfying the so-called Chapman–Jouguet condition, let D be the detonation propagation velocity, w_3 the flow velocity, ρ_3 the density of the gas, a_3 the sound velocity in the gas behind the detonation front, respectively, and w the flow velocity, ρ the gas density, a the sound velocity in the gas behind the rarefaction wave where the gas expands isentropically, and γ the ratio of the specific heats, then the following relations are introduced[147, 148]:

$$w - w_3 = \int_\rho^{\rho_3} a \frac{d\rho}{\rho} \tag{12.42}$$

and as the sound velocity a is expressed by the following equation

$$a = a_3 \left(\frac{\rho}{\rho_3}\right)^{(\gamma-1)/2}, \tag{12.43}$$

then

$$w = w_3 - \frac{2a_3}{\gamma - 1}\left\{1 - \left(\frac{\rho}{\rho_3}\right)^{(\gamma-1)/2}\right\} \tag{12.44}$$

or

$$w = w_3 - \frac{2a_3}{\gamma - 1}\left\{1 - \left(\frac{P}{P_3}\right)^{(\gamma-1)/2\gamma}\right\}. \tag{12.44'}$$

Therefore

$$w + a = w_3 + a_3 - \frac{\gamma+1}{2}(w_3 - w). \tag{12.45}$$

From (11.17) introduced under the Chapman–Jouguet condition where $M_D^2 \gg 1$, in which $M_D = D/a$,

$$w_3 \approx \frac{D}{\gamma + 1}. \tag{12.46}$$

12.3 Shock Tubes Driven by Detonation Waves

Let X be the distance from the detonation initiation point to the detonation front and x that from the detonation front to the rarefaction wave front, $(w_3 + a_3)\,t = -X$ and $(w + a)\,t = -x$, where t is the time in which both the waves propagate the distance X and x, respectively. Here X and x take minus sign, as their coordinates have the opposite direction to the detonation propagation.

As $w_3 + a_3 = D$, the following equation is introduced from (12.45),

$$w + a = D - \frac{\gamma+1}{2}\left\{\frac{D}{\gamma+1} - w\right\} \tag{12.47}$$

then

$$w = \frac{\frac{X}{t} - \frac{2x}{t}}{\gamma+1}, \tag{12.47'}$$

thus

$$w = 0 \text{ where } x = \frac{X}{2}, \tag{12.48}$$

$$w = \frac{D}{\gamma+1} \text{ where } x = 0, \text{ and} \tag{12.49}$$

$$w = -\frac{D}{\gamma+1} \text{ where } x = 1.0 \text{ in an open end tube.} \tag{12.50}$$

Further, following equations can be introduced considering (12.44) and (12.44'):

$$\frac{P}{P_3} = \left\{1 - \frac{\gamma-1}{\gamma}\frac{X-x}{X}\right\}^{2\gamma/(\gamma-1)}, \tag{12.51}$$

$$\frac{T}{T_3} = \left\{1 - \frac{\gamma-1}{\gamma}\frac{X-x}{X}\right\}^{2}, \tag{12.52}$$

where P_3 and T_3 mean the pressure and temperature behind the detonation wave, P and T those behind the rarefaction wave, respectively.

Between $x = X/2$ and $x = 1.0$ in the closed end tube, the gas is at rest, namely the flow velocity w behind the rarefaction wave is always equal to 0 and the pressure ratio keeps a constant value,

$$\frac{P}{P_D} = \left\{1 - \frac{\gamma-1}{2\gamma}\right\}^{2\gamma/(\gamma-1)}. \tag{12.53}$$

Thus, P/P_D and w/D with respect to x/X are expressed in the diagram of Fig. 12.61.

12.3.2 Shock Tube Directly Driven by Detonation Waves[149-151]

In general, after the end of the combustion process, behind a propagation detonation wave, an isentropic expansion takes place propagating together with the

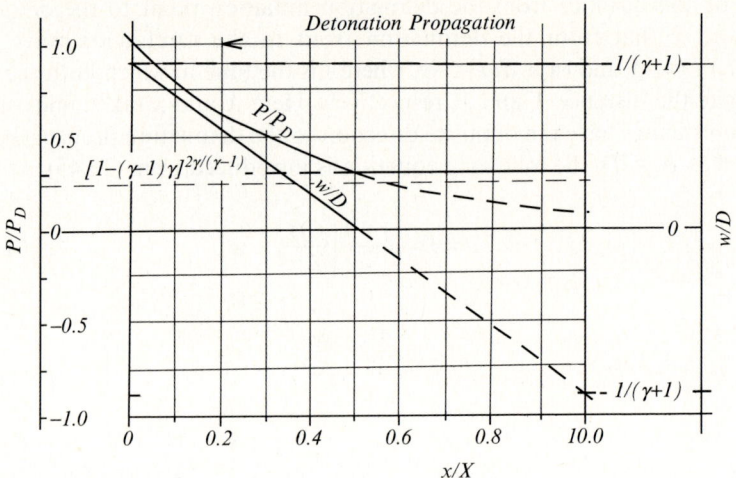

Fig. 12.61. P/P_D and w/D with respect to x/X. P, w: pressure and flow velocity behind rarefaction wave; P_D: pressure behind the detonation front; D: detonation propagation velocity; x: distance from the detonation front to the rarefaction wave front; X: distance of the detonation front to its initiation point. *Solid line*: in the closed end tube, *broken lines*: in the open end tube

detonation, as described above. As the theoretical variations of pressure and flow velocity behind the detonation wave are illustrated in Fig. 12.61, the behavior of the gas behind the detonation wave is just like that behind the high pressure driver gas in a shock tube after the rupture of the diaphragm which drives the shock wave in the low pressure tube accompanying a rarefaction wave.

In an open end tube the gas behind the detonation wave expands approximately isentropically keeping the pressure and flow velocity at the tail of the detonation wave almost constant as shown in the diagram of Fig. 12.61. The flow having a constant pressure and velocity can produce a shock wave propagating in the opposite direction of the detonation propagation with a constant velocity. A brief explanation of the shock tube directly driven by detonation waves is introduced according to the works carried out by Yu, Groenig, and Olivier in Stosswellenlabor der RWTH Aachen University.

An example of the shock tunnels driven by detonation waves developed for the investigation of the hypersonic flow in the Stosswellenlabor, Aachen, is schematically illustrated in Fig. 12.62. In the driver section a hydrogen–oxygen mixture having a rate of $H_2/O_2 = 0.82$ is filled under a pressure of 10 MPa, while air is filled in the driven section under a pressure of 80 kPa. To the end of the driven section a divergent nozzle and test section are connected to examine the hypersonic flow, and a damping section filled by air under a pressure of 0.1 MPa is connected to the driver section. The initial temperature of the gas in each section is 293 K.

In the mixture in the high pressure section a detonation wave is initiated at the right end by the detonation wave propagating from the initiation tube and

12.3 Shock Tubes Driven by Detonation Waves

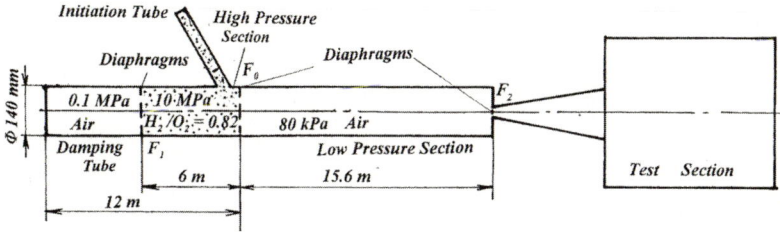

Fig. 12.62. Sketch of the shock tunnel driven by detonation (developed at the Stosswellenlabor/Aachen)

propagates left. The diaphragm F_0 is soon broken and the burned gas flow into the low pressure section, namely the detonation wave propagates along an open rear end, behind which the gas flows into the air in the low pressure tube, keeping the pressure and velocity constant behind the rarefaction wave in an open end, as described above.

Thus, the detonation wave drives a shock wave propagating with a constant velocity under a constant pressure. In Fig. 12.63 a wave diagram obtained theoretically in this shock tube is illustrated.

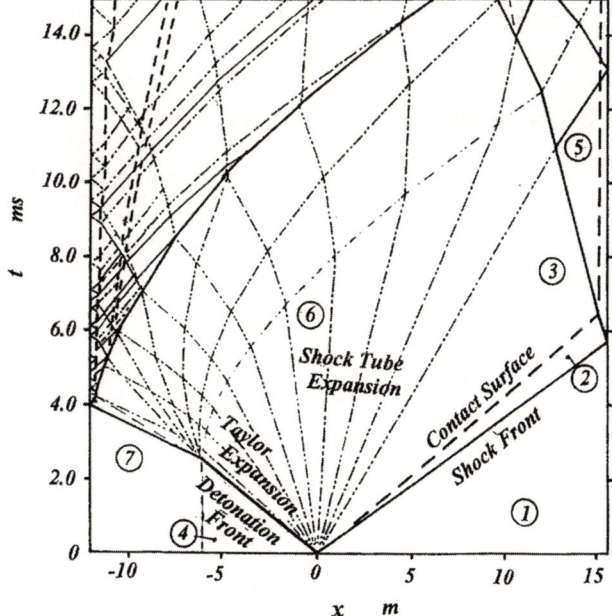

Fig. 12.63. Wave diagram in a shock tube with detonation driver and damping tube. $P_4 = 10$ MPa, $T_4 = 293$ K, $H_2/O_2 = 0.82$, $P_1 = 80$ kPa, $T_1 = 293$ K, driven gas air: $P_7 = 0.1$ MPa, $T_7 = 293$ K, damping gas is air

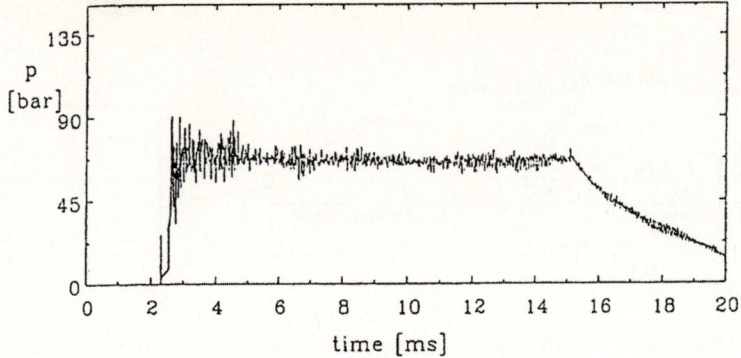

Fig. 12.64. Pressure P in the driven section measured at a point 160 mm from the diaphragm F_1. $P_4 = 1.0$ MPa, $T_4 = 293$ K, $H_2/O_2 = 4$, vacuum in the damping tube and driven section, using a thin diaphragm between the damping tube and high-pressure section

In Fig. 12.64 an example of the pressure variation in the driven section, measured at the position of 160 mm from the diaphragm F_1 of the shock tunnel is illustrated. The pressure behind the rarefaction wave produced behind the detonation wave is kept almost constant during about 12 ms, as the theory explained above.

By this method a shock wave propagating in air theoretically with a Mach number of 8.0, thus, can be obtained.

12.3.3 Shock Tube Using Free Piston Driven by Detonation Waves[152]

A Stalker tube requires a large quantity of high-pressure gas to drive the piston and heavy tubes. Applying detonation waves to drive the piston, we can obtain strong shock waves in a simple shock tube under a low initial pressure of the drive gas.

Apparatus

A light driver gas such as hydrogen or helium should be heated as much as possible to produce strong shock waves in a shock tube. With shock waves, the gas can be heated to a much higher temperature than by isentropic compression, as shown in Fig. 12.65.

Using a free piston driven by a high-pressure gas the shock driver gas can be compressed and heated, as reported by Stollery and Stalker[145]. In order to produce shock waves to heat the driver gas, the free piston must be accelerated to a high velocity corresponding to the sound velocity of the gas. For this purpose a shock tube having a free piston driven by detonation waves is developed. The shock tube is schematically illustrated in Fig. 12.66.

12.3 Shock Tubes Driven by Detonation Waves

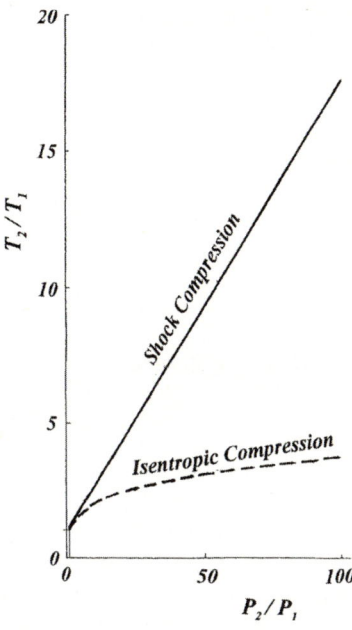

Fig. 12.65. Comparison of the temperature rise by shock compression. Subscript 1 means the initial state, 2 that after the compression

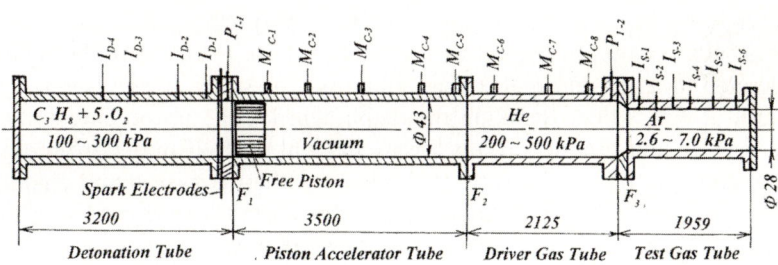

Fig. 12.66. Shock tube using a free piston driven by detonation waves, dimensions in mm

The shock tube consists of four tubes of stainless steel interconnecting along the same center line, i.e., a detonation tube having a length of 1,005 mm, 2,010 mm or 3,200 mm, a piston accelerator tube of 3,500 mm in length, a driver gas tube of 2,125 mm in length and a test tube filled with a low pressure gas of 1,959 mm in length. The detonation tube has an inner-diameter of 43 or 63 mm, while the piston accelerator and driver gas tube have the same inner-diameter of 43 mm, and the test gas tube has 28 mm diameter. These tubes are separated from each other by polyester diaphragms F_1 and F_2 having a thickness of 100 μm and an aluminum or steel diaphragm F_3 having a thickness of 1.5, 2.0, 3.0, or 4.0 mm.

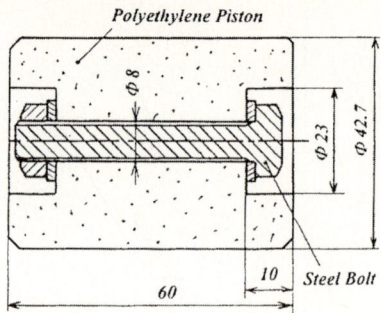

Fig. 12.67. Free piston, dimensions in mm

The free piston set in the piston accelerator tube and contacted with diaphragm F_1 is made of polyethylene having a steel bolt in its center, as shown in Fig. 12.67. The piston has a mass of 102 ± 2 g.

The detonation tube is filled with a stoichiometric propane–oxygen mixture under 200 or 300 kPa and the driver gas tube with He gas under a pressure between 200 and 400 kPa, while the piston accelerator tube is kept almost in vacuum (<100 Pa). The experiments are carried out at room temperature.

The shock tube operates as follows: Following an ignition of the propane–oxygen mixture by an electrical spark at the end of the detonation tube near the diaphragm F_1, a detonation wave propagates to the left, rupturing the diaphragm F_1. The free piston pushed by the high-pressure gas behind the rarefaction wave produced behind the detonation wave is driven to the right and accelerated to a velocity higher than 300 m s^{-1}, as schematically shown in the diagram of Fig. 12.68. Breaking the diaphragm F_2, the piston enters into the driver gas He and produces a shock wave in front of itself in the driver gas.

By reflection repeated several times between the free piston and diaphragm F_3, as shown in Fig. 12.68, a shock wave compresses the driver gas He to a pressure higher than 12 MPa and heats it to a temperature higher than 1,500 K, as described later. When the piston approaches the diaphragm F_3, the diaphragm is broken, producing a shock wave in the argon gas filled in the test tube under an initial pressure between 2.5 and 7.0 kPa.

Experiments

The experiments are carried out under different conditions, that is, different initial pressures of the propane–oxygen mixture, helium and argon, using the diaphragm F_3 of different materials and thickness.

The propagation of the detonation and shock waves are observed using ionization probes set at different positions on the detonation or test gas tube, and shown as I_{D-1}, I_{D-2}, ..., I_{D-4} and I_{S-1}, I_{S-2}, ..., I_{S-6}, respectively, as shown in Fig. 12.66, while the motion of the free piston is observed by measuring the

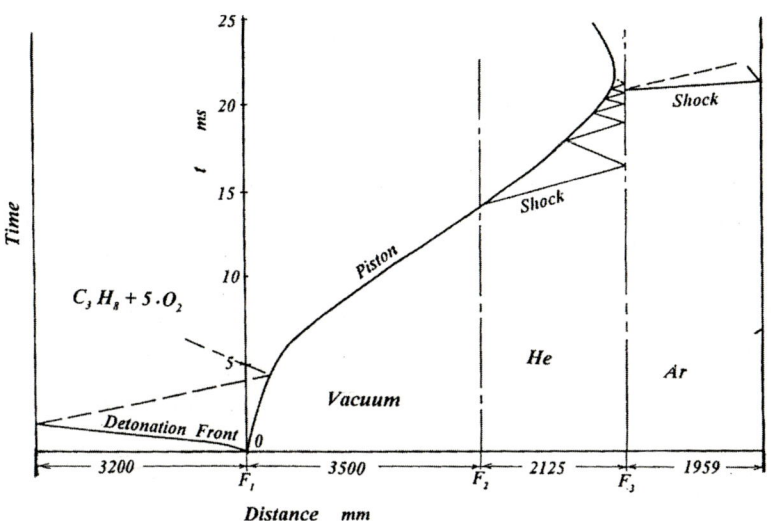

Fig. 12.68. Time–distance diagram of the detonation front, operating piston and shock wave observed in the shock tube shown in Fig. 12.66

current through solenoids wound around permanent magnets set at different positions on the piston accelerator and driver gas tubes, shown as M_{C-1}, M_{C-2}, ... M_{C-8}. The current is induced by the motion of the bolt set in the piston. The pressure of the gas in the detonation tube is measured by a piezoelectric pressure transducer P_{I-1} and that in helium in the driver tube by P_{I-2}.

Experimental Results

Detonation Wave

Figure 12.69 shows two examples of the time–distance diagram of the detonation front propagating in the propane–oxygen mixture under the initial pressure of 300 kPa in the detonation tube having an inner-diameter of 43 mm. t_d (broken line) represents the propagation of the detonation during the diaphragm F_1 is closed, while t_{d-p} (solid line) signifies the propagation when the diaphragm F_1 is broken by the detonation and the free piston is driven by the high pressure burned gas behind the detonation waves.

The propagation velocity of the detonation wave is about 2,160 m s^{-1} at the initial mixture pressure of 300 kPa when F_1 is broken, and is slightly lower than that of 2,290 m s^{-1} observed when both the tube ends are closed. The detonation propagation velocity decreases little with decrease of the initial mixture pressure, and we observe a normal detonation propagation in every case.

In Fig. 12.70 an oscillogram of the pressure history P_D observed by the pressure transducer P_{I-1} at the tube end near the diaphragm F_1 in the detonation tube

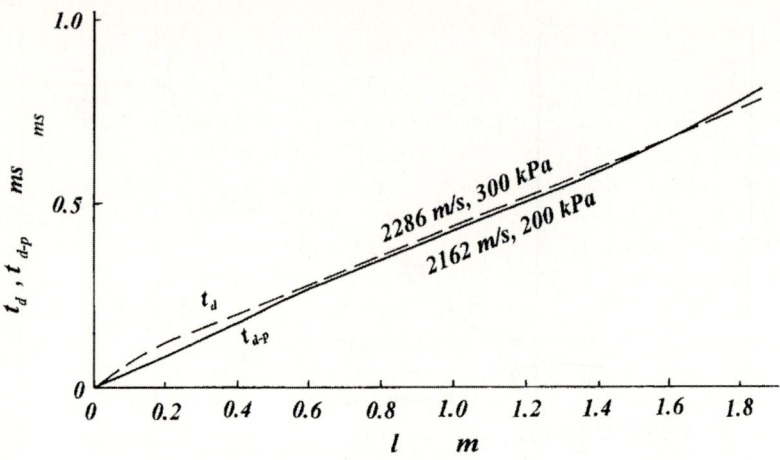

Fig. 12.69. Experimentally obtained time–distance diagram of the detonation front. t_d: when both the ends are closed, t_{d-p}: when F_1 is broken and the free piston is driven. l: distance from F_1 toward the left tube end. Number on each curve is the final propagation velocity and initial pressure of $C_3H_8 + 5O_2$ mixture

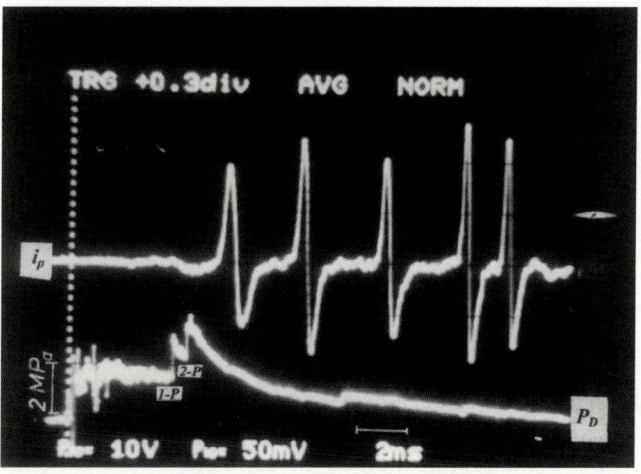

Fig. 12.70. Secondary current i_p (*above*) through solenoids M_{C-1}, M_{C-2}, ..., M_{C-5} induced by the piston motion and pressure P_D (*below*) in the detonation tube at the initial mixture pressure of 200 kPa

is shown. The first pressure jump 1-P is caused by the ignition and detonation, the next jump 2-P by the shock wave produced at the reflection of the detonation wave from the closed tube end and the third one by the shock wave reflected from the piston. The burned gas behind the detonation wave expands almost isentropically, driving the free piston.

12.3 Shock Tubes Driven by Detonation Waves

Acceleration of the Free Piston

Following the rupture of the diaphragm F_1, the free piston is driven and accelerated to a high velocity propagating along the vacuum tube by the flow and high-pressure gas produced in the rarefaction wave behind the detonation wave. The time–distance diagram of the free piston under different conditions is illustrated in Fig. 12.71. The final velocity of the piston reaches a value between 300 and 420 m s^{-1} corresponding to the initial condition.

As long as the same piston is driven along the same vacuum tube, the kinetic energy ε_p of the operating piston is almost proportional to the theoretically calculated work W_p done by the isentropic expansion of the gas behind the detonation wave in the piston accelerator tube, as shown in Fig. 12.72.

In this method, the mixture is ignited near F_1 contacting with the free piston, from which the detonation wave propagates to the other closed end. The free piston, therefore, is pushed by the pressure of the rarefaction wave behind the detonation wave. The pressure is about 1/3 of that at the so-called Chapman–Jouguet point, as explained in Sect. 12.3.1. The mixture can be ignited at the other end, so that the detonation propagates toward the piston with the latter method, however, the running velocity of the piston fluctuates over a wide range of more than ±10%, while that in the former method is kept within ±1.5%. The reason for it is attributed to the higher entropy increase by the latter method than the former because of higher irreversibility, much larger fluctuation is observed, when the high pressure behind the detonation front pushes the piston suddenly, as explained in Sect. 3.1.

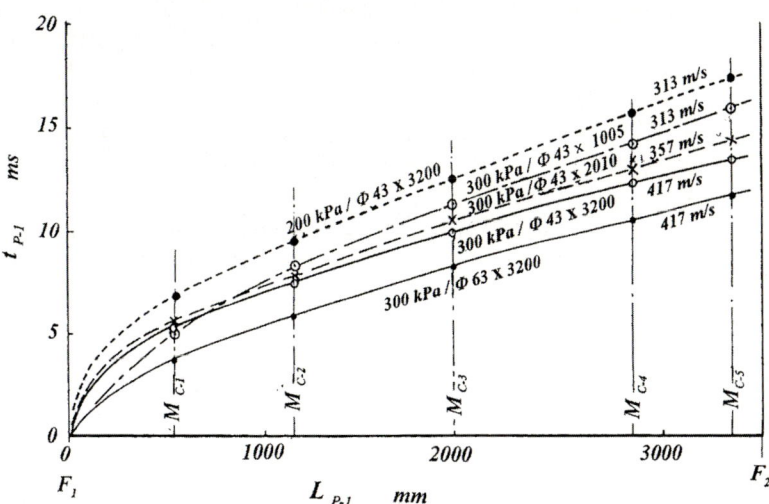

Fig. 12.71. Piston motion diagram. *L*: distance from F_1, Number on each curve: inner diameter (mm) × length of the tube (mm), initial mixture pressure and final velocity

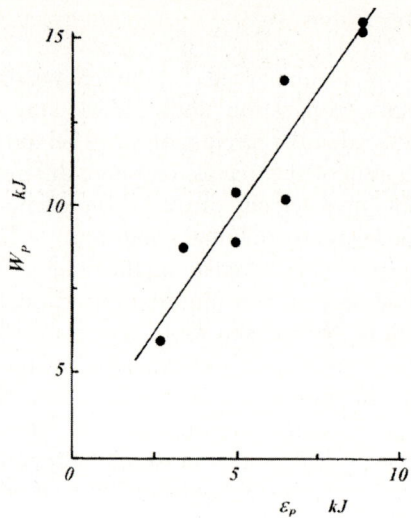

Fig. 12.72. Work W_p performed on the piston by isentropic expansion of the gas behind the detonation with respect to the kinetic energy ε_p of the piston after acceleration

Shock Compression and Driver Gas Heating

The free piston accelerated in the vacuum tube by the rarefaction wave behind the detonation wave enters into the driver gas He having an arbitrary pressure between 200 and 300 kPa, breaking the polyester diaphragm F_2. If the diaphragm F_3 is completely closed, the piston stops near F_3 and returns about 80 cm, but if F_3 is broken, the piston continues to move further and reaches the tube end. Figure 12.73 shows an example of the oscillograms in which the piston motion is observed by measuring the induced current through the solenoids at M_{C-1}, M_{C-2}, ..., M_{C-8} on the driver gas tube, while the pressure variation of He is measured by the pressure transducer P_{1-3} set at the tube end near the diaphragm F_3.

From such oscillograms we obtain time-distance diagrams of the free piston in the driver gas tube from F_2 to F_3, as well as the pressure and temperature variations of He in relation to the piston motion, as shown in Fig. 12.74.

These diagrams suggest that the piston is further accelerated in the first half of the driver gas tube and then decelerated, producing a shock wave in front of the piston with a propagation velocity depending on the piston velocity. A shock wave having a Mach number from 1.10 to 1.40 propagates to the right tube end and is reflected from the diaphragm F_3, as shown in Fig. 12.74. The reflected shock wave propagates left and is again reflected from the piston. With the repetition of such reflections between F_3 and the free piston, He is compressed and heated in a stepwise manner, as shown in Fig. 12.74.

As soon as the He is compressed beyond a certain pressure corresponding to the strength of the diaphragm F_3, the aluminum or steel diaphragm F_3 is broken and He gas flows into the low-pressure argon in the test tube producing a strong

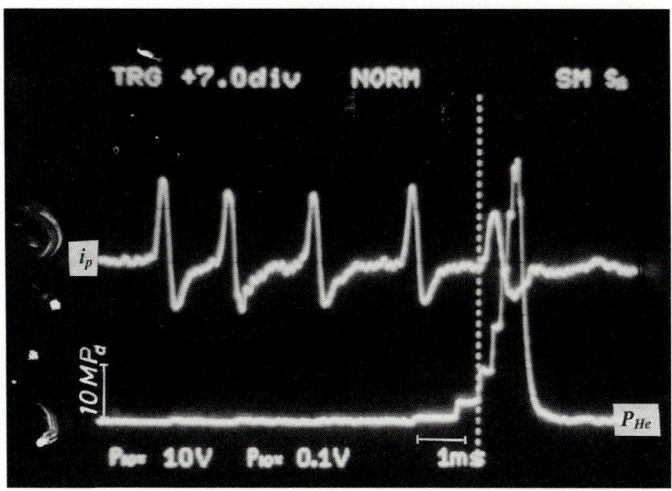

Fig. 12.73. Current i_p through solenoids M_{C-4}, M_{C-5}, ..., M_{C-8} (*above*) and pressure P_{He} of the driver gas He (*below*). Initial pressure of He : 300 kPa, that of argon in the test tube: 2.7 kPa. F_1 comprised two 1.5-mm thick Al plate

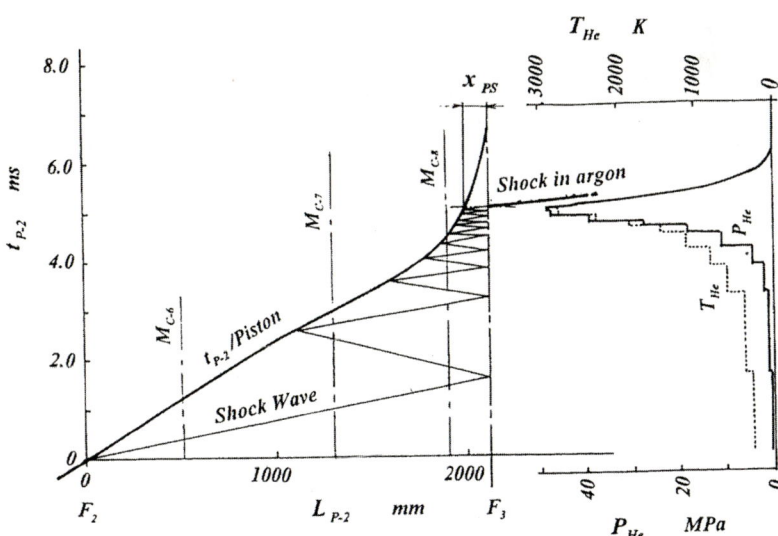

Fig. 12.74. Time–distance diagram of the piston motion as well as the variation of pressure P_{He} and temperature T_{He} of He in the driver gas tube under the same conditions of Fig. 12.73. t_{p-2}: time after the entrance of the piston into the driver tube, L_{p-2}: distance from F_2, x_{ps}: distance from F_3 to the piston at the rupture instant of F_3

shock wave in the test gas behind F_3, while the piston moves to the end of the driver gas tube.

The maximum pressure P_{max} of He in the driver tube as well as the piston position, that is, the distance x_{ps} from the diaphragm F_3 to the piston at the rupture instant of the diaphragm depends on the strength, namely, on the material and thickness of the diaphragm F_3. As shown in Fig. 12.75, P_{max} increases with increase of the diaphragm thickness, while x_{ps} decreases. P_{max}, however, decreases with increase of the initial pressure of He, while x_{ps} increases, as shown in Fig. 12.76, although both P_{max} and x_{ps} vary over a relatively large range.

Production of Strong Shock Waves

Since the driver gas He is compressed to a pressure between 10 and 50 MPa, being heated to a temperature between 1,000 and 3,000 K, as described above, strong shock waves can be produced in argon in the test gas tube. Figure 12.77 shows some examples of the time–distance diagrams of the shock waves driven by the compressed He and propagating in argon in the test gas tube under different conditions, namely, at different initial pressures of He and argon, respectively, and maximum pressure P_{max} of He in the driver tube.

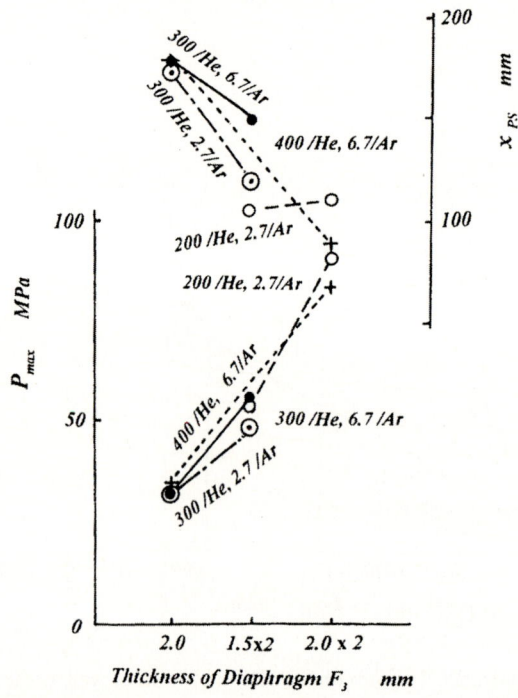

Fig. 12.75. Maximum pressure P_{max} of He and distance x_{ps} from F_3 to the piston at the rupture instant of F_3 Numbers on lines: the initial pressure of He and argon in kPa

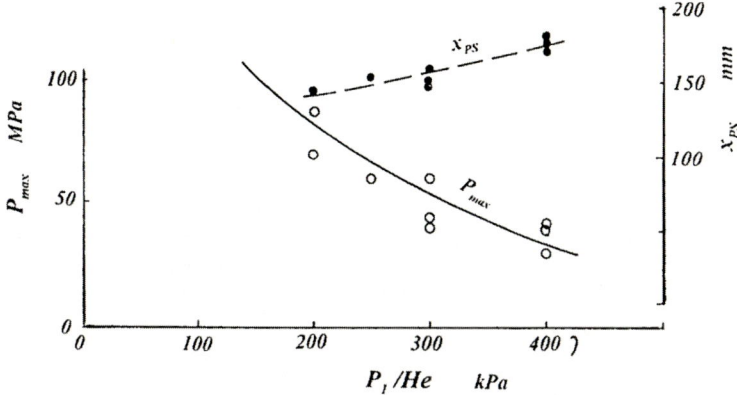

Fig. 12.76. P_{max} and x_{ps} with respect to the initial pressure of He

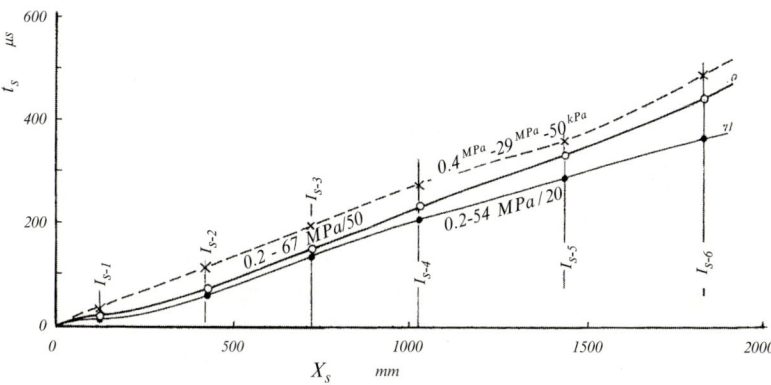

Fig. 12.77. Time–distance diagrams in argon. Numbers on each curve give the initial pressure (MPa), P_{max} (MPa) of the driver gas He and initial pressure of argon (kPa)

From these diagrams, we obtain the Mach number M_s of the shock waves propagating in argon in relation to the distance L_s from the diaphragm F_3, as illustrated in Fig. 12.78. In this diagram each shock wave shows an almost constant propagation velocity after passing a point distant 1,000 mm from the diaphragm F_3.

Figure 12.79 shows the Mach number M_s of the shock waves having an almost constant propagation velocity during the latter half of the test tube in relation to the maximum pressure P_{max} of He in the driver tube. At the initial argon pressure of 2.7 or 6.7 kPa, the experimentally obtained Mach numbers M_s is from 10 to 15 which are always lower than theoretical ones; in particular, the discrepancy becomes larger with increase of the maximum pressure P_{max} of He in the driver gas tube.

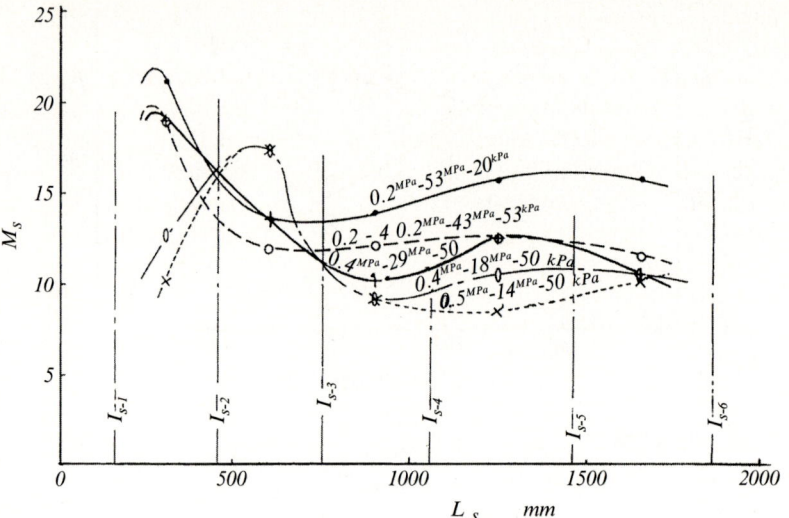

Fig. 12.78. Mach number M_s of shock waves produced in argon with respect to the distance L_s from the diaphragm F_3. Numbers on each curve give the initial pressure (MPa), P_{max} of the driver gas He, and initial pressure of argon (kPa)

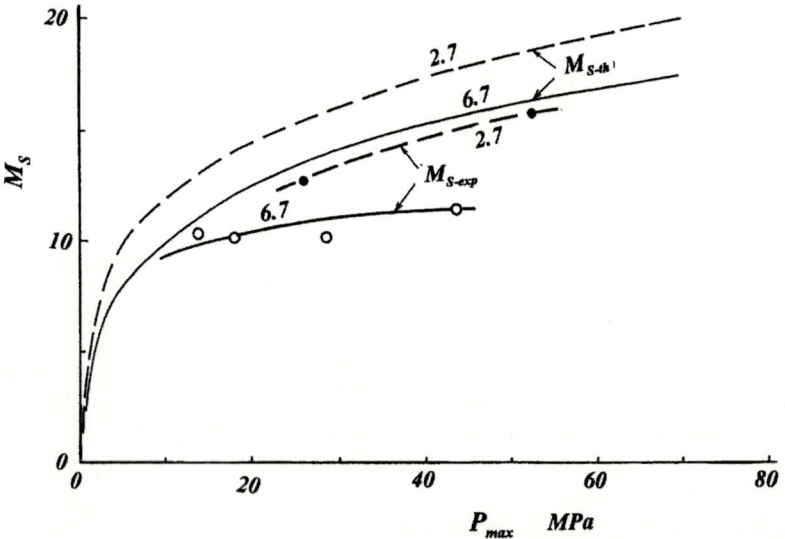

Fig. 12.79. Final Mach number M_s of the shock waves produced in the test tube with respect to the maximum pressure P_{max} of the driver gas He, $M_{S\text{-th}}$: theoretically estimated, $M_{S\text{-exp}}$: experimentally obtained. Number on each curve gives the initial pressure of argon in kPa

Concluding Remarks

In a shock tube using a free piston driven by detonation waves, strong shock waves having a Mach number higher than 10 in the test gas argon can be obtained. Behind reflected shock waves produced by such strong incident shock waves the argon is heated higher than 20,000 K. In such a high temperature the argon is fully ionized and isotropic plasma in relatively large quantity is produced. Such isotropic plasma having a large quantity can be applied to an experimental investigation in plasma physics.

References

1. Becker, R.: *Theorie der Wärme* (Springer, Berlin Heidelberg New York, 1955) Chap. 1A and B
2. Jost, W.: *Explosions-und Verbrennungsvorgänge in Gasen* (Springer, Berlin Heidelberg New York, 1939)
3. Lewis, B. and von Elbe, G.: *Combustion, Flames and Explosions of Gases* (Academic, London, 1959)
4. Kuo, K.K.: *Principles of Combustion* (Wiley, New York, 1986)
5. Jost, W.: *Explosions-und Verbrennungsvorgänge in Gasen* (Springer, Berlin Heidelberg New York, 1939) Chap. 1. p. 3
6. Jost, W.: *Explosions-und Verbrennungsvorgänge in Gasen* (Springer, Berlin Heidelberg New York, 1939) Chap. 8
7. Dainton, F.S.: *Chain Reaction* (Methuen, London, 1956)
8. Semebnoff, N.: Zur Theorie des Verbrennungsprozesses, Z. Phys. **48** (1928) 571
9. Giddings, J.C. and Hirschfeld, J.O.: Flame properties and kinetics of chain-branching reactions, 16th Symp. (Intern.) on Combust. (the Combustion Institute, Pittsburgh, 1965) 295
10. Jost, W.: *Explosions-und Verbrennungsvorgänge in Gasen* (Springer, Berlin Heidelberg New York, 1939) Chap.9.A
11. Lewis, B. and von Elbe, G.: *Combustion, Flames and Explosions of Gases* (Academic, New York, London, 1961) Chaps. II and III
12. Silver, R.S.: Phil Mag. (7) **23** (1947) 633
13. Jost, W.: *Explosions-und Verbrennungsvorgänge in Gasen* (Springer, Berlin Heidelberg New York, 1939) Chaps. 8–2, p. 244
14. Fukutani, S. and Jinno, H.: Mem. Fac. Eng. Kyoto Univ. 7 (1985) No.4
15. Terao, K.: Heterogeneity and stochastic phenomena in the irreversible process, Bull. Fac. Eng. Yokohama Natl. Univ. **33** (1984) 15
16. Becker, R.: *Theorie der Wärme* (Springer, Berlin Heidelberg New York, 1955) Chap. I-5
17. Becker, R.: *Theorie der Wärme* (Springer, Berlin Heidelberg New York, 1955) Chap. I-9
18. Becker, R.: *Theorie der Wärme* (Springer, Berlin Heidelberg New York, 1955) Chap. I-20

19. Becker, R.: *Theorie der Wärme* (Springer, Berlin Heidelberg New York, 1955) Chap. I-10
20. Becker, R.: *Theorie der Wärme* (Springer, Berlin Heidelberg New York, 1955) Chap. I-19
21. Becker, R.: *Theorie der Wärme* (Springer, Berlin Heidelberg New York, 1955) Chap. VII-87
22. Rieckers, A. and Stumpf, H.: *Thermodynamik*, Bd. 2 (Vieweg, Braunschweig, 1977) Chap. 10-b
23. Glandsdorff, P. and Progogne, I.: *Thermodynamic Theory of Structure, Stability and Fluctuations* (Wiley-Interscience, London, New York, Sydney, Toronto, 1971) p. 12
24. Baras, F., Nicolis, G., Malek Mansour, M., and Turner, J.W.: Stochastic theory of adiabatic explosion, J. Statis. Phys. **32** (1983) 1–23
25. Nicolis, G. and Baras, F.: Instric randomness and spontaneous symmetry-breaking in explosive system, J. Statis. Phys. **48** (1987) 1071–1090
26. Zuber, K.: Über die Verzögerungszeit bei der Funkenentladung, Ann. Phys. (1925) 231
27. von Laue, M.: Bemerkung zu K.Zubers Messung der Verzögerungszeiten bei der Funkenentladung, Ann. Phys. (1925) 261
28. Hirata, M.: *Applied Statistics* (Kokuseido, 1949, in Japanese) Chap. 11
29. Terao, N.: Fracture phenomena of solid materials, Butsuri, **9** (1954) 247
30. Becker, R.: *Theorie der Wärme* (Springer, Berlin Heidelberg New York, 1955) Chap. II-34
31. Terao, K.: *Combustion and Detonation Waves as Irreversible Phenomena* (I.P.C. 1991 in Japanese)
32. Hirth, J.P. and Pound, G.M.: *Condensation and Evaporation* (Pergamon, Oxford, 1951)
33. Terao, K., Hatano, F., and Sano, F.: Kinetische Untersuchung über das Sieden der Flüssigkeit, Jpn. J. Appl. Phys. **3** (1964) 728–732
34. Terao, K., Ataka, M., and Kaneko, R.: Das Sieden des überhitzten Wassers unter elektrischer Entladung, Jpn. J. Appl. Phys. **4** (1965) 808–812
35. Mauret, P. and Vorsanger, J.J.: Ėtude cinétique de la vaporization de quelques liquids. Influence de l'association, C.R. Acad. Sci. Paris, **248** (1959) 1808
36. Chalmers, B.: *Principles of Solidification* (Wiley, London, 1964) p. 67
37. Terao, K. and Ohtomo, Y.: Kinetische Untersuchung über die Eisbildung in Wasser, Jpn. J. Appl. Phys. **8** (1969) 545–550
38. *American Institute of Physics Handbook*, (McGraw-Hill, New York, 1963) 4–189
39. Wilson, J.G.: *Principles of Cloud Chamber Technology* (Cambridge University Press, Cambridge, 1951)
40. Glaser, D.A.: Some Effects of ionizing radiation on the formation of bubbles in liquids, Phys. Rev. **87** (1952) 665
41. Glaser, D.A.: Bubble chamber tracks of penetrating cosmic-ray particles, Phys. Rev. **91** (1953) 762
42. Oertel, H.: *Stossrohre* (Springer, Berlin Heidelberg New York, 1966) Chap. C, p. 392
43. Greene, E.F. and Toennies, J.P.: *Chemische Reaktionen in Stosswellen*, (J.P.Dr. Dietrich Streinkopf Verlag, Darmstadt 1959, translated in German by H.Gg. Wagner
44. Martinengo, A.: Untersuchung der Selbstzündung von Kohlenwasserstoff-Luft-gemischen durch adiabatische Verdichtung, Dissertation an der Math-Naturw. Fak. Der Univ. Göttingen, 1958
45. Terao, K.: Explosion limits of hydrogen-oxygen mixture as a stochastic phenomenon, Jpn. J. Appl. Phys. **16** (1977) 29–38

46. Soloukhin, R.I.: *Shock Waves and Detonations in Gases* (Mono-Book Corporation 1966, Baltimore, translated by B.W. Kuvshinoff) Chap. V
47. Terao, K.: Selbstzünmdung des n-hexan-Luft-Gemisches in Stosswellen, J. Phys. Soc. Jpn., **15** (1960) 1113–1122
48. Terao, K.: Selbstzündung von Kohlenwasserstoff-Luftgemsiches in Stosswellen, J. Phys. Soc. Jpn., **15** (1960) 2086–2092
49. Levedahl, W.J.: Multi-stage autoignition of engine fuels, Fifth Symp. (Intern.) Combust. (Reinhold Publishing Corporation New York, 1955)
50. Jungers, J.C.: *Cinétique Chmique Appliquée* (Société des &Ebar;ditions Technip, Paris, 1985)
51. Terao, K., Takahashi, S., and Miyazaki, S. Zündgrenzen des Kohlenwasserstoff-Luft-Gemisches, Jpn. J. Appl. Phys. **2** (1963) 429–432
52. Lewis, B. and von Elbe, G.: *Combustion, Flames and Explosions of Gases* (Academic, New York, London, 1961) Chap. V-6
53. Sturgis, B.M.: Some Concepts of Knock and Antiknock Action, SAE Transaction, **63** (1955) 253–264
54. Terao, K.: Die Wirkung des Bleitetraäthyls auf die Zündung des n-Heptan-Luft-Gemisches in Stosswellen, Jpn. J. Appl. Phys. **2** (1963) 364–369
55. Terao, K. and Liao, Ch.: The most inflammable state of a lean hydrocarbon-air mixture, Jpn. J. Appl. Phys. (1992) 2308–2309
56. Terao, K. and Liao, Ch.: Distribution of ignition probability in a heterogeneous mixture, Jpn. J. Appl. Phys. 31 (1992) 1217–1218
57. Liao, Ch. and Terao, K.: Ignition probability in a fuel spray, Jpn. J. Appl. Phys. **31** (1992) 2299–2303
58. Jost, W.: *Explosions-und Verbrennungsvorgänge in Gasen* (Springer, Berlin Heidelberg New York, 1939) Chap. 10
59. Lewis, B. and von Elbe, G.: *Combustion, Flames and Explosions of Gases* (Academic, New York, London, 1961) p. 13
60. Terao, K.: Die Wirkung der Kapazitäts-und Induktanzkomponente auf die Aktivierung des Gemisches bei der Funkenzündung in Gasen, Jpn. J. Appl. Phys. **1** (1962) 295–301
61. Terao, K., Shimoda, K., and Nakano, H.: Quantitätseffekt des elektrischen Funken auf die Funkenzündung in Gasen, Jpn. Appl. Phys. **2** (1963) 578–583
62. Miyashiro, S. and Terao, K.: Zündung brennbarer Gemische, Research Reports of the Anan Technical College, **14** (1978) 1–8
63. Unites States Patent Office 3,538,372, Wide Gap Discharge Plug
64. List, H.: *Thermodynamik der Verbrennungskraftmaschine* (Springer, Berlin Heidelberg New York, 1939) Chap. B-c
65. Terao, K. and Sato, T.: Der Zusammenhang zwischen der Fortpflanzungs-geschwindigkeit und Ionendichte der Flamme, Jpn. J. Appl. Phys. **7** (1968) 563–564
66. Jost, W.: *Explosions-und Verbrennungsvorgänge in Gasen* (Springer, Berlin Heidelberg New York, 1939) Chap. 7
67. Justi, E.: *Spezifische Wärme, Enthalpie, Entropie, Dissoziation technischer Gase* (Springer, Berlin Heidelberg New York, 1938)
68. Weast, R.C.: *CRC Handbook of Chemistry and Physics* (CRC, West Palm Beach, 1962)
69. Lawson, J. and Weinberg, F.J.: *Electrical Aspect of Combustion* (Clarendon, Oxford, 1969)
70. Thomson, W.B.: *An Introduction to Plasma Physics* (Pergamon, Oxford, London, 1962) Chap. 3

71. Calcote, H.F.: Ion and electron profiles in flames, Ninth Symp. (Intern.) Combust. (Academic, New York, London, 1963) p. 622
72. Chen, F.F.: Plasma `Physics, Probe Techniques (Summer Institute, Princeton University, 1962) 113–119
73. Travers, B.E.I. and Williams, H.: The use of electrical probes in flame plasma, Tenth Symp. (Intern.) Combust. (the Combustion Institute, Pittsburgh, 1965) p. 657
74. Mukherjee, N.R., Fueno, T., Eyring, H., and Ree, T.: Ions in flames, Eighth Symp. (Intern.) Combust. (Williams & Wilkins, Baltimore, 1961) p. 1
75. Terao, K.: Flame temperature measurement using the double probe method, Jpn. J. Appl. Phys. **3** (1964) 169–170
76. Terao, K.: Die Temperature aktivierter Teilchen bei der Verbrennung, Jpn. J. Appl. Phys. **3** (1964) 486–489
77. Terao, K.: Untersuchung der Flamme des Methan-Luft-Gemisches mittels Doppelsondenmethode, Jpn. J. Appl. Phys. **4** (1965) 64–68
78. Terao, K., Nishida, Y., and Sato, T.: Doppelsondenmethode mit einer stufenförmigen Hochfrequenzstossspannung, Jpn. J. Appl. Phys. **6** (1967) 1154–1162
79. Ohtake, K.: Simultaneous measurement of temperature and velocity in turbulent diffusion flames by Rayleigh scattering and LDV, *Laser Diagnostics and Modeling of Combustion* (edited by Iinuma, K., Asanuma, T., Ohsawa, T., and Doi, J., Springer, Berlin Heidelberg New York, 1987) p. 29
80. Salpeter, E.E.: Electron density fluctuations in a plasma, Phys. Rev. **120** (1965) 1528–1535
81. Kunze, H.J.: Messungen der lokalen Elektronentemperatur und Elektronendichte in einem θ-Pinch mittels der Streuung eines Laserstrahls, Z. Naturforschg, **20a** (19765) 801–813
82. Kunze, H.J.: *Plasma Diagnostics* (edited by Lochte-Holtgreven, Northholland Publishing Co., Amsterdam, 1968) Chap. 9
83. Kegel, W.H.: Kurven zur Bestimmung von Plasmaparameter durch Lichtstreuexperimente, IIP 6/34 (Max-Planck-Institut für Plasmaphysik, Garching, 1965)
84. Yamamoto, K. and Terao, K.: Experimental study on a propagating flame by a laser light scattering method, Jpn. J. Appl. Phys. **27** (1988) 1262–1267
85. Terao, K., Hozaka, M., and Kaitoh, H.: Experimental study on the ionization of argon behind reflected shock waves, Jpn. J. Appl. Phys. **22** (1983) 735–741
86. Penner, S.S.: *Quantitative Molecular Spectroscopy and Gas Emissivities* (Addison-Wesley, London, 1959)
87. Lochte-Holtgreven, W.: *Plasma Diagnostics* (North Holland Publ. Co. Amsterdam, 1968) Chap. 3
88. Desai, S.V. and Corcoran, W.H.: J. Quant. Spectrosc. Radiat. Transfer, **8** (1968) 1721
89. Lewis, B. and von Elbe, G.: *Combustion, Flames and Explosions of Gases* (Academic, New York, London, 1961) Chap. V
90. Lovachev, L.A.: The theory of flame propagation in systems involving branching chain reactions, Eighth Symp. (Intern.) Combust. (Williams & Wilkins, Baltimore, 1962) p. 411
91. Povienelli, L.A. and Fuchs, A.H.: The special theory of turbulent flame propagation, Eighth Symp. (Intern.) Combust. (Williams & Wilkins, Baltimore, 1962) p. 554
92. Hoffmann-Berling, E., Günther, R., and Leuckel, W.: The Effect of the flame front structure on flame propagation in premixed gases, 19th Symp. (Intern.) Combust. (the Combustion Institute, Pittsburgh, 1981)

93. Terao, K. and Nagata, K.: Die Flammenfortpflanzungsgeschwindigkeit im laminaren und turbulenten Gebiet, Bull. Fac. Eng. Yokohma Natl. Univ. **14** (1965) 1–7
94. Abouseif, G.E., Toong, T., and Converti, J.: Acoustic- and Shock-Kinetic Interaction in Non-Equilibrium H_2-Cl_2 Reactions, 17th Symp. (Intern.) on Combust. (The Combustion Institute, Pittsburgh, 1978) p. 1341
95. Terao, K. and Inagaki, T.: Interaction between combustion and shock waves, Jpn. J. Appl. Phys. **28** (1989) 1226–1234
96. Schmidt, P.: *Periodisch wiederholte Zündungen durch Stosswellen* (Arbeitsgemeinschaft für Forschung des Landes Nordrhein-Westfalen, Westdeutscher Verlag, Köln, Opladen, 1958)
97. Lembcke, H.: Das Schmidtrohr, Z. VDI Bd. **96** (1952) No. 31, p. 1005
98. Putnam, A.A., Belleser, F.E., and Kentfield, J.A.C.: Pulse combustion, Prog. Energy Combust. Sci. **12** (1986) 43–79
99. Nakano, T., Setoguchi, T., Zeutzius, M., and Terao, K.: Effect of a divergent tail pipe on the performance of puls jet engines, ISABE-2005–1303, 17th Intern. Symp. Airbreath. Engines, München, September 2005
100. Lewis, B. and von Elbe, G.: *Combustion, Flames and Esplosions of Gases* (Academic, New York, London, 1961) Chap. VIII
101. Shchelkin, K.I. and Troshin, Ya.K.: *Gasdynamics of Combustion* (Mono Book Corp., Baltimore, 1965, translated by B.W. Kunvshinoff and L. Holtschlag) Chap. 1
102. Nettleton, M.A.: *Gaseous Detonations* (Chapman and Hall, London, New York, 1987)
103. Soloukhin, R.I.: *Shock Waves and Detonations in Gases* (Mono Book Corp. translated by B.W. Kuvshinoff, 1966) Chap. VI
104. Takai, R., Yoneda, K., and Hikita, T.: Study of detonation wave structure, 15th Symp. Combust. (1974) p. 69
105. Libouton, J.C., Dormal, M., and van Tiggelen, P.J.: The role of chemical kinetics on structure of detonation waves, 15th Symp. Combust. (1974) p. 79
106. Brode, H.L., Glass, I.I., and Oppenheim, A.K.: Shock Tube Research (Chapman and Hall, London, 1971) p. 2
107. Edwards, D.H.: A survey of recent work on the structure of detonation waves, 12th Symp. (Intern.) Combust. (The Combustion Institute, Pittsburgh, 1969) p. 819
108. Burcat, A. and Hassen, A.: Deflagration to detonation transition in hexane and heptane mixtures with oxygen, 19th Symp. (Intern.) Combust. (The Combustion Institute, Pittsburgh, 1982) p. 625
109. Terao, K., Ishikawa, Y., Shiraishi, H., Ishii, K., and Miyashiro, S.: Interaction between converging shock and combustion waves, 22th Intern. Symp. Shock Waves, Imperial College, London, July 18–23, 1999
110. Miyashiro, S. and Grönig, H.: Low jitter reliable nanosecond spark source for optical short-duration measurement, Exp. Fluids **3** (1989) 71–75
111. Terao, K. and Azumatei, T.: Cellular pattern formation in detonation waves as a stochastic phenomenon, Jpn. J. Appl. Phys. **28** (1989) 723–728
112. Terao, K.,Yoshida, T., Kishi, K., and Ishii, K.: Interaction between shock and detonation waves, ICDERS, Seattle, 2001
113. Shchelkin, K.I. and Troshin, Ya.K.: *Gasdynamics of Combustion* (Mono Book Corp. Baltimore, 1965, translated by B.W. Knuvshinoff) Chap. 4
114. Soloukhin, R.I.: *Shock Waves and Detonations in Gases* (Mono Book Corp., Baltimore, 1966) Chap. V

115. Nettleton, M.A.: *Gaseous Detonations* (Chapman and Hall, London, New York, 1987) Chap. 5
116. Terao, K. and Sawada, R.: Stochastic aspect of the transition from combustion to detonation, Jpn. J.Appl. **18** (1979) 1463–1470
117. Nettleton, M.A.: *Gaseous Detonations* (Chapmn and Hall, London, New York 1987) Chap. 2
118. Terao, K. and Motoyama, Y.: Propagation velocity of detonation waves in a high temperature mixture, Proc. Int. Symp. Shock Waves, Sendai, July 12–16, 1991
119. Terao, K. and Kobayashi, H.: Experimental study on suppression of detonation waves, Jpn. J. Appl. Phys. **21** (1982) 1577–1579
120. Terao, K., Neda, T., Nakano, K., and Sato, T.: Ionisation in detonationswellen, Jpn. J. Appl. Phys. **8** (1969) 834–843
121. Yamamoto, K.: Experimental Investigation of Flame Using a Laser Light Scattering Method, Thesis of Dr.-Eng., Yokohama Natl. Univ. (1987) p. 97
122. Perry, R.W. and Kantrowitz, W.N.: The production of converging shock waves, J. Appl. Phys. **22** (1951) 878–886
123. Guderley, G.: Starke kugelige und zylindrische Verdichtungsstösse in der Nähe des Kugelmittelpunktes bzw. Zylinderachse, Luftfahrtforschung **19** (1942) 302–312
124. Chisnel, R.F.: The motion of a shock wave in a channel with application to cylindrical and spherical shock waves, J. Fluid Mech. **2** (1957) 286–298
125. Whitham, G.B.: On the propagation of shock waves through regions of non-uniform area of flow, J. Fluid. Mech. **4** (1958) 337–360
126. Lee, B.H.K.: Nonuniform propagation of imploding shocks and detonations, AIAA J. **5** (1967) 1997–2003
127. Huni, H.J.: A study of cylindrical imploding detonations, Ph. D. Thesis, Univ. British Columbia, 1970
128. Terao, K.: Experimental study on cylindrical and spherical implosions, Jpn. J. Appl. Phys. **22** (1983) 446–453
129. Terao, K.: Temperature in imploding detonation waves, Jpn. J. Appl. Phys. **23** (1984) 27–33
130. Terao, K. and Wagner, H.Gg.: Experimental study on spherically imploding detonation waves, Shock Waves, **1** (1991) 27–34
131. Terao, K., Akaba, H., and Shiraishi, H.: Spherically imploding detonation waves initiated by two-step divergent detonation, Shock Waves, **4** (1995) 187–193
132. Lawson, J.D.: Some criteria for power producing thermonuclear reactor, Proc. Phys. Soc. **3** (1952) 6–10
133. Nettleton, M.A.: Recent work on gaseous detonations, Shock Waves, **12** (2002) 3–12
134. Knystautas, R. and Lee, J.H.: Experiments on the stability of converging cylindrical detonations, Combust. Flame, **16** (1971) 61–73
135. Terao, K.: Nuclear fusion reactor ignited by imploding detonation waves, AIAA 2000–2965, 35th Intersociety Energy Conversion Engineering Conference, Las Vegas, Nevada, July 24–28, 2000
136. Linhart, J.G., *Plasma Physics* (North-Holland, Amsterdam, 1960) Chap. 1. and 7
137. Ikeda, Y., Yoshimoto, U., Maekawa, F., Smith, D.L., Gomes, I.C., Ward, R.C., Filatenkov, A.A.: An investigation of the activation of water by D-T fusion neutron and some implications for fusion reactor technology, Fusion Eng. Des. **37** (1997) 107–150
138. Terao, K. and Ishii, K.: Nuclear fusion rocket engine applying spherically imploding detonation waves, AIAA 2003–4885, 39th AIAA/ASME/SAE/ASEE Joint Propulsion Conference and Exhibit, Hunstville, Alabama, July 20–23, 2003

139. Terao, K., Ishii, K., Totsuka, T., and Ishikawa, Y.: An experimental investigation of hypersonic combustion for ram jet engine applying detonation waves, 11th AIAA/AAAF International Conference Space Planes and Hypersonic Systems and Technologies, Paper 2002–5164, 29 Sept–4 Oct. 2002, Orleans, France
140. Currun, E.T. and Murthy, S.N.B.: *Scramjet Propulsion*, AIAA, Virginia, 2000, Chap. 13 "Detonation Wave Ramjets" 823–841
141. Kailasanath, K.: Review of propulsion applications of detonation waves, AIAA J. **38** (2000) 1698–1708
142. Hiraiwa, T., Kanda, T., and Enomoto, Y.: Experiments on a scram jet engine with ramp-compression inlet at Mach 8 condition, AIAA paper 2002–4129, 48th AIAA/ASME/SAE/ASEE Joint Propulsion Conference and Exhibit, Indianapolis, 7–10 July 2002
143. Oertel, H.: *Stossrohre* (Springer, Berlin Heidelberg New York, 1966) Chap. E2 "Stossrohr mit divergierender Düse", 540–564
144. Oosthuzen, P.H. and Carscallen, W.E.: *Compressible Fluid Flow* (McGraw-Hill, New York, 1977) Chap. 6. Sec.2 "Oblique Shock Wave Relations" 120–128
145. Stollery, J.L. and Stalker, R.J.: The development and use of free piston wind tunnels, Proc. 14th Int. Symp. Shock Tubes Waves (Sydney, 1983) p. 41
146. Kendall, M.A., Morgan, R.G., and Jacobs, P.A.: A compact shock-assisted free-piston driver for impulse facilities, Shock Waves, **7** (1997) 219–230
147. Soloukhin, R.I.: *Shock Waves and Detonations in Gases* (Mono Book Corp., Baltimore, 1966) Chap. IV, p. 121
148. Nettleton, M.A.: *Gaseous Detonations* (Chapman and Hall, London, New York, 1987) Chap. 2.8. p. 37
149. Yu, H., Esser, B., Lenarts, M., and Grönig, H.: Gaseous detonation driver for a shock tunnel, Shock Waves, **2** (1992) 245–254
150. Habermann, M., Olivier, H., and Grönig, H.: Operation of a high performance detonation driver in upstream propagation mode for a hypersonic shock tunnel, 22nd Intern. Symp. Shock Waves, London, July 18–23, 1999
151. Bleilebens, M. and Olivier, H.: Druckmessungen hoher räumlicher Auflösung an einem aufgeheizten Rampenmodell im Stosswellenkanal, Deutscher Luft-und Raumfahrtkongress 2003, München, Nov. 2003, DGLR-2003-129, Jahrbuch 2003
152. Terao, K.: Shock tube using free piston driven by detonation waves, Jpn. J. Appl. Phys. **33** (1994) 2811–2816

Author Index

Abouseif, G.E. 199[94]
Akaba, H. 343[131]
Ataka, M. 29, 36[34]
Azumatei, T. 241[111]

Baras, F. 21[24][25]
Becker, R. 1[1], 18[16], 18[17][18][19], 20[20], 21[21], 21, 23, 24[30]
Belleser, F.E. 216[98]
Bleilebens, M. 373[151]
Brode, H.L. 230[106]
Burcat, A. 232[108]

Calcote, H.F. 148[71]
Carscallen, W.E. 366[144]
Chalmers, B. 36[36]
Chen, F.F. 149[72]
Chisnel, R.F. 311[124]
Converti, J. 199[94]
Corcoran, W.H. 180[88]
Currun, E.T. 362[140]

Dainton, F.S. 8[7]
Desai, S.V. 180[88]
Dormal, M. 230[105]

Edwards, D.H. 230[107]
Enomoto, Y. 362[142]
Esser, B. 373[149]
Eyring, H. 149[74]

Filatenkov, A.A. 358[137]
Fuchs, A.H. 190[91]

Fueno, T. 149[74]
Fukutani, S. 16[14]

Gidding, J.C. 8[9]
Glandsdorff, P. 21[23]
Glaser, D.A. 42[40][41]
Glass, 230[106]
Gomes, I.C. 358[137]
Greene, E.F. 45[43]
Grönig, H. 236[110], 373[149][150]
Guderley, G. 308[123]
Günther, R. 191[92]

Habermann, M. 373[150]
Hassen, A. 232[108]
Hatano, F. 29, 36[33]
Hikita, T. 228[104]
Hiraiwa, T. 362[142]
Hirata, M. 21, 24[28]
Hirschfeld, J.O. 8[9]
Hirth, J.P. 27, 34[32]
Hoffmann-Berling, E. 191[92]
Hozaka, M. 174, 175[85]
Huni, H.J. 313[127]

Ikeda, Y. 358[137]
Inagaki, T. 199[95]
Ishii, K. 233[109], 251[112], 358[138], 361, 362[139]
Ishikawa, Y. 233[109], 361, 362[139]

Jacobs, P.A. 372[146]
Jinno, H. 16[14]

Jost, W. 3[2][5], 8[6], 13[10], 15[13], 114, 125[58], 139[66]
Jungers, J.C. 51, 87[50]
Justi, E. 140[67]

Kailasanath, K. 362[141]
Kaitoh, H. 174, 175[85]
Kanda, T. 362[142]
Kaneko, R. 29, 36[34]
Kantrowitz, W.N. 304[121]
Kegel, W.H. 162[83]
Kendall, M.A. 372[146]
Kentfield, J.A.C. 216[98]
Kishi, K. 251[112]
Knystautas, R. 354[134]
Kobayashi, H. 289[119]
Kuo, K.K 3[4]
Kunze, H.J. 161[81][82]

Lawson, J. 146[69]
Lawson, J.D. 353[132]
Lee, B.H.K. 313[126]
Lee, J.H. 354[134]
Lembke, H. 216[97]
Lenarts, M. 373[149]
Leuckel, W. 191[92]
Levedahl, W.J. 80[49]
Lewis, B. 3[3], 13[11], 87[52], 114, 115[59], 187[89], 223[100]
Liao, Ch. 99, 103[55], 103[56][57]
Libouton, J.C. 230[105]
Linhart, J.G. 350[136]
List, H. 135[64]
Lochte-Holtgreven, W. 174, 179[87]
Lovachev, L.A. 187[90]

Maekawa, F. 358[137]
Martinengo, A. 57, 80[44]
Malek Mansor, M. 21[24]
Mauret, P. 34[35]
Miyashiro, S. 132[62], 233[109]
Miyazaki, S. 84[51]
Morgan, R.G. 372[146]
Motoyama, Y. 283[118]
Mukherjee, N.R. 149[74]
Murthy, S.N.B. 362[140]

Nagata, K. 192[93]
Nakano, H. 117[61]

Nakano, K. 297[120]
Nakano, T. 216[99]
Neda, T. 297[120]
Nettleton, M.A. 223[102], 255[114], 282[117], 354[133], 372[148]
Nicolis, G. 21[24][25]
Nishida, Y. 157[78]

Oertel, H. 45[42], 365[143]
Ohtake, K. 160[79]
Ohtomo, Y. 36[37]
Oliveier, H. 373[150][151]
Oosthuzen, P.H. 366[144]
Oppenheim, A.K. 230[106]

Penner, S.S. 174, 179[86]
Perry, R.W. 304[121]
Pound, G.M. 27, 34[32]
Povienelli, L.A. 190[91]
Progogne, I. 21[23]
Putnam, A.A. 216[98]

Ree, T. 149[74]
Rieckers, A. 21[22]

Salpeter, E.E. 161[80]
Sano, F. 29, 36[33]
Sato, T. 137[65], 157[78], 297[120]
Sawada, R. 255[115]
Schmidt, P. 216[96]
Setoguchi, T. 216[99]
Semenoff, N. 8[8]
Shchelkin, K.I. 223, 282[101], 251[112]
Shimoda, K. 117[61]
Shiraishi, H. 233[109], 343[131]
Silber, R.S. 14[12]
Soloukhin, R.I. 61[46], 228[103], 255[113], 372[147]
Smith, D.L. 358[137]
Stalker, R.J. 372[145]
Stollery, J.L. 372[145]
Sturgis, B.M. 87[53]
Stumpf, H. 21[22]

Takahashi, S. 84[51]
Takai, R. 228[104]
Terao, K. 18[15], 21, 24[31], 29, 36[33][34], 36[37], 57[45], 74[47][48], 84[51], 90[54], 99, 103[55], 103[56][57],

117[60][61], 132[62], 137[65], 152[75], 153[76][77], 157[78], 165[84], 174, 175[85], 192[93], 199[95], 216[99], 233[109], 241[111], 251[112], 257[116], 283[118], 289[119], 297[120], 314, 319[128], 328[129], 336[130], 343[131], 354[135], 358[138], 361, 362[139], 376[152]
Terao, N. 21, 24[29]
Thomson, W.B. 147, 178[70]
Toennies, J.P. 45[43]
Toong, T. 169[94]
Totsuka, T. 361, 362[139]
Travers, B.E.I. 149[73]
Troshin, Ya.K. 223, 282[101], 255[113]
Turner, J.W. 21[24]

van Tiggelen, P.J. 230[105]
von Elbe, G. 3[3], 13[11], 87[52], 114, 115[59], 187[89], 223[100]

von Laue, M. 21, 24[27]
Vorsanger, J.J. 34[35]

Wagner, H.Gg. 336[130]
Ward, R.C. 350[136]
Weast, R.C. 143[68]
Weinberg, F.J. 146[69]
Whitham, G.B. 311[125]
Williams, H. 149[73]
Wilson, J.G. 42[39]

Yamamoto, K. 165[84], 304[121]
Yoneda, K. 228[104]
Yoshimoto, U. 358[137]
Yoshida, T. 251[112]
Yu, H. 373[149]

Zeutzius, M. 216[99]
Zuber, K. 21, 24[26]

Subject Index

Acetone 29, 33, 34
Activated atoms, radicals, O_2 10, 11, 13, 14, 91, 191
Activation energies 4, 7, 23, 25, 26, 75, 79, 82–83, 94, 101, 137, 262
  effective 71, 117, 192, 194, 195, 249, 257
  for initiation, development 81
  of apex point formation 246, 249
  of ebullition 32, 34
  of ice formation 40, 41, 43
  of reaction with lead 94
Activation energies of spontaneous ignition in hydrogen-oxygen mixture 62, 65, 66, 68, 71
  of hydrocarbon-air mixtures 78, 79, 85, 95
Active atoms, radicals, chain carrier 94
Adiabatic
  combustion temperature 139–142, 169, 173, 202, 210, 212
  compression 14, 57, 58, 361
  process 18
Aldehyde 87
Aluminum diaphragm 62
Antiknock ability, effect 87, 88
  additive 88, 96
α-particle 358
Apex point 244, 247, 250, 251, 254, 263, 264, 290, 316, 317
Argon 174–180, 328
Argus-Schmidtrohr 221
ArI-line 179, 328–330, 341, 342

Arrhenius' equation, formula, relation 8, 23, 62, 71, 116, 262, 286
Associated molecule in water 34
Autoignition 89, 90
Avogadro's number 23

Bell-like (Gaussian) curve, profile 170
Blanket 357–359
Blast wave 250
Boltzmann constant 7, 18, 21, 28, 146, 161, 174, 209
Boundary layer 62, 72, 250
Breakdown
  by electric spark, 113–115, 120
  by laser beam focusing 164, 233
Bremsstrahlung 355
Bubble chamber 42
Bulk modulus of water 38
Bunsen burner 152, 188, 189, 192
Buoyancy 234, 307, 314, 319, 354, 361

Calcium oxide (CaO) 246, 247
Capacity component of electrical discharge 114, 115, 117, 120–124, 129
Capsule of $D+T$ 356, 357
Cellular pattern, structure (formation)) 229–231, 241–251, 253, 254, 260, 262, 263, 266, 272–276, 296, 316, 322, 332
  on a incombustible powder film 247
CFR-engine 88

Subject Index

Chain-branching explosion, kinetics, reaction 3, 7, 8, 10, 11, 13, 14, 61, 72, 87, 88, 97
Chain-breaking reaction 10
Chapman-Jouguet (C-J) condition, point, velocity 227, 231, 233, 253, 266, 274–278, 282–286, 295, 296, 335
Characteristic length (plasma) 163, 164
  parameter 161, 163, 167, 177
Charged particle 206, 304
Chemical bond C-C, C-H 81, 82
  binding energy 81
Chemical ionization 146
  kinetics 8, 9, 16, 74
  potential 28
  reaction process 4, 8, 10, 139, 142
Circumference, circular, cylindrical, elliptical, 235, 238, 239, 240, 314, 315, 321, 338, 344
  in detonation chamber 320
Classical theories 15, 17, 29, 72, 76, 192
  of detonation 223, 282, 295
  of ignition 3, 61, 74, 76, 113
  of nucleation 27, 35, 41
Cloud chamber 47
Coefficient of energy loss 68
Collision
  between particles 32, 68, 72, 81, 137, 185
  between shock and flame 199–216, 233–240
  between shock and detonation 251–255
Collision cross-section 355
  number, frequency between fuel and oxygen molecule 25, 68, 81
Combustion chamber 87, 88, 165, 166, 216, 217, 220, 234, 235
  rate 136
  velocity 96, 137, 138, 191, 192, 197, 198
  wave 187, 191, 199, 223, 226, 232, 234, 240, 258, 259, 260, 268, 282, 296
  zone behind detonation front 282
Compression ratio 58, 87, 96, 133
Compression wave produced by combustion 224, 259, 260, 267
Concave mirror (schlieren method) 235, 236

Condenser 114, 122, 123, 124, 166, 235
  lens (schlieren method) 236
Conical combustion wave 189, 270
  flame, front, surface 181, 188, 190, 198
  shock 259, 260, 266, 267
Contact surface (shock tube) 51, 59, 284
Convergence (implosion) 322, 323, 345, 346, 353, 354
Convergent nozzle 362, 364, 369
Cool flame 74, 84–90, 95–98
Cooling effect of electrodes 115, 116, 155, 160
  probe 153, 159, 160
  system (rocket engine) 361
Cosmic ray (nucleation) 29
Critical nucleus 24, 28, 34, 35
  radius and size 27, 28, 34, 41
Cross section area 60, 75, 87, 100, 160, 190, 193
Cylindrically imploding shock or detonation waves 308–314, 319–328

Debye length (plasma) 161
Deflagration wave 226, 232, 255
Detonation chamber (for implosion) 320, 321, 337, 339, 344, 355, 356, 359
Detonation waves 186, 189, 215, 216, 223, 224, 226, 228, 232, 233, 241, 307, 315, 362, 379
Detonation tube 242, 378
Detour spark plug 133–136
Deuterium 355–361
Developing period, process 23, 24, 25
  reaction 73
Development from initiation to detonation waves 263, 264
Deviation of wavelength 161, 329
Diffuse density 166, 168, 170, 171
Diffusion, zone 110, 185, 190, 191
Dissociation 140, 141, 185, 227, 282, 363
  degree 140, 141
Distance of gap
  between two electrodes 116, 125–129
  between two successive apex points 244, 254, 292, 331
Distilled water 30, 34, 37
Distribution of apex point 317

Distribution of ignition position 99
   of ignition probability 103, 108, 110, 111
Divergent-convergent space 343–345, 356
Divergent detonation, chamber 314–319
Divergent nozzle 220, 364, 369
Doppler effect 160, 161
Double probe, method 137, 147, 149,
   152–154, 157, 201, 202, 203, 206,
   296, 297, 331
Driver gas 50, 52, 59, 175, 270
$D+T$ nuclear fusion reaction 355–361

Ebullition of liquid 27, 29
   mechanism 35
Effective activation energy 4, 15, 32,
   72, 125, 192, 246, 248, 255, 257,
   262, 265, 293
Effect of lead (anti-knock) 94, 95
Electrical conductivity of water 30, 38
Electrical discharge circuit 114, 115
   delay 22
   current 129
   potential, tension 35, 113, 114, 115,
      119, 120, 121, 129, 134
   power 124, 126, 129
Electron component 161, 163, 166, 167,
   168, 350
Electron density 137, 148, 161, 162, 164,
   169, 173, 177–179, 183, 305, 356
   mass 148, 151, 161, 162, 356
   correlated to ions 161
Electron temperature 146, 148, 150–152,
   154, 155, 159–161, 163, 164, 168,
   173–178, 183, 184, 207, 209–214,
   298–301, 303–305, 328, 330, 333
Elementary charge 146, 151, 161, 209, 355
Elementary reaction 8, 10, 15, 16
Elliptical combustion chamber 234
   foci, space 234, 235, 238, 239
Endothermic process, reaction 36, 173
Energy level (spectroscopic) 174, 181
Energy loss coefficient (at collision) 68, 173
Energy process (stochastic) 25
   (nucleation) 35, 42
   (ignition) 72, 77
Energy released by initiation reaction (W)
   25, 32, 65, 68, 101, 126
   from external energy source 117
   from spark 121–124, 129, 130

Enthalpy 47, 145
Entropy decrease 20
   increase 18
Envelope curve crossing at ignition point
   250, 251
Equilibrium 21, 27, 38, 140, 187, 240
   constant 141, 142, 143
Equivalence ratio 75–78, 80, 81, 95, 96,
   99, 100, 102, 103, 123, 134, 135, 137,
   194–198
Establishment of detonation 259, 263,
   264, 266, 267
Ethane-air mixture 74, 77, 79
Ethylene-air mixture 78, 79
Exhaust gas composition 88, 89
Exothermic process, reaction 36, 215,
   216, 222
Explosion 3, 5, 7, 45, 230, 231
   limits of hydrogen-oxygen mixture 12,
      13, 61, 68–70, 98
   limit at thermal explosion 7
   limit having an inverse S-curve 7, 12,
      69
   peninsula 8, 12, 61, 69, 70
   reaction model of hydrogen-oxygen
      mixture 73

Filter 84, 164, 165
Fine structure of detonation 223
First limit of explosion 12, 13
First law of thermodynamics 17, 18
First wall (fusion reactor) 357–359
Flame front 198, 203, 205
   propagation 188, 191, 192, 194, 195,
      262
   tail, tip 203, 262
Floating potential (plasma) 149, 150,
   209
   double probe 150
Flow leak method 293
Flow velocity 369
   at exit of divergent nozzle 369
   before arrival of shock 319
   behind combustion wave 188
   behind detonation front 369
   behind rarefaction wave 372–376
   in divergent-convergent nozzle 369
Fluctuating phenomena 17, 20, 21, 23
Fluctuation 57, 61, 99, 117, 120, 257

Focusing instant 235
Focus 234, 255, 355
Formaldehyde 88
Formation enthalpy 144, 145
Free electron 137, 138, 147, 148, 160, 161, 174, 183, 185, 191, 240
Free enthalpy 28
Free piston driven
  by high pressure gas 376
  by detonation waves 378, 381, 382
Frequency factor 25, 34, 35, 41, 65, 79, 84, 117, 127, 245, 246, 249, 255, 257, 265
Fuel concentration 101, 123
  consumption 123, 219, 220
  injection, nozzle 106
  injector 106, 217
  particle velocity 96
  spray 99, 103, 106, 107, 111

Gap distance between electrodes (spark ignition) 115–117, 125, 126, 128–130, 233
Gas dynamics 224, 231–233
Gasdynamic temperature 176, 178, 181–184, 202, 332
γ-ray 358
Glass disk 116, 129
  plug 346
Graphite, reflector, surface 307–310
Growth of nucleus 33, 34
G-type detonation chamber 336–343, 352
Guderley's formula 310

Halfwidth 162, 163, 170–172, 177
He gas 62, 376
Heat loss by conduction 5
  transfer 191
  transferred outside 4, 5
Hemispherical space, vessel 315, 343
Heterogeneity 62, 229
Heterogeneous mixture 99
  state 17, 20, 113, 137, 174, 183, 185, 186, 192
High frequency multistep potential 157–159, 207, 331
Histogram 24
  of apex point distance 244, 253, 254, 264, 292

of induction period 22
of ice formation 38, 39
of ebullition induction period 31
of emission induction period 182
of ignition induction period 59, 63, 64, 75, 103, 108, 120, 121
of ignition position 63, 65
of induction distance 260
of interval between two successive apex points 244, 254, 264
Homogeneous state 4, 27, 57, 61, 192
Hot ignition 74, 84–87, 85, 91–98
Hugoniot-curve 48, 49, 188, 189, 223, 225, 227, 282
Hydrocarbon 74, 140
Hydrocarbon-air mixture 74, 90
Hydrogen-oxygen mixture 61, 62, 64, 362, 363
-peroxides 88, 89
Hypersonic flow 362, 364
  combustion 307, 361
  counter flow 312

Ice formation 36–43
  probability 37, 39–41
  under radiation 42, 43
Ignition 3, 45, 59, 99, 119, 186, 220, 224
  ability 68, 99
  mechanism 13, 14, 15, 74, 82
  in a fuel spray 99, 110
  in Schmidtrohr 222
  by electric spark 98, 101, 112
  energy 116, 125
  by an external energy 3, 113
  delay, 3, 14
  probability 74, 125, 126
  in a fuel spray 108, 109, 110
Ignition limit of hydrocarbon-fuel-air mixtures 84, 86
Ignition point, position 60, 63, 229
Ignition probability
  obtained from induction period 60, 64, 75, 97
  obtained from ignition position 65
  in fuel spray 103
Ignition region 97
  suppression effect 96
Image intensifier 165, 166, 167

Imploding shock wave, front 308–313
Imploding detonation wave 307, 313, 314, 319, 343, 355
Implosion center, focus 308, 322, 346
  size 346
  temperature 349
Incident shock wave 50–53, 57, 63, 183, 229
Induced current through coil 30, 383
Inductance component of electric discharge 114, 115, 117, 120
Induction coil 114, 118
Induction distance of detonation initiation 259–261, 269, 270, 277
Induction period 3, 8, 13, 15
Induction period of cool flame initiation 84, 85, 95, 97
Induction period of ebullition 29–31
  of hot ignition 84, 89, 95, 97
  of ice formation 37, 38, 39
  of ignition 14, 59, 61, 84, 99, 104, 107, 117, 120
  of light emission 182
Industrial applications of detonation waves 307
Initiation of detonation 229, 255, 259, 263, 264, 266, 267, 269, 272
Initiation process 23, 25
  of fusion reaction 353
Intensity of spectral line 163, 169, 174, 175, 341
Interaction between combustion and pressure or shock waves 186, 199, 202–216, 222
  between converging shock and combustion wave 233–240
  between shock and detonation waves 251–255
Interaction mode (collision of shock with combustion waves) 199, 214
Interfacial energy 28
Intermediates, products 15, 70, 95
Initiation process, reaction 24, 25, 73
Injection period 106, 219
  pressure 106
Inverse S-form curve 7, 8, 12, 69, 97
Ion, 137, 138, 240
  C, H, O 170, 180
  component 161, 169
  current 148–153
  density 137, 138, 155, 156, 161, 164, 210, 211, 298–301, 356
Ionization (behind combustion, detonation, shock and retonation waves) 138, 152, 153, 165, 173, 174, 185, 206, 296
  degree 178–180, 183, 213, 214
  potential 178, 213, 214
Ion mass 149, 151, 163, 169, 210
Ion probe 315, 321
Ion sheath 147
Ion temperature 146, 151, 154, 161, 164, 168, 172, 173, 210, 304, 328, 351
Ion density 137, 138, 155, 156, 206, 210, 212, 213, 214, 299, 302, 303, 351
  weight 156, 210
Irreversible phenomenon, process 13, 15, 17, 18, 21, 23, 27, 57, 113, 116, 137, 173, 187, 191, 192, 233
Isentropic change, curve 227, 377
Isobaric combustion 141
Isochoric combustion 141, 220, 222, 250
Iso-octane-air mixture 57, 58
Isotope 43, 358

Joule's heat 113

Kegel's graphs 163, 164, 177
Ketone 88
Knocking, phenomenon 13, 87, 88, 233

Laminar flame, area, flow 186, 187, 190, 191, 195
Langmuir probe 147, 148
Laser beam, light 164, 171
Laser light scattering method 138, 160, 161, 165, 169, 176, 296, 304
Latent heat of evaporation 34
  of solidification 41
Launch of rocket 361
Law of mass action 9
Lawson's criterion 353
Lead 92, 94, 95
Leak method 293
Light emission 182, 200
Light velocity 162, 174, 356
Liquid hydrogen 359

Logarithmic relation between
  probability and reciprocal mixture temperature 33, 40, 41, 67, 76–78, 91–93, 101, 102, 123, 126, 132, 183, 245, 248
  pressure and radial distance 326, 327, 341, 348
  temperature and radial distance 333, 343, 352
Luminescence of flame, ignition, spark 59, 63, 84, 118, 154, 166, 182, 203

Mach angle 366
  number 49, 50, 53, 55, 63, 176, 177, 183, 200, 202–204, 206, 211, 212, 228, 242, 243, 252, 262, 268, 272–282, 311, 361, 363, 369
  reflection 230, 246
  stem 229, 230
Maclaurin's series 92
Macroscopic structure of detonation waves 224, 228
Macroscopic change (thermodynamics) 18
Main shock wave (detonation) 230
Major axis of ellipse 234, 238, 239, 240
Mass action 9, 140
Maximum entropy 8, 18
Maxwell-Boltzmann, Maxwell's, Boltzmann's distribution 8, 23, 147
Mean induction period 26, 71, 75, 80, 95, 96, 102, 121, 122
Mean power of capacity component 126
  of inductance component 126, 129
Measurable size 25, 32, 34, 39, 117
Mechanical noise 323
Methane-air mixture 74, 78, 153, 154
Microphotometer 165, 166
Microscopic structure of detonation 228
Minor axis of ellipse 234
Minimum entropy 18, 23
Minimum ignition energy (spark) 116, 128, 129
Mixture density 69, 191, 192
  quantity 71, 76, 128, 132
  volume excited by a spark 126–128
Molar (mole) number 31, 60, 75, 117, 139, 140–143
Molar volume 60, 101, 253, 255, 257

Molecular dynamics 21
Monochromator 165, 166, 177, 180, 329, 349
Multisteppotential 157, 206–208, 296–299

Nanospark apparatus 107, 236, 367
Neutral particle 172, 178, 185
Neutron, neutron multiplier 357, 358
Newton's principle 309
n-hexane-air mixture 74, 76, 113–125, 131, 152, 193
n-heptane-air mixture 84, 88–93, 95
n-octane, spray 99, 100–102, 107
Non-equilibrium 17, 20, 95, 113, 137, 165, 173, 174, 183, 185, 186, 216, 233, 240, 296
Non-ignition region 97
Normalized
  flame propagation velocity 192, 194, 195–198
Nuclear fusion, reactor, reaction 307, 355–358
Nucleus growing period 25, 26, 33, 34, 36, 41, 60, 64, 67, 75, 76–78, 90–93, 100–102, 120, 122, 123, 132, 182, 183
Nucleation, factor 17, 26, 27, 29, 39
Number of C-atom 81
Neumann-Zeldovich-Doering (NZD)-model of detonation wave 222, 224, 227

Oblique shock 229, 240, 366, 367
Obstacle method 289
Overall (whole) ignition probability 33, 39, 40, 69, 71, 121, 123, 126, 131
Overdriven detonation 277, 281
Oxygen, concentration 79, 82

Parabola of n-th order 310
Paraffin hydrocarbon 74
Partial ignition probability 103, 104, 109, 111
Peak (highest)cycle temperature 330, 355
  (highest) pressure 205, 206, 324, 325, 355
Performance curve 135, 218, 220
Planck's constant 174, 356

Subject Index

Plasma, physics 137, 138, 147, 164, 307
  diagnostic 160
Pressure (variation) diagram, history 85, 136, 203–205, 218, 219, 266, 276–278, 287, 323, 324, 325, 338, 340, 346, 347, 372
  -density diagram 226, 279, 280, 334, 335
  -temperature diagram 7, 8, 13, 70, 87
  transducer (piezoelectric) 63, 84, 105, 201, 239, 252, 258, 265, 272, 315
  wave 199, 218, 232
Probability density 24, 32, 59, 64, 100, 117, 120, 269
Probability 23, 24
  of apex point formation or apex point formation probability 244, 245, 255, 262, 263, 265, 266, 290, 293, 331–333
  of chain-breaking, chain-branching 12
  of cool flame initiation 85, 90–93
  of ignition 57, 59, 60, 64, 67, 75–78, 99, 100–103, 248
Probability of phenomenon, 24, 25
  for initiation, development 25
  of detonation initiation 257, 260, 261, 269
  of hot ignition 85, 90–93
  of spark ignition 117, 120, 122, 123, 131, 132, 134
  of light emission (argon behind shock) 182, 183
  of nucleation 29, 36
  of ebullition 28, 29, 31, 33
Probability of state 21
Probe method 138, 146, 147
Probe characteristic 149, 150, 154, 158, 159
  current 148–150, 154, 158, 207–209, 298, 299–302, 303
  potential 148–152, 154, 158
Production of hypersonic flow 365
Propagating flame 153, 165, 166, 168, 199
Propagation velocity of combustion wave or flame 186, 188, 189, 190, 296
  of pressure, shock or reflected shock 48, 60, 64, 202, 252, 267
  or pressure wave, 60, 63, 238, 258, 266, 272–278

  of detonation wave 224, 226, 237, 252, 253, 267, 276–278, 282–289, 291, 294, 295, 296, 298, 307, 317, 318, 324, 347, 348, 357, 369, 389
Propane-air mixture 3, 199
  -oxygen mixture 142, 158, 165, 233, 234, 248, 252, 258, 270, 271–279, 282–297, 314, 317, 321, 330, 344, 378
Pseudo-detonation 267, 268

Q-switched giant pulse ruby laser 165, 176, 349
Quantity effect 71, 130, 192, 195, 196

Radial distance from implosion center 315, 323, 325, 327, 348
Radially divergent detonation 314–319
Radioactivity, radioactive isotope 42, 43, 358
Radius of spark column 127
Ram jet engine 361
Rankine-Hugoniot equations 47, 268, 282
Rarefaction wave 51, 52, 59, 217, 218, 220, 227, 382
Rate coefficient of reaction 9
Rayleigh line 48, 278, 282
  scattering 160, 172
Reaction heat 4, 141, 215, 225
  Mechanism, order 42, 192
  first order (single molecule) 9, 11
  process 140
  rate (H_2-O_2 mixture) 9, 13
  second order (two molecule) 4, 9
Rectangular (square) cross section 228, 242, 250
Reducing agent, reducer 96, 97
Reed valve 217
Reflected shock wave, front 50, 52, 59, 60, 75, 84, 96, 100, 175, 199, 230
Relaxation time 185
Resonance, pulse jet engine 199, 216, 218, 222
Retardation of detonation wave 289–293, 370
Reversible phenomenon, process 15, 17, 61
Reynolds number 190, 191–198
Rhomboidal section 344
Ruby laser 163, 165

Saha's equation 178, 213
Saturated electron, ion probe current 149, 151–153
Sawtooth probe potential 152–154
Scattering cross section 160
  angle 161–164, 166
Scattered light 160, 166, 167
Schlieren photograph 234, 235–237, 240, 366
Schmidtrohr 216–218, 221, 222
Scram jet engine 361
Second law of thermodynamics 17, 18
  explosion limit 3, 7, 8, 13, 70, 74, 87
Second order reaction 6, 9
Self-similarity method, of flow 308
Self-sustained detonation wave 227, 242, 263, 282, 315, 362
Shadow photograph 107
Shock (wave) diagram 52, 59, 100, 105, 200, 240, 271, 284
  collision 205, 240, 253
  -induced detonation wave 241, 270–280
  intersection 269, 270
  reflection 318
Shock tube 17, 45, 49, 62, 75, 89, 100, 104, 175, 199, 242, 270, 271, 284
  connecting with detonation tube 271
  driven by detonation wave 371, 373
  using a free piston 376
Shock tunnel 375
Shock waves 45, 46, 50, 52, 57, 174, 175, 182, 183, 199, 202, 203, 206, 217, 223, 224, 232, 234, 268, 282, 296, 307, 386
  at detonation front 242, 245, 250
  produced by propagating flame 259, 267
Slow-down detonation 277
Sodium sulfate 96
Soot film trace of combustion wave 235, 239, 241, 259, 268, 269, 322
  of detonation waves 229, 242, 243, 252, 254, 272–275, 291, 295, 321
  at transition from combustion wave 259, 262, 263, 264, 269
  of shock induced detonation 273–275, 288
  of imploding detonation 332
Sound velocity 45, 47, 223, 225, 228, 253, 308, 376

Spark ignition 115–117, 119, 193, 200, 234, 252, 258, 265, 268, 378
  in a flow 132
Spark-ignition engine 13, 87, 88, 133, 135, 136, 233, 234
  -plug 133, 314, 315, 321
Special electrode 47
Spectra, spectral distribution, profile 160, 163, 165, 167–172, 174, 177
Spectroscopic temperature 174, 175, 178–181, 183, 184, 328, 329, 330, 333, 340, 342
Spherically converging shock wave 310–313
  imploding detonation waves 319–355
Spherical implosion 319–355
Spinning detonation 270
Spontaneous ignition 3, 13, 17, 45, 57, 58, 61, 74, 192, 223, 230, 232, 248, 250, 268, 270
Stalker tube 372
Standing detonation wave 361, 362, 364, 370
  flame 152, 193, 194
Statistical weight 174
Stepwise probe current 157, 158, 206–208
Stimulation of combustion 199, 206, 215
Stochastic ignition theory 17, 23, 51, 74, 84, 90, 95, 97, 112, 116, 117, 233
  nucleation theory 29, 36
  phenomena 9, 182, 183, 241, 257, 314
  theory 27
  for apex point formation in detonation waves 242–251
Strong shock 232
Sturgis' work 87
Superheated, supersaturated state 30, 36, 38
Supersaturation pressure 31, 32
Subsonic flow 230
Supersonic flow 230
Surface tension 27
Symmetric double probe 151
System dividing 19

Subject Index 409

Taylor expansion 372
Temperature behind combustion waves 155, 159, 168
  behind detonation waves 266, 297–305
  behind imploding detonation waves 317, 328–334, 341, 342
  behind shock waves 175–181
  in the implosion focus 349–352
Temperature ratio at C-J point 283 rise by adiabatic or shock compression 377
Tetra-ethyl lead 13, 87–95
Theoretical temperature of $D+T$ at fusion reaction 356
Thermal explosion theory 3, 97, 98
Third explosion limit 13
Thomson scattering 162, 172
Three conservation laws, equations 186, 224, 309
Time-distance diagram along shock tube 52, 59, 105, 200, 252, 271, 271, 379
  of combustion wave and detonation front 259, 291, 302
  of piston 379, 381, 382, 383
  of shock and detonation 284, 285, 302, 316, 323, 324, 339, 345, 347, 348, 375, 380, 385
  of shock and flame propagation 238, 239
Top part of conical block (implosion) 353

Translation velocity 162
Transition from combustion wave (or propagating flame) to detonation wave 199, 255, 257, 259, 260, 262
  from deflagration to detonation wave 229, 232
  from laminar flame to turbulent flame 198
  from shock to detonation wave 270–281
Transition probability 174, 190
Transverse wave 230, 231, 241, 246
Triple point, shock 229–231
Tritium 355
Turbulent area, flame, flow 186, 187, 190, 191, 195
Two-step divergent detonation 343–355
  expansion method 366

Unsaturates 89

Vaporization 30–35
Vapor pressure 27, 30
Vibrating combustion 220, 221
Volumetric efficiency 221
Vortex 230, 231, 246

Wavelength (of spectral line) 161, 163, 165, 174, 177, 328, 341
Wave number vector 161, 162
Work on piston by gas expansion 382

Printing: Krips bv, Meppel
Binding: Stürtz, Würzburg